LONDON MATHEMATICAL SOCIETY LECTURE NOTE SERIES

Managing Editor:
Professor M. Reid, Mathematics Institute, University of Warwick, Coventry CV4 7AL, United Kingdom

The titles below are available from booksellers, or from Cambridge University Press at
http://www.cambridge.org/mathematics

London Mathematical Society Lecture Note Series: 416

Reversibility in Dynamics and Group Theory

ANTHONY G. O'FARRELL
National University of Ireland, Maynooth

IAN SHORT
The Open University

CAMBRIDGE
UNIVERSITY PRESS

CAMBRIDGE
UNIVERSITY PRESS

University Printing House, Cambridge CB2 8BS, United Kingdom

One Liberty Plaza, 20th Floor, New York, NY 10006, USA

477 Williamstown Road, Port Melbourne, VIC 3207, Australia

4843/24, 2nd Floor, Ansari Road, Daryaganj, Delhi - 110002, India

79 Anson Road, #06-04/06, Singapore 079906

Cambridge University Press is part of the University of Cambridge.

It furthers the University's mission by disseminating knowledge in the pursuit of education, learning and research at the highest international levels of excellence.

www.cambridge.org
Information on this title: www.cambridge.org/9781107442887

First published 2014

A catalogue record for this publication is available from the British Library

ISBN 978-1-107-44288-7 Paperback

Contents

Preface

Reversibility is the study of those elements of a group that are conjugate to their own inverses. Reversible maps appear naturally in classical dynamics; for instance, in the pendulum, the n-body problem, and billiards. They also arise in less obvious ways in connection with problems in geometry, complex analysis, approximation, and functional equations. When a problem has a connection to a reversible map, this opens it to attack using dynamical ideas, such as ergodic theory and the theory of flows. Reversibility has its origins in work of Birkhoff, Arnol'd, Voronin, Sevryuk, Siegel, Moser, Smale, and Devaney, among others, mainly in the context of continuous dynamical systems [8, 9, 10, 11, 31, 32, 66, 131, 211, 214, 237]. Devaney initiated the formal study of smooth reversible systems, not necessarily derived from a Hamiltonian, and there has been considerable work on such systems. The main focus has been on higher dimensions, and the systematic study of discrete reversible systems in low dimensions is more recent. We concentrate here on the discrete system theory, and on developments since the turn of the century.

The subject relates to involutions, conjugacy problems, and automorphism groups. The reversible elements of a group are those elements that are conjugate to their own inverses, and the strongly-reversible elements are those elements that are conjugate to their own inverses by involutions. Both types of element have been studied in many contexts. For finite groups, the terms *real* and *strongly real* are used instead of *reversible* and *strongly reversible* [117, Section 9.1] because of the connections with real characters. Questions of reversibility for classical groups have been addressed in works such as [82, 150, 151, 153]. The authors of these papers use the term *bireflectional* to describe a group comprised entirely of strongly-reversible maps. There is a rich modern literature on reversibility in dynamical systems, which includes [16, 66, 159, 160, 162, 163, 164, 165, 198, 199, 205, 211]. Some authors in this field describe reversible elements as *weakly reversible*, and describe strongly-reversible elements as *reversible*.

The book opens with a brief account of the origins of the subject in the theory of reversible systems in physics, in finite group theory, and in topics where the dynamics of a reversible map prove useful in tackling problems that have no apparent dynamic connection. Then we proceed to a rapid review of general facts about the reversible elements in a group, and the reversers of these elements. The remainder of the text is a survey of (mostly) recent work on reversibility in classes of groups in which there are often attractive geometric properties. The groups we examine are finite groups, classical groups, compact groups, isometry groups, certain groups of integer matrices, and larger groups: the homeomorphism groups of the line and circle, the diffeomorphism group of the line, formal power series groups, and groups of biholomorphic germs in one variable.

The choice of topics reflects the expertise of the authors, and there are substantial results about some groups that we omit in order to keep this work within reasonable limits. These include groups of polynomial automorphisms (see for instance [15, 18, 103, 142]) and area preserving and symplectic maps (see [17, 161, 204] and, in particular, the survey by Lamb and Roberts [164]). We also neglect the long history of reversibility in ergodic theory. Arguably, the relationship between reversibility and ergodic theory began with the work of Halmos and von Neumann [127], who proved that in the group of invertible measure-preserving transformations of a Borel probability space, those transformations that are ergodic and have a discrete spectrum are strongly reversible. Halmos and von Neumann suggested that perhaps every element of this group is reversible; however, this was shown not to be so by Anzai [7]. This work was continued by Goodson, del Junco, Lema'nczyk, and Rudolph [108] who found remarkably weak conditions in the group of invertible measure-preserving transformations that ensure that the conjugating map of a reversible transformation is an involution (so the transformation is strongly reversible). For more on this topic, consult the work of Goodson [107, 108, 109, 110, 111, 112, 115], and for some recent applications of reversibility in ergodic theory, see [1, 91]. Another significant collection of groups that we omit are the finite simple groups. The finite simple groups that consist entirely of reversible (real) elements were classified by Tiep and Zalesski in [231], and it has recently been shown [22, 81, 93, 121, 122, 153, 154, 202, 233] that these finite simple groups composed entirely of reversible elements are in fact composed entirely of strongly-reversible elements.

We only scratch the surface of discrete reversible systems, and give a taste of their applications. There are several excellent accounts in books and surveys that are almost completely disjoint from ours, such as Sevryuk [211]. The survey of Lamb and Roberts [164] includes a substantial bibliography for the period up to 1998. A good deal of the work is primarily concerned with physics, such as that of Hawking, Lahiri, MacKay, Roy, Wigner, and some of Penrose. We have little or nothing to say about this, nor about the equally interesting philosophical aspects of reversibility, as reviewed and discussed, for instance, in Nickel's contribution to [83]. The subject of reversibility is massive, and we cannot include all references. Major contributions have been omitted. We include a large bibliography with sources most relevant to the material we cover, and the reader should refer to the above texts for references to other works.

The main dependencies between the chapters are described by the directed graph below. One chapter depends on another if and only if there is a directed path from the second of these chapters to the first. Chapter 2 (represented by '2. Basics' in the graph) contains a small number of definitions that are used throughout the text, but for the most part, notation is local to chapters.

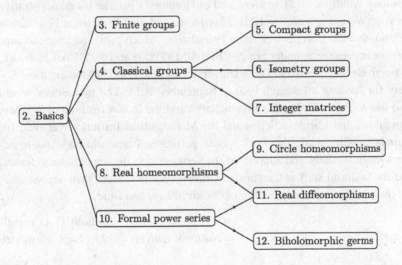

Each chapter finishes first with a Notes section, which includes references and further material, and then an Open problems section. The Open problems section contains unresolved issues about reversibility from that chapter, and we hope this section will prove useful as a source of research problems, particularly for doctoral students.

In general, when we do not give the proof of a proposition, we indicate this by putting the usual QED symbol □ right after the statement. If a mature mathematician, the reader may take this as an indication that the proof is straightforward. If a student, he or she may take it as a suggestion for an exercise. In case we are quoting a substantial result from another source, without proof, we will give the reference. We write the composition of elements f and g of a group by fg. Sometimes we compose elements of a group with functions that do not belong to the group, in which case we often use the symbol $\circ$ to denote composition, for clarity. Occasionally we use the symbol $\cdot$ for multiplication, when there is a chance that multiplication may be confused with group composition.

The content of this book owes much to the advice and help of our research collaborators, especially Patrick Ahern, Nick Gill, Roman Lávička, Frédéric Le Roux, Maria Roginskaya, and Dmitri Zaitsev. We are also grateful to sometime members of the Reversible Maps Group (Mary Boyce, Mary Hanley, Ying Hou, Simon Joyce, Dennis O'Brien, Jesús San Martin, David Walsh, Richard Watson) and other participants in our seminars, who helped us to refine some of the ideas presented here. Special thanks are due to Javier Aramayona, Stefan Bechtluft-Sachs, Kurt Falk, Xianghong Gong, John Murray, Azadeh Nikou, and Claas Röver. It is a pleasure to acknowledge the support of Janice Love and Anthony Waldron, on IT matters, and of Gráinne O'Rourke for administrative backup. We are grateful to NUI, Maynooth, and also to Science Foundation Ireland, and the European Science Foundation, which provided financial support for our research, under grants RFP/05/MAT0003 and the HCAA Network, respectively. We thank the Open University and the London Mathematical Society for funding a research visit in September 2011. The first author would also like to thank Tirthankar Bhattacharya and the Indian Institute of Science, Bangalore, and Caroline Series and the Mathematical Institute, Warwick, for their hospitality while the work was in gestation. We would both like to acknowledge the help and support of the series editor, the anonymous referees, and the editorial staff at Cambridge University Press. More than anyone else, we owe an unmeasurable debt to our beloved Lise and Ellie.

Anthony G. O'Farrell
National University of Ireland, Maynooth

Ian Short
The Open University

July 2014

1

Origins

To motivate our study of reversibility, we describe how the concept originates in dynamical systems, finite group theory, and in a subject known as *hidden dynamics*. Full details of these topics are beyond the scope of this book, and none of the material in this chapter is needed later on.

1.1 Origins in dynamical systems

Here we discuss several examples of reversibility in the study of conservative dynamical systems.

1.1.1 The harmonic oscillator

The simple pendulum is approximately modelled by the harmonic oscillator: the system in which a particle on the real line $\mathbb{R}$ is attracted to the origin by a force directly proportional to its distance from the origin. This system also models a weight suspended from a spring, oscillating about its equilibrium position (in which case the relationship between the force and distance is given by Hooke's law). Newton's second law states that the rate of change of the momentum of a body is equal to the force applied to it. Momentum is mass times velocity, so this gives the differential equations

$$\begin{aligned}
\frac{dp}{dt} &= -\kappa q, \\
\frac{dq}{dt} &= \frac{p}{m},
\end{aligned} \qquad (1.1)$$

where q represents the position of the particle, p its momentum (both p and q are functions of time t), κ is the constant of proportionality between the force and the distance to the origin, and m is the particle's mass.

It follows at once that the quantity

$$H(q,p) = \frac{p^2}{2m} + \frac{\kappa q^2}{2},$$

called its *Hamiltonian* (which is, physically, the energy of the system, given by the sum of its kinetic and potential energy), has derivative zero with respect to time, and hence is constant along trajectories. It follows that the trajectories are the concentric ellipses $H(q,p) = E$, for constant $E \geqslant 0$.

Consider the map $\tau : \mathbb{R}^2 \to \mathbb{R}^2$ defined by $\tau(q,p) = (q,-p)$. Evidently, $\tau \circ \tau = \mathbb{1}$, the identity map. A simple calculation establishes the following result.

Lemma 1.1 *If $(q(t),p(t))$ is a solution of the differential equations* (1.1), *then so is $\tau(q(-t),p(-t))$.* $\qquad\qquad\square$

This lemma is usually expressed as saying that τ is a *time-reversal symmetry* of the system.

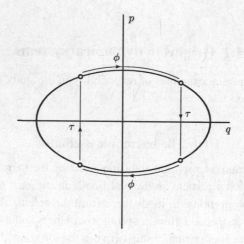

Figure 1.1 Time-reversal symmetry of the harmonic oscillator

Let $t \mapsto (q(t),p(t))$ represent the solution of (1.1) subject to the initial conditions $(q(0),p(0)) = (q_0,p_0)$, where (q_0,p_0) is some pair in $\mathbb{R}^2$. We define $\phi : \mathbb{R}^2 \to \mathbb{R}^2$ to be the *time-one step of the system*, given by $\phi(q_0,p_0) = (q(1),p(1))$. Then

$$\phi \circ \tau \circ \phi \circ \tau = \mathbb{1},$$

or

$$\tau \circ \phi \circ \tau = \phi^{-1}, \qquad\qquad (1.2)$$

the inverse map of ϕ (see Figure 1.1).

1.1.2 The n-body problem

The above behaviour is not particular to the harmonic oscillator. We can make similar observations whenever the Hamiltonian $H(q,p)$ of a dynamical system is quadratic in the momentum variable p.

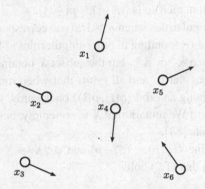

Figure 1.2 The n-body problem

Consider, for instance, the problem of n point bodies moving under their mutual gravitational attraction, illustrated in Figure 1.2. If we denote the masses by m_i and the positions by $x_i : \mathbb{R} \to \mathbb{R}^3$ $(i = 1, \ldots, n)$, then in Newtonian form the equations of motion are

$$\frac{d}{dt}\left(m_i \frac{dx_i}{dt}\right) = \sum_{\substack{r=1 \\ r \neq i}}^{n} \frac{Gm_i m_r}{|x_r - x_i|^2}\left(\frac{x_r - x_i}{|x_r - x_i|}\right),$$

where G is the gravitational constant. Let $x_i = (x_{i1}, x_{i2}, x_{i3})$ for $i = 1, \ldots, n$ and, for $j = 1, 2, 3$, let

$$\mu_{3i-3+j} = m_i, \quad q_{3i-3+j} = x_{ij}, \quad p_{3i-3+j} = m_i \frac{dx_{ij}}{dt}.$$

We also define

$$K(p) = \sum_{r=1}^{3n} \frac{p_r^2}{2\mu_r}, \quad V(q) = -\frac{1}{2}\sum_{\substack{r,s=1 \\ r \neq s}}^{n} \frac{Gm_r m_s}{|x_r - x_s|},$$

$$H(q,p) = K(p) + V(q),$$

where $p = (p_1, \ldots, p_{3n})$ and $q = (q_1, \ldots, q_{3n})$. Then the equations of motion become

$$\frac{dq_k}{dt} = \frac{\partial H}{\partial p_k},$$
$$\frac{dp_k}{dt} = -\frac{\partial H}{\partial q_k},$$

for $k = 1, \ldots, 3n$. We have, as before, that $H(q, -p) = H(q, p)$, and that if $(q(t), p(t))$ is a solution, then so is $(q(-t), -p(-t))$.

This system has singularities when $n > 1$, some corresponding to collisions, and, for $n \geqslant 4$, some corresponding to other singularities [248]. Let us consider not the full phase space $\mathbb{R}^{3n} \times \mathbb{R}^{3n}$, but the subset X obtained by removing all orbits that end in a singularity, and all orbits that when run backwards end in a singularity. (By running an orbit $(q(t), p(t))$ backwards, we mean taking the orbit $(q(-t), -p(-t))$.) We remark that X is nonempty, but its structure is not fully understood to date [84].

Again, we can define $\tau(q, p) = (q, -p)$ and $\phi : X \to X$ to be the time-one step of the system, so that (1.2) holds.

1.1.3 Billiards

Consider billiards on an arbitrary smoothly-bounded, strictly-convex table without pockets. Let Γ denote the boundary. We ignore the motion in which the ball

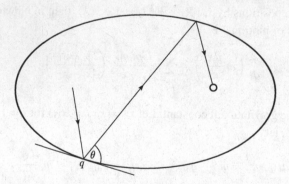

Figure 1.3 Trajectory of a billiard ball

rolls around the cushion, considering only trajectories in which it bounces to and fro. We assume that it moves in a straight line between bounces, and that at each bounce the line of incidence and the line of departure make equal angles with the normal to the boundary at the point of impact.

We may parametrise the set of states at which the ball leaves the cushion by two parameters q and θ, where q is the point in Γ at which the ball leaves, and θ is the angle between the line it departs on and the tangent to Γ, in the counterclockwise direction (as shown in Figure 1.3). Thus the state space is $X = \Gamma \times (0, \pi)$, and the dynamical step $\phi : X \to X$ is the map that takes (q_0, θ_0) to (q_1, θ_1), where (q_1, θ_1) parametrises the state that results one bounce after the state parametrised by (q_0, θ_0).

If we denote by τ the bijection of X defined by $\tau(q, \theta) = (q, \pi - \theta)$, which reverses the direction of travel, then we have $\tau \circ \tau = \mathbb{1}$, and

$$(\phi \circ \tau \circ \phi)(q_0, \theta_0) = (\phi \circ \tau)(q_1, \theta_1) = \phi(q_1, \pi - \theta_1) = (q_0, \pi - \theta_0) = \tau(q_0, \theta_0)$$

so that, again, equation (1.2) holds.

Henceforth we omit the symbol $\circ$ from equations such as $\tau \circ \phi \circ \tau = \phi^{-1}$, unless the omission is likely to cause confusion.

1.1.4 Significance of the equation $\tau \phi \tau = \phi^{-1}$

Equation (1.2) says that the dynamical step ϕ is conjugate to its own inverse by the involutive map τ. Birkhoff was probably the first to point out the significance of this equation for a dynamical system [31, page 311]. It implies that the dynamics of the system are essentially the same as the dynamics of the inverse system. This has strong consequences. For instance, if a periodic point is fixed by the involution, then it cannot be attracting. Billiards always has periodic points of all orders [31, page 328ff]. This is a consequence of the famous *Last Geometric Theorem of Poincaré*, conjectured by Poincaré and proven eventually by Birkhoff. Period-two points correspond to orbits that bounce back and forth between two boundary points, and obviously each of the two states is fixed by the map τ that reverses the direction of travel. Thus these orbits are necessarily neither attracting nor repelling.

As we shall see in Chapter 3, the very same equation comes up when one considers the symmetry group of a regular polygon, a *dihedral group*, and we shall meet the equation later in many other situations in geometry, algebra, and analysis.

1.2 Origins in finite group theory

Reversibility is significant in the theory of finite groups because of its connections with representation theory. The next theorem, which is proven in Chapter 3, is the primordial result about reversibility in finite group theory.

Theorem 1.2 *An element g of a finite group G is conjugate to g^{-1} if and only if $\chi(g)$ is real for each complex character χ of G.* □

All the representations we consider in this section are complex representations.

It is because of this theorem that elements of finite groups that are conjugate to their own inverses are called *real* elements. However, we prefer to use the term *reversible* instead of *real* because most of the groups we consider later are infinite groups, for which there is no such characterisation of reversible elements using real characters.

In Chapter 3 we also prove that the number of conjugacy classes of reversible elements in a finite group is equal to the number of real-valued irreducible characters.

Theorem 1.2 tells us that we can identify the reversible elements of a finite group by studying the group's character table. We can obtain another result of the same type, using a similar (if slightly harder) proof.

Theorem 1.3 *An element g of a finite group G is conjugate to g^m for each integer m coprime to $|G|$ if and only if $\chi(g)$ is rational for each character χ of G.* □

Theorems 1.2 and 1.3 indicate that the structure of a finite group is closely related to the values taken by the group's characters. To investigate this relationship more thoroughly, group theorists use the *Schur index* of a character, which we describe briefly.

Let χ be an irreducible complex character of G. Given a subfield k of $\mathbb{C}$, we say that χ can be *realised over k* if there is an irreducible representation ϕ that has character χ such that the matrix entries of $\phi(g)$ lie in k, for every element g of G. Given a field F such that $\mathbb{Q} \subset F \subset \mathbb{C}$, we define $F(\chi)$ be the smallest subfield of $\mathbb{C}$ that contains F and all the values $\chi(g)$, for $g \in G$. The *Schur index of χ over F*, denoted $m_F(\chi)$, is the smallest degree of an extension of $F(\chi)$ over which χ can be realised. If $F = \mathbb{R}$, then we call $m_{\mathbb{R}}(\chi)$ the *Schur index of χ*.

The only possible values of $m_{\mathbb{R}}(\chi)$ are 1 and 2 because the only algebraic extensions of $\mathbb{R}$ are $\mathbb{R}$ and $\mathbb{C}$. The next theorem, the Brauer–Speiser theorem [85], invests this trivial observation with a great deal of value.

Theorem 1.4 *If χ is a real-valued irreducible character of a finite group G, then $m_{\mathbb{Q}}(\chi)$ is either 1 or 2.* □

In particular, if $m_{\mathbb{R}}(\chi)$ is 2, then because $\mathbb{Q}(\chi) \subset \mathbb{R}(\chi)$ it follows that $m_{\mathbb{Q}}(\chi)$ is also 2.

An important tool for calculating $m_{\mathbb{R}}(\chi)$ is the *Frobenius–Schur indicator* $v_2(\chi)$, which is defined by

$$v_2(\chi) = \frac{1}{|G|} \sum_{g \in G} \chi(g^2).$$

The connection between the Frobenius–Schur indicator and the Schur index is explained by the following result [138, page 58].

Theorem 1.5 *Given an irreducible character χ, the Frobenius–Schur indicator $v_2(\chi)$ is either 0, 1 or 2. Furthermore,*

$$v_2(\chi) = \begin{cases} 0, & \text{if } \chi \text{ is not real-valued,} \\ 1, & \text{if } \chi \text{ is realised over } \mathbb{R}, \\ -1, & \text{if } \chi \text{ is real-valued, but cannot be realised over } \mathbb{R}. \end{cases}$$

$\square$

Clearly, if $v_2(\chi)$ is 0 or 1, then $m_{\mathbb{R}}(\chi) = 1$, and if $v_2(\chi) = -1$, then $m_{\mathbb{R}}(\chi) = 2$.

Theorem 1.5 has an important application in counting square roots of elements. It is proven in [138, page 49] and [140, Corollary 23.17] that the number of square roots in G of an element g is given by

$$\sum v_2(\chi)\chi(g),$$

where the sum is taken over all irreducible characters of G. This formula can be used to help prove the Brauer–Fowler theorem about centralisers of involutions in finite simple groups (see [140, Chapter 23]), which is a pivotal result in the classification of the finite simple groups.

The standard proof of Theorem 1.5 throws up an interesting subsidiary result [138, Theorem 4.14] (in which we denote the transpose of a matrix A by A^t).

Theorem 1.6 *Suppose that ϕ is an irreducible representation with a real-valued character χ. Then there exists a nonzero square matrix M such that*

$$\phi(g)^t M \phi(g) = M$$

for all elements g of G. Furthermore, for any such matrix M, we have $M^t = v_2(\chi)M$. $\square$

It follows from Theorem 1.6 that the image of the representation $\phi : G \to \mathrm{GL}(V)$ lies inside the isometry group of the bilinear form

$$\beta : V \times V \to \mathbb{C}, \quad (u,v) \mapsto u^t M v.$$

If $v_2(\chi) = 1$, then $M^t = M$, and β is a symmetric bilinear form; in this case ϕ is called an *orthogonal representation*. If $v_2(\chi) = -1$, then $M^t = -M$, and β is a skew-symmetric bilinear form; in this case ϕ is called a *symplectic representation*.

It is helpful to summarise parts of Theorems 1.5 and 1.6 in the following corollary.

Corollary 1.7 *Let χ be an irreducible character. The following are equivalent:*

(i) $m_{\mathbb{R}}(\chi) = 2$
(ii) $v_2(\chi) = -1$
(iii) *χ is real-valued and is realised by a symplectic representation.*   □

Brauer [36, Problem 14] asked for a group theoretic description of the number of irreducible characters of a finite group with Schur index 1. This question has been answered by Gow [119] when the Sylow 2-subgroups of G are nontrivial and cyclic, and Gow's answer is given in terms of the number of reversible conjugacy classes of G.

A slightly easier question than Brauer's asks for a characterisation of those groups G such that *all* real-valued irreducible characters χ have Schur index 1. A partial answer to this was given by Gow in [119, Corollary 1]. To state Gow's result, we recall that a reversible element of a group is *strongly reversible* if it is conjugate to its inverse by an involution.

Theorem 1.8 *Let G be a finite group whose Sylow 2-subgroup is abelian. Then all real-valued irreducible characters χ of G have Schur index 1 if and only if all reversible elements of G are strongly reversible.*

See [118, 120] for related work of Gow.

Gow notes that often (but not always) the existence of real-valued irreducible characters with Schur index 2 is accompanied by the existence of reversible elements that are not strongly reversible. A conjecture in [146], which the authors of [146] attribute to Tiep, makes this more precise. To understand the conjecture, remember that the Schur index $m_{\mathbb{R}}(\chi) = 1$ if and only if the Frobenius–Schur indicator $v_2(\chi)$ is 0 or 1. Finite groups for which all irreducible characters χ satisfy $v_2(\chi) = 1$ are called *totally orthogonal* because, as we have seen, all their representations are orthogonal.

Conjecture 1.9 *A finite simple group is totally orthogonal if and only if all its elements are strongly reversible.*

It is therefore of interest to determine all those finite simple groups whose elements are all strongly reversible. This very question appeared in the famous Kourovka notebook [178], as Problem 14.82, posed by Sozutov (but described as a "well-known problem"). In 2005, Tiep and Zalesski [231] classified all the simple (and quasi-simple) finite groups in which all elements are reversible. Results of [22, 81, 93, 121, 122, 153, 154, 202, 233] imply that each of these finite simple groups that consists entirely of reversible elements in fact consists entirely of strongly-reversible elements (therefore Problem 14.82 is solved). To prove Tiep's conjecture, then, one must calculate $v_2(\chi)$ for all irreducible characters χ of each of the finite simple groups.

Recently, Kaur and Kulsherstha [146] have shown that the conjecture is false if the finite groups are no longer required to be simple. They construct infinite families of special 2-groups that are totally orthogonal but do not consist entirely of strongly-reversible elements, and they also construct infinite families of special 2-groups that consist entirely of strongly-reversible elements but are not totally orthogonal.

1.3 Origins in hidden dynamics

To further motivate our formal study of reversibility, we briefly describe some examples in which the dynamics of a reversible map play a key role in resolving a problem with no apparent dynamic connection; this phenomenon is known as *hidden dynamics*. Our first example is elementary, our second example is from complex analysis, and our third example is from approximation theory.

1.3.1 Small fibres

A map $f : X \to Y$ is said to have *small fibres* if $f^{-1}(f(x))$ has cardinality at most two, for each x in X. Whenever f has small fibres, we may define an associated involution $\tau : X \to X$ by requiring that

$$f^{-1}(f(x)) = \{x, \tau(x)\}, \quad \text{for } x \text{ in } X.$$

In other words, τ swaps the preimages of each point, if there are two, and fixes the preimage, if there is only one.

Many interesting involutions arise in this way. (In fact, all involutions arise in this way: given an involution τ on a set X, we may define Y as the space of orbits of τ and $f : X \to Y$ as the quotient map.) For example, the quadratic map

$$f : \mathbb{C} \to \mathbb{C}, \quad z \mapsto z^2 + bz,$$

induces the involution $z \mapsto -z - b$. The cubic map

$$f : \mathbb{C} \to \mathbb{C}, \quad z \mapsto z^3 + z^2,$$

has three preimages for most points, but if restricted to a small enough neighbourhood of the origin, it has at most two-point fibres, and the induced involution τ is actually holomorphic on a neighbourhood of 0, with

$$\tau(z) = -z - z^2 + \cdots.$$

It may happen that a problem leads us to two maps $f_1 : X \to Y_1$ and $f_2 : X \to Y_2$, each having small fibres. From the induced involutions τ_1 and τ_2 we can define $\phi = \tau_1 \tau_2$, a reversible bijective map of X. This scenario arises in a significant number of cases, and often the dynamics of ϕ prove useful.

1.3.2 Two-valued reflections

Webster [240, 241, 242] exploited the small fibres idea in connection with the concept of two-valued reflections. Two-valued reflections occur when the familiar local antiholomorphic reflection across a real-analytic curve γ on a Riemann surface Σ happens to have precisely two global extensions. This happens, for instance, for reflection across an ellipse in the sphere, and also for certain particular quartic lemniscates.

The abstract situation involves a compact Riemann surface $\hat{\Gamma}$ (related to the complexification Γ of γ) and maps π_1, π_2, ρ, and ρ', where π_1 and π_2 are each two-fold branched covers of Σ, ρ is an antiholomorphic involution of $\hat{\Gamma}$, and ρ' is an antiholomorphic involution of Σ, such that the following diagram commutes.

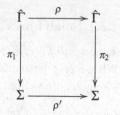

When Σ is the Riemann sphere $\hat{\mathbb{C}}$, it turns out that the existence of both π_1 and π_2 implies (by the Riemann–Roch theorem) that $\hat{\Gamma}$ is the Riemann sphere or a torus. Webster considered these cases, and discovered which Γ correspond to various curves, by an argument that produced explicit parametrisations of Γ and formulas for the reflections. The point is that these formulas are found by a natural process of discovery; we do not need a stroke of genius to find them; nothing has to be pulled out of the air.

Webster realised that a similar outcome should occur for the other curves that admit two-valued reflection. He showed that only tori with certain specific moduli can arise as the complexification of such a curve, and gave parametrizations in terms of the Weierstrass $\wp$-function.

Webster used two-valued reflections to construct the conformal map of the interior E of an ellipse to the disc $\mathbb{D}$. Intuitively, the reason this is possible is that the Riemann map $\mathbb{D} \to E$ sends a curve with a single-valued reflection to a curve with a two-valued reflection, so this map ought to be somehow reflectable in two ways, and then reflectable back, producing symmetries. Functions having many symmetries can often be nailed down.

Webster went on to "rediscover" the conformal maps of the other conic sections. These were known since the time of H. A. Schwarz. He also proposed a programme of imitating this construction for the other curves that admit a two-valued reflection, and he carried it out in detail in one example, obtaining a new formula.

He also considered double-valued reflections for ellipsoid in $\mathbb{C}^n$, and their use in calculating extremal disks for the Kobayashi metric, and the closely-related Lempert disks. He sketched this programme in [241] and partially executed it in [242]. As far as we know, the full programme has not been carried out.

Earlier, in 1988, Bullett [43] studied the dynamics of quadratic correspondences, and in particular those called *maps of pairs*. These are two-valued relations on the sphere $\hat{\mathbb{C}}$ that can be transformed to the form $z \mapsto z'$, where $q(z) = r(z')$, for some quadratic rational maps q and r. They are obviously invariant under related involutions of the source $\hat{\mathbb{C}}$ and the target $\hat{\mathbb{C}}$. They include as special cases the Möbius transformations, the quadratic maps, and the inverses of quadratic maps. They also include, for example, the correspondence defined by

$$\frac{z^2 + 2z}{z+1} = \frac{z'^2 - 1}{z'},$$

which can be rewritten as

$$z' = z + 1 \quad \text{or} \quad z' = \frac{-1}{z+1}.$$

When one considers the iteration of this correspondence, it appears that the full two-sided orbits are just the orbits of the usual representation of the modular group $\mathrm{PSL}(2, \mathbb{Z})$, because the maps $z \mapsto z + 1$ and $z \mapsto -1/(z+1)$ generate the modular group. More generally, one may associate a map of pairs to any Kleinian group which is generated by two elements g_1, g_2 such that $g_2^{-1} g_1$ is an involution. So Bullett's theory provides a common framework in which to study the Julia sets of quadratic maps and the limit sets of certain Kleinian groups.

1.3.3 Sums of functions of fewer variables

The question of approximating a function of two real variables $h(x,y)$ by sums $h_1(x) + h_2(y)$ arises in several applications.

One application is the art of nomography. A nomograph is a planar diagram or chart used for calculating some function (or solving some equation). There are nomographs for a vast range of purposes in engineering, science, commerce, and (of course) warfare. One finds nomographs associated to the use of sundials. One may even regard sundials and astrolabes as particular nomographs. Usually, one uses a nomograph by positioning a straight edge to pass through specific points on two scales (a scale is a curve labelled with values), and reads the corresponding value off a third scale. One may, however, have any number of scales on the chart. The systematic study of nomographs was initiated by Philbert Maurice d'Ocagne (1862–1938) in 1880. Hilbert's 13th problem was inspired by the question of whether every function, of any number of variables, could be calculated by using a suitable collection of scales. This led Kolmogorov directly to the consideration of $h(x,y) = h_1(x) + h_2(y)$. Other areas of application occur in numerical analysis [69, 181], problems about data storage of data in computer memory, statistics, economics [148], and functional equations [42]. Sometimes, the application involves functions defined on the whole plane, or on a rectangle $[a,b] \times [c,d]$. Sometimes, it involves functions defined on a special subset of the plane, such as a finite union of smooth curves.

Let us now approach the problem formally. Define

$$\pi_1 : (x,y) \mapsto x \quad \text{and} \quad \pi_2 : (x,y) \mapsto y$$

to be the two coordinate projections $\mathbb{R}^2 \to \mathbb{R}$. For the purposes of this section, denote by $\Pi(x)$ the set of all continuous real-valued functions on $\mathbb{R}^2$ that depend only on the first coordinate x; that is, functions of the form $h \circ \pi_1$, where $h : \mathbb{R} \to \mathbb{R}$ is continuous. Similarly, let $\Pi(y)$ denote the continuous real-valued functions on $\mathbb{R}^2$ that depend only on y.

Given a topological space Y, let $C(Y)$ denote the set of continuous real-valued functions on Y. We focus on $C(X)$, where X is a compact subset of $\mathbb{R}^2$. Our objective is to approximate functions in $C(X)$, uniformly on X, by elements of the vector space sum $\Pi(x) + \Pi(y)$. That is, we wish to minimise the *distance*

$$\text{dist}(f, \Pi(x) + \Pi(y)) = \inf_{h_1, h_2 \in C(\mathbb{R})} \sup_{(x,y) \in X} |f(x,y) - h_1(x) - h_2(y)|.$$

The problem of when this distance can be made arbitrarily small was solved by combining methods of unctional analysis (the Hahn–Banach theorem, Riesz representation theorem, and Krein–Milman theorem) with ergodic theory. The key to the solution is the idea of a *lightning bolt*. The concept of a lightning bolt originated with V. I. Arnol'd, but was independently invented later by many people, and given various names, such as *zig-zag* and *trip*. A lightning bolt is a sequence $p_1, p_2, \ldots$ of points in the plane such that either

$$\pi_1(p_1) = \pi_1(p_2), \ \pi_2(p_2) = \pi_2(p_3), \ \pi_1(p_3) = \pi_1(p_4), \ \pi_2(p_4) = \pi_2(p_5), \ldots$$

(a type I bolt) or

$$\pi_2(p_1) = \pi_2(p_2), \ \pi_1(p_2) = \pi_1(p_3), \ \pi_2(p_3) = \pi_2(p_4), \ \pi_1(p_4) = \pi_1(p_5), \ldots$$

(a type II bolt). See Figure 1.4.

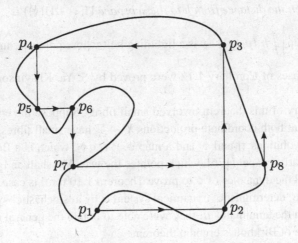

Figure 1.4 A type II lightning bolt in a planar set

Suppose that the sequence p of points $p_1, p_2, \ldots$ lies in some compact subset X of $\mathbb{R}^2$. We say that the lightning bolt generates the measure μ on X if the sequence of combinations of point masses

$$\mu_n(p) = \frac{1}{n} \sum_{j=1}^{n} (-1)^{j+1} \delta_{p_j}$$

converges weak-star to μ in $C(X)^*$.

It is easy to see that each measure generated by a lightning bolt is an annihilating measure of norm at most 1. In the other direction, we have the following theorem from [177].

Theorem 1.10 *Let X be a compact subset of* $\mathbb{R}^2$*. Then each extreme norm 1 annihilating measure for* $\Pi(x) + \Pi(y)$ *on X is generated by some lightning bolt whose points all belong to X.*

The theorem has two immediate corollaries.

Corollary 1.11 *Let X be a compact subset of* $\mathbb{R}^2$*. Then the following two statements are equivalent.*

(i) *Each continuous function* $h : X \to \mathbb{R}$ *is the uniform limit on X of a sequence of functions drawn from* $\Pi(x) + \Pi(y)$*.*
(ii) *For each lightning bolt p whose points all lie in X, the measures* $\mu_n(p)$ *converge weak-star to 0.* $\qquad\square$

Corollary 1.12 *Let X be a compact subset of* $\mathbb{R}^2$*, and let* $h : X \to \mathbb{R}$ *be continuous. Then the distance from h to the subspace* $\Pi(x) + \Pi(y))$ *is*

$$\max\left\{ \limsup_n \left| \int h \, d\mu_n(p) \right| : p \text{ is a lightning bolt whose points lie in } X \right\}. \quad \square$$

Special cases of Corollary 1.12 were proved by S. Ya. Khavinson and others [148].

The history of this theorem involved small fibres. Suppose the compact set X is such that both coordinate projections $X \to \mathbb{R}$ have small fibres. Then we have two involutions τ_1 and τ_2 and a map $\phi = \tau_1 \circ \tau_2$, which is a Borel bijection. In an earlier paper [176], which pointed the way, Marshall and one of the authors used the dynamics of ϕ to prove Theorem 1.10 for this case. Later, by replacing the 'deterministic' dynamical system ϕ by a 'stochastic' system that depended on the annihilator μ, they were able to obtain the general result as a consequence of Birkhoff's ergodic theorem.

As a simple example, consider the problem of approximating functions on the unit circle by elements of the set S of all sums of the form

$$h_1(\alpha_1 x + \beta_1 y) + h_2(\alpha_2 x + \beta_2 y), \tag{1.3}$$

where h_1 and h_2 are real-valued continuous functions. (This is essentially equivalent to approximating functions on the ellipse parametrised by

$$x = \alpha_1 \cos\theta + \beta_1 \sin\theta$$
$$y = \alpha_2 \cos\theta + \beta_2 \sin\theta$$

using elements of $\Pi(x) + \Pi(y)$.) Applying Corollary 1.11, one obtains the following proposition from [176].

Proposition 1.13   *The set S is a dense subset of $C(X)$ if and only if the angle between the level sets is an irrational multiple of π.*   □

It is worth remarking that these results do not extend to sums such as $f(x) + g(y) + h(z)$ on subsets of $\mathbb{R}^3$. There is no way to generate the annihilators of a sum of three algebras using any analogue of the lightning bolts. This is related to the fact that ergodic theory is essentially limited to actions of groups that have subexponential growth (that is, the number of elements representable by words of length at most n in the generators grows less than exponentially with n). There is a remarkable example [221] of a subset X of $\mathbb{R}^3$, such that each continuous real-valued function on X can be written as $f(x) + g(y) + h(z)$, with f, g, and h bounded, but cannot be written in this way with f, g, and h continuous (or even Borel). This shows that there are essentially new phenomena that arise with the sum of three subalgebras of $C(X)$, as opposed to two.

In fact, it was this striking difference that first focussed the attention of one of us on the special character of dihedral groups of maps, and led us to study reversibility.

1.4 The reversibility problem

Motivated by the importance of reversibility in conservative dynamical systems, finite group theory, and hidden dynamics this book classifies the reversible elements (elements conjugate to their own inverses) in some of the most commonly encountered and important groups in mathematics. We describe the problem of classifying the reversible elements of a group as the *reversibility problem*. This is a special case of the *conjugacy problem*: the problem of determining whether any pair of elements from a group are conjugate. For groups described by a presentation, the reversibility problem (and likewise the conjugacy problem) can be interpreted formally as a decision problem. Many of the large groups that we consider have unmanageable presentations, and so we take a less formal view of the reversibility problem (and conjugacy problem) in that we seek only to describe the reversible elements of a group in understandable mathematical terms.

Usually we are able to solve the conjugacy problem, and therefore also solve the reversibility problem, but this is not always so. For example, conjugacy in the group of diffeomorphisms of the real line is not fully understood, whereas we are able to give a complete solution to the reversibility problem, in Chapter 11. Often the reversibility problem can be solved in much simpler terms than the conjugacy problem. For example, in the group of formal power series over the complex numbers (with formal composition) studied in Chapter 10, we see that a power series $x + a_2 x^2 + \cdots$ is reversible if and only if it is conjugate to the expression

$$\frac{x}{(1 + \lambda x^p)^{1/p}},$$

which you can expand as a series in the usual way, for some positive integer p and complex number λ.

For each of our groups, we also address certain other questions related to the reversibility problem. Primarily, we determine the elements of the group that are *strongly reversible*; that is, those elements that are reversible and for which the conjugating map can be chosen to be an involution. It is an elementary fact, proven in Chapter 2, that an element of a group is strongly reversible if and only if it can be expressed as a product of two involutions. Extending this, we also classify those elements of the group that can be expressed as a product of n involutions, for $n = 3, 4, \ldots$, and we determine the normal subgroup generated by all involutions.

We remark that all the properties discussed so far, such as the reversible or strongly-reversible properties, depend on the group under consideration. For example, in the rotation group of a circle, the rotation ϕ by $\pi/2$ is not reversible. In the larger group of rotations and reflections of the circle, however, ϕ is strongly reversible, because $\tau \phi \tau = \phi^{-1}$ for any reflection τ.

To finish this chapter, let us discuss some attractive aspects of reversibility in a selection of familiar groups. Each of these groups will be studied in more detail later in the text.

1.4.1 The symmetric group

Our first example is from finite group theory: we consider the group S_n of permutations of the set $\{1, 2, \ldots, n\}$. Finite groups will be examined in detail in Chapter 3. Using disjoint cycle notation, we let f be the permutation $(123 \ldots m)$ in S_n, for some $m \leqslant n$. Plot the m points $1, 2, \ldots, m$ with equal spacing around a circle. Then f corresponds to a rotation by angle $2\pi/m$, and f is conjugate to f^{-1} by a reflection τ that preserves the m points. The action of one such reflection τ is shown in Figure 1.5.

Because each member of S_n is expressible as a product of disjoint cycles, we see that every element in S_n is strongly reversible. Similar reasoning applies to

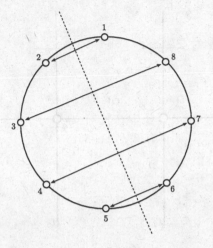

Figure 1.5 The action of the reflection τ

infinite symmetric groups. On the other hand, there are elements of alternating groups that are not reversible, such as (123) in A_3.

The smallest group in which the classes of reversible and strongly reversible elements do not coincide is the group of unit quaternions $Q = \{\pm 1, \pm i, \pm j, \pm k\}$ (where $ijk = -1$ and $i^2 = j^2 = k^2 = -1$). All elements of this group are reversible: ± 1 are reversed by 1, $\pm i$ are reversed by j, $\pm j$ are reversed by k, and $\pm k$ are reversed by i. In contrast, since the only involutions are ± 1, and these lie in the centre of Q, we see that ± 1 are the only strongly-reversible elements.

1.4.2 Euclidean isometries

Our second example concerns the group $\mathrm{Isom}^+(\mathbb{R}^2)$ of orientation-preserving Euclidean isometries of the plane $\mathbb{R}^2$. This group consists of translations and rotations. The nontrivial involutions in $\mathrm{Isom}^+(\mathbb{R}^2)$ are the rotations by π. Suppose that f is a strongly-reversible element of $\mathrm{Isom}^+(\mathbb{R}^2)$ that is not an involution. Then $f = \sigma\tau$, where σ is a rotation by π about a point p, and τ is a rotation by π about a point q, and $p \neq q$. Let ℓ_1 denote the line through p and q. Let ℓ_2 denote the line perpendicular to ℓ_1 that contains p, and let ℓ_3 denote the line perpendicular to ℓ_1 that contains q. Denote reflection in the line ℓ_i by r_i. Then $\sigma = r_2 r_1$ and $\tau = r_1 r_3$. Hence $f = r_2 r_3$. Since ℓ_2 and ℓ_3 are parallel, we see that f is a translation. Conversely, by reversing this argument, we see that all translations are strongly reversible. Figure 1.6 illustrates this geometry.

In summary, an element of $\mathrm{Isom}^+(\mathbb{R}^2)$ is strongly reversible if and only if

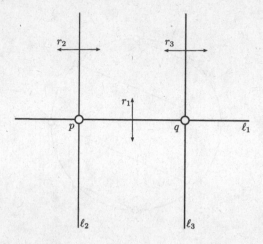

Figure 1.6 A translation composed of two rotations by π

it is a translation or an involution. By considering fixed points, one can also show that the rotations that are not involutions – which, as we have just seen, are not *strongly* reversible – are not even reversible.

We turn now to the group $\mathrm{Isom}^+(\mathbb{R}^3)$ of orientation-preserving isometries of three-dimensional Euclidean space $\mathbb{R}^3$. Each isometry g consists of a rotation ρ about a Euclidean line ℓ, followed by a translation t in the direction of ℓ. We can choose two planes Π_1 and Π_2 that intersect in the line ℓ, and that satisfy $\rho = r_2 r_1$, where r_i denotes reflection in Π_i. We can also choose two parallel planes Π_3 and Π_4 that are orthogonal to ℓ, and that satisfy $t = r_4 r_3$, where, again, r_i denotes reflection in Π_i. See Figure 1.7. Since Π_2 and Π_3 are

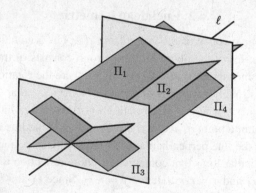

Figure 1.7 Planes used to show that a Euclidean isometry is strongly reversible

orthogonal, the reflections r_2 and r_3 commute. Therefore

$$f = t\rho = r_4 r_3 r_2 r_1 = (r_4 r_2)(r_3 r_1).$$

Because Π_2 and Π_4 are orthogonal, and likewise Π_1 and Π_3 are orthogonal, the two maps $r_4 r_2$ and $r_3 r_1$ are each rotations by π. Hence f is strongly reversible.

In summary, every map in $\text{Isom}^+(\mathbb{R}^3)$ is strongly reversible.

The same techniques we have used for Euclidean geometry can also be applied to hyperbolic geometry. The group of orientation-preserving isometries of three-dimensional hyperbolic space is the group of complex Möbius transformations; namely the group of maps of the form $z \mapsto (az+b)/(cz+d)$, where $ad - bc \neq 0$. In this case the reversibility problem has a particularly simple algebraic solution. The isometry groups of Euclidean and hyperbolic space are considered in detail in Chapter 6.

1.4.3 Homeomorphisms of the real line

For our third example we consider the group $\text{Homeo}(\mathbb{R})$ of all homeomorphisms of the real line $\mathbb{R}$. The nontrivial involutions in this group are order-reversing homeomorphisms, and form a single conjugacy class in $\text{Homeo}(\mathbb{R})$. In particular, they are all conjugate to the map $\sigma(x) = -x$ (see Proposition 8.2). Let $\Gamma(h)$ denote the graph of a homeomorphism h; that is, $\Gamma(h)$ is the subset $\{(x, h(x)) : x \in \mathbb{R}\}$ of $\mathbb{R}^2$. Let ρ denote a rotation by π about $(0,0)$, let r denote reflection in the line $y = x$, and let s denote reflection in the line $y = -x$. Then $\rho = sr$. It is well known, and easy to prove, that *a homeomorphism f is an involution if and only if its graph is symmetric about the line $y = x$.* That is, if and only if $r(\Gamma(f)) = \Gamma(f)$.

Suppose now that f is a product of two nontrivial involutions (it is strongly reversible). After replacing f with a conjugate we can assume that $\sigma f \sigma = f^{-1}$ (where, as above, $\sigma(x) = -x$). Now, one can easily verify that for any homeomorphism h, $\Gamma(\sigma h \sigma) = \rho(\Gamma(h))$ and $\Gamma(h^{-1}) = r(\Gamma(h))$, which implies that

$$\Gamma(f) = \rho(\Gamma(\sigma f \sigma)) = \rho(\Gamma(f^{-1})) = \rho r(\Gamma(f)) = s(\Gamma(f)).$$

In other words, *a homeomorphism f is a product of two nontrivial involutions if and only if it is conjugate to a homeomorphism whose graph is symmetric about the line $y = -x$.* Graphs of an involution and a strongly-reversible homeomorphism are shown in Figure 1.8.

The group $\text{Homeo}(\mathbb{R})$ is explored in detail in Chapter 8. The larger group of homeomorphisms of the circle has a richer reversible structure, because it has

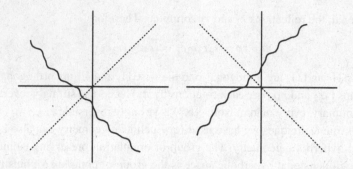

Figure 1.8 Graphs of an involution (left) and a strongly-reversible homeomorphism (right)

sense-preserving involutions (rotations) as well as sense-reversing involutions (reflections). This group is discussed in Chapter 9.

1.4.4 Real-analytic arcs

Our fourth example is about the problem of classifying pairs of real-analytic arcs that meet at a point, up to biholomorphic equivalence. (This is a special case of the general Poincaré problem of biholomorphic equivalence, which is discussed in the notes of Chapter 12.) A real-analytic arc is the image of an interval on the real axis under a biholomorphic map f of some neighbourhood of the interval, so it has an associated reflection map τ, obtained by conjugating the reflection across $\mathbb{R}$ by f:

$$\tau(z) = f\left(\overline{f^{-1}(z)}\right),$$

where the bar denotes complex conjugation. The map τ is an antiholomorphic involution. We met such involutions earlier (albeit informally) in the discussion of two-valued reflections.

Given two arcs γ_1 and γ_2 through a point p there are associated two antiholomorphic involutions τ_1 and τ_2 that fix p. See Figure 1.9.

The composition $\phi = \tau_1 \tau_2$ is a biholomorphic map in a neighbourhood of p that can be conjugated to its inverse using either of the maps τ_1 or τ_2. In Chapter 12 we describe how Écalle–Voronin theory allows us to understand reversibility in the group of biholomorphic germs in one variable. Ahern and Gong [2] used Écalle–Voronin theory to classify pairs of arcs with tangential intersections; in such cases ϕ is of the form $z + az^{k+1} + \cdots$, and its dynamics are described in terms of a Leau flower, such as that illustrated in Figure 1.10.

This is another example of hidden dynamics: the complex dynamics of the

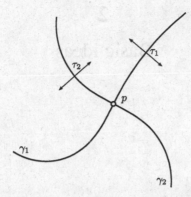

Figure 1.9 Two reflections in real-analytic arcs

Figure 1.10 A Leau flower

map ϕ play a key role in resolving a problem with no apparent dynamical connection.

2

Basic ideas

2.1 Reversibility

The natural context for the study of reversibility is the theory of groups. In this chapter we review the basics of reversibility, borrowing heavily from work of Aschbacher, Baake, Goodson, Lamb, and Roberts [12, 19, 110, 159, 164]. We assume throughout that G is a group with identity $\mathbb{1}$.

2.1.1 Reversible elements

An element g of a group G is said to be *reversible* in G if there is another element h of G such that

$$hgh^{-1} = g^{-1}.$$

In finite groups, the reversible elements are called *real elements* because of their relationship with real characters (see Chapter 3, page 49). We denote by $R(G)$, or just R if the context is clear, the set of reversible elements in G. The *reversibility problem* in G is the problem of characterising the reversible elements of G.

If the equation $hgh^{-1} = g^{-1}$ is satisfied, then we say that h *reverses* g and h is a *reverser* for g. We denote by $R_g(G)$, or just R_g, the set of reversers for g, which may of course be empty. We note that h reverses g if and only if $ghg = h$; that is, the reversers of g are the fixed points of the bijection $x \mapsto gxg$.

We denote the set of involutions in G (elements of order at most two) by $I(G)$, or just I. Thus

$$I = \{g \in G : g^2 = \mathbb{1}\}.$$

Example 2.1

(i) The identity $\mathbb{1}$ is reversible, and every element of G reverses $\mathbb{1}$.

22

(ii) An element of G is an involution if and only if it is reversed by $\mathbb{1}$.

(iii) If G is an abelian group, then $R = I$ and R is a subgroup of G. As a special case, if G is a cyclic group of even order, then it has two reversible elements: the identity and the element of order two.

(iv) More generally, denoting, as is usual, the *centre* of G by

$$Z(G) = \{g \in G : gh = hg, \text{ for all } h \in G\},$$

we have $R(G) \cap Z(G) = I(G) \cap Z(G)$.

An immediate and interesting consequence of the reversibility equation

$$hgh^{-1} = g^{-1}$$

is that the three similar equations

$$h^{-1}gh = g^{-1}, \quad hg^{-1}h^{-1} = g, \quad \text{and} \quad h^{-1}g^{-1}h = g$$

are all true. These are recorded along with other properties in the next pair of propositions.

Proposition 2.2 *Let $g, h \in G$, and suppose that h reverses g. Then h and h^{-1} both reverse g^n, for each integer n.* $\square$

Proposition 2.3 *Let $h \in G$. Then*

(i) *h reverses $\mathbb{1}$*

(ii) *if h reverses g, then h reverses g^{-1}*

(iii) *if h reverses f and g, then h also reverses fgf.* $\square$

It follows from Proposition 2.3 that the set of elements reversed by h forms a *twisted subgroup* of G (see [12, 90]). A twisted subgroup of G is a subset T of G that satisfies

(i) $\mathbb{1} \in T$

(ii) T is closed under taking inverses

(iii) if $x, y \in T$, then $xyx \in T$.

The theory of twisted subgroups has developed only recently, particularly in the context of finite groups, and although we do not explicitly study twisted subgroups in this text, there are close links with reversibility, not least because the set $R(G)$ of reversible elements in a group is a union of twisted subgroups. In fact, $R(G)$ has structure similar to that of a subgroup of G. By Proposition 2.3, it contains the identity and is closed under taking inverses. It is not necessarily closed under composition; however, if $hfh^{-1} = f^{-1}$ and

$hgh^{-1} = g^{-1}$ (that is, $f, g \in R(G)$ and both are reversed by the *same* element h of G) then

$$(hg)(fg)(hg)^{-1} = (hgh^{-1})(hfh^{-1})(hgh^{-1})(hg^{-1}h^{-1})$$
$$= g^{-1}f^{-1}g^{-1}g$$
$$= (fg)^{-1}.$$

Hence $fg \in R(G)$, and indee29+8d $fg \in R_{hg}$.

2.1.2 Conjugation

Two elements f and g of a group G are said to be *conjugate* in G if there is a third element h such that

$$f = hgh^{-1}.$$

The process of forming hgh^{-1} from g is called *conjugating* g by h. The group G can be partitioned into classes each made up of mutually conjugate elements. These equivalence classes are known as *conjugacy classes*. The task of characterising whether any pair of elements in a group are conjugate – or equivalently, the task of characterising the conjugacy classes of the group – is known as the *conjugacy problem*. Clearly this is a more general problem than the reversibility problem.

We often use the language of conjugacy to discuss reversibility. In particular, we can describe the reversible elements of a group as those elements that are conjugate to their own inverses. The property of being reversible is preserved under conjugation, in the sense that if g is a reversible element of G, then so is hgh^{-1} for any element h of G. It follows that each conjugacy class of G either consists entirely of reversible elements or otherwise contains no reversible elements. Those conjugacy classes comprised of reversible elements are called *reversible classes*.

We sometimes write

$$g^h = h^{-1}gh,$$

for elements g and h of G.

2.1.3 Automorphism groups

Suppose that a group G embeds in a larger group H. Within H the conjugacy classes of G may fuse, which implies that an element of G that is not reversible may become reversible in H. These conjugacy classes may also contain elements of $H \setminus G$. We can regard G as a group of permutations of a set X, and choose H to be the full symmetry group $S(X)$ of X. We shall see in Proposition 3.4 that each element of $S(X)$ is reversible by an involution, and hence every element of G becomes reversible when G is embedded in a sufficiently large group.

More interesting perhaps is the situation when G embeds as a normal subgroup of H. The conjugacy classes of G may fuse within H, but now, because of normality, these conjugacy classes will not be supplemented by elements of $H \setminus G$. To classify the possible groups H, we need to study the automorphism group $\mathrm{Aut}(G)$ of G. Given h in G, there is an automorphism ϕ_h of G given by $\phi_h(g) = g^h$. It is called the *inner automorphism* associated to h. The map $h \mapsto \phi_h$ from G to $\mathrm{Aut}(G)$ is a homomorphism with kernel equal to the centre of G, denoted $Z(G)$. Its image is a normal subgroup of $\mathrm{Aut}(G)$, denoted $\mathrm{Inn}(G)$. Automorphisms of G that are not inner automorphisms are known as *outer automorphisms*.

Suppose $Z(G)$ is trivial. Then G is isomorphic to $\mathrm{Inn}(G)$, and thus G embeds as a normal subgroup of $\mathrm{Aut}(G)$. Now suppose that G is a normal subgroup of some other group H. The maps ϕ_h, for h in H, are all automorphisms of G. We define

$$C_H(G) = \{h \in H : hg = gh \text{ for all } g \in G\};$$

that is, $C_H(G)$ consists of those elements of H for which ϕ_h is trivial. It is a normal subgroup of H, and G is a normal subgroup of the quotient group $H/C_H(G)$. This quotient group fuses conjugacy classes in G in the same way that H does, and it is a normal subgroup of $\mathrm{Aut}(G)$. In particular, if an element g of G is reversible in H, then it is also reversible in $\mathrm{Aut}(G)$.

In the preceding discussion we assumed that $Z(G)$ was trivial. When $Z(G)$ is not trivial, it is necessary to work with the quotient group $G/Z(G)$ instead. We should be careful now, because although, by Proposition 2.14, reversible elements of G are necessarily reversible in $G/Z(G)$, the converse does not hold.

We must also consider additional structure of the group G; for example, G may be a topological group or a Lie group. The automorphisms of most interest are those that preserve this additional structure. In future when discussing automorphisms, we assume that they are group automorphisms that preserve any topological or differential structure associated with the group.

Given the importance of automorphisms to reversibility, we make the following definition.

Definition 2.4   An element g of G is *reversed by an automorphism* if there exists an automorphism of G that maps g to g^{-1}.

Example 2.5   We saw in Example 2.1 that only the involutions in an abelian group G are reversible. On the other hand, the map $\theta(x) = x^{-1}$ is an automorphism of G, which implies that *every* element of G is reversed by an automorphism. Refer to [130] for more on automorphisms of abelian groups.

The next proposition is an immediate consequence of Proposition 2.14.

Proposition 2.6 *If ϕ is an automorphism of a group G, then ϕ maps $I(G)$ bijectively onto itself, and also maps $R(G)$ bijectively onto itself.*   □

Corollary 2.7 *Suppose that H is a normal subgroup of a group G. If g and h are elements of H that are conjugate in G, then*

(i) *$g \in I(H)$ if and only if $h \in I(H)$*
(ii) *$g \in R(H)$ if and only if $h \in R(H)$.*   □

Proposition 2.6 and Corollary 2.7 indicate that involutions and reversibles are useful in studying automorphism groups. This is indeed the case, and a strong example of such use is Dieudonné's treatise on the automorphisms of the classical groups [67], in which he determines automorphism groups by studying the action of automorphisms on involutions. In this work we do not consider automorphism groups in depth, but instead confine ourselves to select remarks about elements of groups that are reversed by automorphisms.

2.1.4 Centralisers and reversers

We denote the *centraliser* of an element g of a group G by

$$C_g(G) = \{h \in G : hg = gh\},$$

or just by C_g if the group G is clear from the context. We use this notation (rather than the usual notation $C_G(g)$ in which G and g are interchanged) because it ties in with our notation for reversible and reversing elements. The centraliser is a subgroup of G.

Proposition 2.8 *Let g be reversible in G, and suppose that $h \in R_g$. Then*

$$R_g = C_g h = hC_g.$$

That is, the set of reversers of g is both a right and a left coset of the centraliser of g.

Proof   If $f \in C_g$, then

$$(hf)g(hf)^{-1} = hfgf^{-1}h^{-1} = hgh^{-1} = g^{-1}.$$

Thus $hC_g \subset R_g$.

Conversely, if $k \in R_g$ then

$$(h^{-1}k)g(h^{-1}k)^{-1} = h^{-1}(kgk^{-1})h = h^{-1}g^{-1}h = g.$$

Therefore $h^{-1}k \in C_g$, which implies that $k \in hC_g$, so $R_g \subset hC_g$.
The identity $R_g = C_g h$ is proved similarly. $\square$

This proposition motivates the definition of the *extended centraliser*

$$E_g(G) = \{h \in G : g^h = g \text{ or } g^h = g^{-1}\} = C_g \cup R_g.$$

This is also a subgroup, in which C_g has index 1 or 2.

(In the physics literature, the centraliser is sometimes called the *symmetry group*, and the extended centraliser is called the *reversing symmetry group*.)

Proposition 2.9   *Let g be an element of a group G. Then exactly one of the following holds:*

(i)   *g is not reversible, and $E_g = C_g$ and $R_g = \emptyset$*
(ii) *g is an involution, and $E_g = C_g = R_g$*
(iii) *g is reversible but not an involution, and $[E_g : C_g] = 2$.* $\square$

We also have the following basic properties of centralisers and reversers, easily proven:

Proposition 2.10   *Given elements g and h of a group G we have*

(i) $C_{g^{-1}} = C_g$
(ii) $R_{g^{-1}} = R_g$
(iii) $C_{hgh^{-1}} = hC_g h^{-1}$
(iv) $R_{hgh^{-1}} = hR_g h^{-1}$ $\square$

2.1.5 Strongly-reversible elements

In the dynamical examples of Section 1.1, we can consider the associated group to be the set of bijections of the space X under the operation of composition. We saw that the equation $\tau\phi\tau = \phi^{-1}$ is satisfied in each example. Motivated by the fact that these maps τ are involutions, we make the following definition.

Definition 2.11   We say that an element g of a group G is *strongly reversible* if it is reversed by an involution τ from G. In other words, g is strongly reversible if there is an involution τ such that

$$\tau g \tau = g^{-1}.$$

The property of being strongly reversible is preserved under conjugation, as one can easily verify. There is a pleasing alternative method of characterising the strongly reversible elements of a group, suggested by the following proposition.

Proposition 2.12 *An element g of a group G is strongly reversible if and only if there are involutions a and b in G such that $g = ab$.*

Proof If g is strongly reversible then there is an involution a such that $aga = g^{-1}$. It follows that $g = ab$, where $b = g^{-1}a$. This element b is an involution because

$$b = g^{-1}a = ag = b^{-1}.$$

Conversely, if there are two involutions a and b such that $g = ab$, then

$$aga = a(ab)a = ba = g^{-1},$$

so g is strongly reversible.   □

Notice that if $g = ab$ for involutions a and b, then g is reversed by *both* a and b. (For instance, in the case of the harmonic oscillator of Section 1.1.1, one can check that the second involution $\tau\phi$ is an oblique reflection in the (q, p)-plane.) The elements a and b generate a (possibly infinite) dihedral group, and in fact any element of this group is reversed by both a and b. This is because any reduced word in a and b is either a palindrome, in which case it is an involution, or else it is a power of ab, in which case it is reversed by both a and b. We return to dihedral groups in Section 3.2.

Some authors refer to strongly-reversible elements as *2-reflectional* or *bireflectional*. The idea is that involutions can be thought of as abstract reflections. In finite groups, strongly-reversible elements are usually described as *strongly real* and occasionally described as *strictly real*. Strongly-real elements of finite groups will be studied in Chapter 3.

2.1.6 Products of involutions and reversibles

Given any group G, we define, for each positive integer n,

$$I^n(G) = \{g_1 g_2 \cdots g_n : g_i \in I(G), i = 1, 2, \ldots, n\},$$

$$I^\infty(G) = \bigcup_{n=1}^{\infty} I^n(G),$$

$$R^n(G) = \{g_1 g_2 \cdots g_n : g_i \in R(G), i = 1, 2, \ldots, n\},$$

$$R^\infty(G) = \bigcup_{n=1}^{\infty} R^n(G).$$

We often write I^n and R^n instead of $I^n(G)$ and $R^n(G)$, when the group G is clear from the context. With this notation, Proposition 2.12 says that I^2 is the set of strongly-reversible elements.

The set I^∞ is the subgroup of G generated by involutions, and likewise R^∞ is the subgroup of G generated by reversible elements.

Proposition 2.13

(i) *For each positive integer n, $I^n \subset I^{n+1}$ and $I^{2n} \subset R^n \subset R^{n+1}$.*
(ii) *The sets I^n and R^n are invariant under conjugation.*
(iii) *The groups I^∞ and R^∞ are both normal subgroups of G.*

Proof Proposition 2.12 tells us that $I^2 \subset R$, and statement (i) follows immediately.

For statement (ii), notice that the sets I^n and R^n are invariant under *all* automorphisms of G, including conjugation. Therefore the groups I^∞ and R^∞ are also invariant under all automorphisms of G; that is, they are *characteristic subgroups* of G. In particular, as they are invariant under conjugation they are normal subgroups of G, which is statement (iii). □

In many of the examples in this book we find that $I^n = I^\infty$ for some positive integer n; in fact, often $n = 4$ is sufficient.

Let us now look at two propositions that can be used to help determine the classes I^n and R^n.

Proposition 2.14 *If $\phi : G \to H$ is a group homomorphism, then*

$$\phi(I^n(G)) \subset I^n(H) \quad and \quad \phi(R^n(G)) \subset R^n(H). \qquad □$$

To see how this proposition can be useful, suppose, for example, that H is a quotient group of G, and ϕ is the quotient map. Suppose also that every element of G is strongly reversible. Then $I^2(G) = G$, so

$$I^2(H) \supset \phi(I^2(G)) = \phi(G) = H.$$

Therefore $I^2(H) = H$; that is, every element of H is strongly reversible too. The next elegant observation is about products of involutions.

Proposition 2.15 *Suppose that g is an element of the centre of a group G.*

(i) *If $g = ab$ for involutions a and b in G, then $g^2 = 1$. Thus $Z \cap I^2 \subset I$.*
(ii) *If $g = abc$ for involutions a, b, and c in G, then $g^4 = 1$.*

Proof   Statement (i) is true because

$$g^2 = gab = agb = a(ab)b = 1.$$

Statement (ii) is true because

$$g^4 = agbgcg = (ag)b(gc)g = (bc)b(ab)g = bg^{-1}bg = 1.$$

□

Suppose that g is an element of the centre of a group G, and there are involutions a, b, c, and d in G such that $g = abcd$. Despite the pattern that seems to emerge from Proposition 2.15, it is *not* necessarily true that $g^8 = 1$. For example, consider the matrix

$$g = \begin{pmatrix} \omega & & \\ & \omega & \\ & & \omega \end{pmatrix},$$

where ω is a complex third root of unity, which belongs to the centre of the general linear group $GL(3, \mathbb{C})$. It can be expressed as a composite of four involutions as follows:

$$g = \begin{pmatrix} \overline{\omega} & \omega & \\ & 1 & \\ & & 1 \end{pmatrix} \begin{pmatrix} & 1 & \\ 1 & & \\ & & 1 \end{pmatrix} \begin{pmatrix} 1 & & \\ & \overline{\omega} & \\ & & \omega \end{pmatrix} \begin{pmatrix} 1 & & \\ & & 1 \\ & 1 & \end{pmatrix},$$

and yet $g^8 \neq 1$.

2.1.7 Products and limits

The following proposition, easily proved, records what happens to reversibility in direct products.

Proposition 2.16 *Let G and H be groups. Then*

(i) $R(G \times H) = R(G) \times R(H)$,
(ii) $R_{(g,h)}(G \times H) = R_g(G) \times R_h(H)$.

□

Proposition 2.16 fails when the direct products are replaced by semidirect products. Consider, for example, the dihedral group D_n, which is a semidirect product of C_n and C_2. We noted in Section 2.1.5 that each element of D_n is reversible, and so are both elements of C_2. On the other hand, none of the noninvolutive members of C_n are reversible. See Section 3.2 for more on the dihedral group.

For another example, consider the group G of two-by-two invertible complex matrices with determinant 1 or -1. The group $\mathrm{SL}(2,\mathbb{C})$ is a normal subgroup of G of index 2, and G is a semidirect product of $\mathrm{SL}(2,\mathbb{C})$ and the copy of C_2 generated by the involution

$$A = \begin{pmatrix} -1 & 0 \\ 0 & 1 \end{pmatrix}.$$

The matrix

$$B = \begin{pmatrix} -i & 0 \\ 0 & i \end{pmatrix}.$$

is reversible in $\mathrm{SL}(2,\mathbb{C})$ because $CBC^{-1} = B^{-1}$, where

$$C = \begin{pmatrix} 0 & i \\ i & 0 \end{pmatrix}.$$

Likewise A is reversible in C_2; however,

$$AB = \begin{pmatrix} i & 0 \\ 0 & i \end{pmatrix},$$

and this matrix lies in the centre of G, so, because it is not an involution, it is not reversible. We return to reversibility in matrix groups in Chapter 4.

We finish this section with some brief remarks on group limits. Given groups G_i and injective homomorphisms $G_i \to G_{i+1}$ for $i = 1, 2, \ldots$, a reversible element of one of the groups G_i remains reversible when embedded in the direct limit $\varinjlim G_i$. Next, given groups H_i and surjective homomorphisms $H_{i+1} \to H_i$ for $i = 1, 2, \ldots$, a reversible element of the inverse limit $\varprojlim H_i$ projects to reversible elements in each of the groups H_i.

2.1.8 Group actions

An *action* of a group G on a set X is a function

$$\theta : G \times X \to X$$

such that $\theta(gh, x) = \theta(g, \theta(h, x))$ and $\theta(\mathbb{1}, x) = x$ for all elements g and h of G and points x in X. We can think of the group G as a subgroup of the group of permutations of the set X; with this in mind, we write $g(x)$ instead of $\theta(g,x)$. The condition $\theta(gh,x) = \theta(g,\theta(h,x))$ then becomes $gh(x) = g(h(x))$. Many of the groups we consider are defined in terms of their actions on particular sets; for example, we look at the group of permutations of $\{1,2,\ldots,n\}$, the group of automorphisms of a finite-dimensional vector space, the group of isometries of the hyperbolic plane, the group of diffeomorphisms of the real line, and the group of homeomorphisms of the circle.

We make good use of the next lemma and corollary.

Lemma 2.17 *Suppose that a group G acts on a set X. Given elements g and h of G, if either $hgh^{-1} = g$ or $hgh^{-1} = g^{-1}$, then h permutes the fixed points of g.*

Proof Let x be a fixed point of g. If $hgh^{-1} = g$, then

$$gh(x) = hg(x) = h(x),$$

so $h(x)$ is a fixed point of g. Similarly we see that $h^{-1}(x)$ is a fixed point of g. Therefore h permutes the fixed points of g. The case when $hgh^{-1} = g^{-1}$ is proved in much the same way. ☐

Corollary 2.18 *Suppose that a group G acts on a set X. Let g and h be elements of G such that g has a unique fixed point. If either $hgh^{-1} = g$ or $hgh^{-1} = g^{-1}$, then h also fixes the fixed point of g.* ☐

A group action is said to be *faithful* if for any two distinct elements f and g of G there is a point x in X such that $f(x) \neq g(x)$. We often have recourse to the following proposition.

Proposition 2.19 *Suppose that a group G acts faithfully on a set X, and for each point x in X there is an element of G that fixes x but fixes no other points in X. Then G has trivial centre.*

Proof Choose an element h of the centre of G. Let x be a point in X, and let g be an element of G that fixes x but fixes no other points in X. Since g and h commute, Corollary 2.18 tells us that $h(x) = x$. It follows that h fixes every point in X, and since the action is faithful, we deduce that h is the identity element. ☐

2.2 Reformulations

In this section, based on work of Lamb [159], we present some equivalent formulations of the properties of being reversible and being a reverser.

Lemma 2.20 *Let f, g, and h be elements of a group G. Then $hgh^{-1} = f$ if and only if there exists an element k of G such that $f = hk$ and $g = kh$.*

Proof Suppose first that $hgh^{-1} = f$. Let $k = h^{-1}f$. Then $f = hk$ and

$$g = h^{-1}fh = kh.$$

Conversely, suppose there is an element k with $f = hk$ and $g = kh$. Then

$$hgh^{-1} = h(kh)h^{-1} = hk = f.$$

$\square$

Corollary 2.21 *Let f and g be elements of a group G. Then f is conjugate to g if and only if there exist elements h and k of G such that $f = hk$ and $g = kh$.* $\square$

The next result is a consequence of Lemma 2.20.

Proposition 2.22 *Let g and h be elements of a group G. The following are equivalent:*

(i) $hgh^{-1} = g^{-1}$
(ii) *there is an element k of G such that $g = kh$ and $g^{-1} = hk$*
(iii) *there is an element f of G such that $h^2 = f^2$ and $g = f^{-1}h$.*

Proof That (i) implies (ii) follows from Lemma 2.20.
To see that (ii) implies (iii), define $f = k^{-1}$, then $g = kh = f^{-1}h$ and

$$h^2 = (k^{-1}g)(g^{-1}k^{-1}) = k^{-2} = f^2.$$

Finally, assuming (iii) we see that

$$hgh^{-1} = h(gh^{-1}) = hf^{-1} = h^{-1}(h^2f^{-2})f = h^{-1}f = g^{-1},$$

which is statement (i). $\square$

We remark that in (iii), we may replace $f^{-1}h$ by any of fh^{-1}, $h^{-1}f$, or hf^{-1}, without affecting the validity of the result.

Corollary 2.23 *Let g be an element of a group G. The following are equivalent:*

(i) *g is reversible*
(ii) *there exist elements h and k of G such that $g = kh$ and $g^{-1} = hk$*
(iii) *there exist elements h and f of G such that $h^2 = f^2$ and $g = f^{-1}h$.* $\square$

Observe that in (ii) and (iii), the elements h, k, and f reverse g, if they exist. Also, h may be taken to be any reverser of g.

An element h of a group G is said to be a *square root* of another element g if $g = h^2$. Corollary 2.23 shows that reversibility relates to the existence of multiple square roots of elements.

2.3 Signed groups

Some groups have a 'sign', or 'orientation' homomorphism, mapping them onto the cyclic group $C_2 = (\{\pm 1\}, \times)$. The next result, relating reversibles and involutions, is sometimes useful in such groups.

Lemma 2.24 *Let* $\theta : G \to C_2$ *be a homomorphism for which* $R(\ker(\theta)) = \{\mathbb{1}\}$. *Then*

$$R(G) \setminus \ker(\theta) \subset I(G).$$

Proof Let $g \in R(G) \setminus \ker(\theta)$. Then there is an element h of G that reverses g. We may assume that $h \in \ker(\theta)$, for if $\theta(h) \neq 1$, then we may replace h by hg, which also reverses g and satisfies $\theta(hg) = 1$.

Then h also reverses g^2, and both h and g^2 belong to $\ker(\theta)$, so $g^2 = \mathbb{1}$, as required. $\square$

2.4 Whither next?

We shall go on to consider reversibility in more specific contexts, beginning with finite groups, and then moving on to various classes of infinite groups. In general, we shall focus on the reversibility problem; that is, the problem of characterising the reversible elements of a group G. We also consider the following subsidiary questions.

(i) The conjugacy problem: provide a procedure to determine whether or not two given elements f and g of G are conjugate. If such a procedure is available, and easy to use, then it automatically provides a solution to the reversibility problem. However, it may well happen that the reversibility problem is substantially easier than the conjugacy problem. This happens, for instance, with the diffeomorphism group of the real line, discussed in Chapter 11. In that case, we take a short cut, and just deal with the reversible elements. It may also happen that a solution to the conjugacy problem is not so explicit as to be readily used for the reversibility problem. An example is the group of biholomorphic germs, discussed in Chapter 12.

(ii) What is $I^2(G)$? This question, the identification of the strongly-reversible elements, has received more attention in the literature than any other aspect of reversibility, on account of its wide application.

(iii) What are $I^n(G)$ and $R^n(G)$, for each positive integer n, and when, if ever, is $I^n = G$ or $R^n = G$? It is not essential to resolve these questions in order to understand reversibility, but we often find ourselves in shooting distance of solutions to them, once we have understood the reversibility problem, and in that case we usually shoot. It is fascinating how, even in some very large groups, the product of a small number of involutions or reversibles suffices to express the general element. For instance, four involutions suffice in most of our big groups.

(iv) Describe the centraliser C_g of each element g of G.

(v) Describe the set of reversers R_g of each element g of G. In view of Proposition 2.8, providing we know C_g, it suffices to find a single element of R_g, if there is one.

(vi) Classify those elements of G that become reversible once G is embedded in a larger group H. We have seen that this question is most interesting when G embeds as a normal subgroup of H, and in that case, providing G has trivial centre, we should consider $H = \text{Aut}(G)$. Thus we seek the elements of G reversed by an automorphism.

In view of this, we shall treat collections of subgroups of a single large group together in a single chapter. For instance, we consider various classical matrix groups together in Chapter 4. For the same reason, we treat reversibility in large groups before their subgroups, and sometimes the story is sufficiently complex that these require their own chapters. For instance, before dealing with the diffeomorphism group of the line, we have separate chapters about the homeomorphism group (Chapter 8) and about the formal power series groups (Chapter 10). The latter chapter does double duty, as we also apply its results when we study the group of biholomorphic germs.

(vii) What is $R(G/H)$, for a given normal subgroup H of G? Again, this is an easier problem than the reversibility problem, and again, it makes sense to consider a group along with its quotients. For instance, we consider $\text{GL}(2, \mathbb{Z})$ along with its subgroup $\text{SL}(2, \mathbb{Z})$ and the projective quotients $\text{PGL}(2, \mathbb{Z})$ and $\text{PSL}(2, \mathbb{Z})$ in Chapter 7.

Notes

Sources

The main purpose of this chapter has been to fix terminology and notation. The background group theory can be found in a standard text such as [206]. Many of the basic results on reversibility are laid out in work of Aschbacher, Baake, Goodson, Lamb, and Roberts [12, 19, 110, 159, 164], and the reader should consult those references for further similar material. An accessible introduction to twisted subgroups can be found in [90].

Automorphism towers

Let G be a group with a trivial centre. Recall from Section 2.1.3 that, after identifying G with the inner automorphisms of G, we have $G \lhd \text{Aut}(G)$. Notice that $\text{Aut}(G)$ also has a trivial centre. To see this, suppose that $\phi \in \text{Aut}(G)$, and ϕ commutes with all other automorphisms of G. In particular, ϕ commutes with all inner automorphisms, so given h in G we have

$$h\phi(g)h^{-1} = \phi(hgh^{-1}) = \phi(h)\phi(g)\phi(h)^{-1}$$

for all g in G. Since ϕ is an automorphism, and G has trivial centre, we see that $\phi(h) = h$. As the map h was chosen arbitrarily, we conclude that ϕ is the identity.

We can now define a chain of automorphism groups $G_0 \lhd G_1 \lhd G_2 \lhd \cdots$, where $G_i = \text{Aut}(G_{i-1})$ for $i = 1, 2, \ldots$. This chain can be continued transfinitely by defining, for a limit ordinal λ, $G_\lambda = \bigcup_{\alpha < \lambda} G_\alpha$. The resulting chain is called the *automorphism tower* of G. Automorphism towers can also be defined for groups that have nontrivial centres; see [226]. Thomas [227] proved that the tower eventually terminates, in the sense that there is an ordinal α with $G_\alpha = G_\beta$ whenever $\beta > \alpha$, and later Hamkins [128] proved the same result without the assumption that G has a trivial centre. It is a classic result of Wielandt that the automorphism tower of a *finite* group with trivial centre terminates after finitely many steps [206, 13.5.4]. On the other hand, the automorphism towers of both the finite dihedral group D_4 with eight elements (which has a nontrivial centre) and the infinite dihedral group D_∞ (which has trivial centre) terminate after $\omega + 1$ steps [226, Theorems 1.4.6 and 5.2.1].

Squares and commutators

The *commutator* $[a,b]$ of two elements a and b of a group G is given by $[a,b] = aba^{-1}b^{-1}$. From the identity

$$[a,b] = (aba^{-1})^2 a^2 (a^{-1}b^{-1})^2$$

we see that each commutator can be expressed as a product of three squares. On the other hand, if g is reversible in G, that is, $hgh^{-1} = g^{-1}$ for some element h, then $g^2 = h^{-1}g^{-1}hg$, and hence g^2 can be expressed as a commutator.

The *commutator subgroup* $[G,G]$ of G is the (normal) subgroup generated by commutators. We see from above that if every element of G is reversible then $[G,G]$ is equal to the subgroup of G generated by squares.

Conjugacy of an element to its square

Goodson [113, 116, 114] has studied elements of a group G that are conjugate to their own squares. That is, he studied elements g of G such that $hgh^{-1} = g^2$ for some element h of G. It follows from $hgh^{-1} = g^2$ that $g = [h,g]$. Thus if every element of G is conjugate to its square then G is *perfect*; that is, $[G,G] = G$.

Another consequence of the equation $hgh^{-1} = g^2$ is that $h^n g h^{-n} = g^{2^n}$ for each positive integer n. Hence $g = (h^{-n}gh^n)^{2^n}$. Thus g is conjugate to its 2^nth power, and it has a 2^nth root to which it is also conjugate.

Reversibility of powers of elements

It may be that an element g of a group G is not reversible, but some power g^k is reversible. In the dynamical systems literature, an element g for which g^k is reversible is said to possess a *k-reversing symmetry*. Properties of maps with k-reversing symmetries are studied in, for example, [160, 162, 163].

Open problems

This book just scratches the surface of reversibility. Answers to the questions posed in Section 2.4 are known in some groups that we do not touch, but for the most part the questions are open.

Each subsequent chapter finishes with a section on open problems and suggestions for future research. Most of these open problems are special cases of the main problem below.

The main problem

We encourage the reader to take his or her favourite group, and determine I, R, I^n, R^n, $\inf\{n : I^n = I^\infty\}$, $\inf\{n : R^n = R^\infty\}$, and R_g, for each element g. Let us know how you get on!

3
Finite groups

In this chapter, we denote the cardinality of a set F by $|F|$. Thus the order of a group G is denoted $|G|$. We recall that $R(G)$, or just R, denotes the set of reversible elements of G, and $R_g(G)$, or just R_g, denotes the set of elements h that satisfy $hgh^{-1} = g^{-1}$. Also $I(G)$, or I, denotes the set of involutions in G, and $I^n = \{\tau_1 \cdots \tau_n : \tau_i \in I\}$.

3.1 Reversers of finite order

Proposition 3.1 *Suppose that elements g and h of a group G satisfy $hgh^{-1} = g^{-1}$. If h has finite order, then either g is an involution or else the order of h is even. In the latter case, g is also be reversed by an element whose order is a power of 2.*

Proof To prove the first assertion, suppose that the order of h is odd. Since each odd power of h reverses g, it follows that the identity $\mathbb{1}$ reverses g. Therefore $g = g^{-1}$, so g is an involution.

Suppose now that h has even order $2^k q$ for positive integers k and q, where q is odd. Then h^q also reverses g, and this element has order 2^k. This proves the second assertion. □

Corollary 3.2 *A finite group G contains a nontrivial reversible element if and only if G has even order. In fact, if G contains a nontrivial reversible element, then it contains a nontrivial involution.*

Proof Suppose first that G has a nontrivial element g that is reversed by another element h. By Proposition 3.1, one of g or h has even order, so G also has even order.

Suppose now that G has even order. By partitioning G into subsets $\{g, g^{-1}\}$

we see that G contains a nontrivial involution. Since involutions are reversible, all parts of the corollary have been accounted for. $\square$

3.2 Dihedral groups

There is one dihedral group, D_n, of order $2n$, for each positive integer n. It may be described as the symmetry group of the regular n-gon, or as the group

$$\langle a,b : a^2 = b^2 = (ab)^n = \mathbb{1} \rangle.$$

There is also one infinite dihedral group D_∞, which may be described as

$$\langle a,b : a^2 = b^2 = \mathbb{1} \rangle,$$

or as the isometry group of $\mathbb{Z}$ with the usual metric d given by $d(n,m) = |n-m|$, or as a semidirect product $(\mathbb{Z},+) \ltimes C_2$, where C_2 is the group of order 2.

Proposition 3.3 *In any dihedral group (finite or infinite), each element is strongly reversible.*

Proof By Proposition 2.14, it suffices to prove the assertion for D_∞, since all finite dihedral groups are quotients of this group. Let us choose the presentation $\langle a,b : a^2 = b^2 = \mathbb{1} \rangle$ for D_∞. A word using this presentation consists of the letters a and b, alternating. Those words that begin and end in the same letter are involutions, and those words that begin and end with different letters are reversed by the involution a. It follows that every element of D_∞ is strongly reversible. $\square$

3.3 Symmetric groups

The *symmetric group* $S(X)$ of a set X is the group of all bijections of X onto itself, under composition. Elements of $S(X)$ are sometimes called *permutations* of X. We will prove shortly that every element of $S(X)$ is strongly reversible. Before that, let us introduce a couple of familiar concepts about group actions.

Let G be a group that acts on a set X. An *orbit* of G is a subset of X of the form

$$\{g(x) : g \in G\}$$

for some element x of X. The orbits of G partition X. The action of the group G is said to be *transitive* on X if there is only one orbit. Equivalently, the action of G is transitive if for any pair of points x and y in X there is an element g of G such that $g(x) = y$.

We denote by $\langle g \rangle$ the cyclic group generated by a permutation g of X. The group $\langle g \rangle$ also acts on X. Any orbit has the form $\{g^n(x) : n \in \mathbb{Z}\}$ for some point x in X. If $\langle g \rangle$ acts transitively on X, then for any point x in X, every other point of X can be expressed as $g^n(x)$ for some integer n.

Proposition 3.4 *Each element of the symmetric group of a set X is strongly reversible.*

Proof Let $g \in S(X)$. We wish to find involutions h and k in $S(X)$ with $g = hk$.

The orbits of $\langle g \rangle$ partition X, so it suffices to find permutations h and k of a single orbit such that $g = hk$ when restricted to that orbit. In other words, it suffices to consider the case in which $\langle g \rangle$ acts transitively on X. But then, up to isomorphism, we may assume that g is a translation by 1 of $\mathbb{Z}$ (when the order of g is infinite), or g is a rotation by $2\pi/n$ about the centre of the vertices of a regular n-gon (when the order of g is n). In either case, g is an element of a dihedral group, and hence is the product of two involutions, by Proposition 3.3. $\qquad\square$

When $X = \{1, \ldots, n\}$, we denote, as is usual, $S(X)$ by S_n. The conjugacy classes of S_n are in bijective correspondence with the partitions of the integer n. Each element g of S_n, when written as a product

$$g = c_1 \cdots c_m$$

of pairwise-disjoint cycles (including the one-element cycles), determines a partition $p(g)$, given by

$$n = o(c_1) + \cdots + o(c_m),$$

(where $o(c)$ denotes the order of the element c) and the map $g \mapsto p(g)$ induces a well-defined bijection from conjugacy classes onto partitions. This provides another way to see the above proposition when $X = \{1, \ldots, n\}$, because $p(g) = p(g^{-1})$.

Each group G may be regarded as a group of bijections, by representing it (for instance) as acting on itself by right-multiplication. The proposition shows that as "unstructured mappings", all bijections are reversible. This does not mean that all elements of each group G are reversible *in* G. It means that they are reversible in $S(G)$. Therefore the interesting aspects of reversibility are not a matter of set theory: the interesting questions concern groups of maps that have additional structure of some kind.

Take, for instance, the homeomorphism group $G = \mathrm{Homeo}(X)$, where X is a topological space, and the group consists of all the homeomorphisms of X onto itself. The above proposition shows that, given a homeomorphism g, on each orbit of $\langle g \rangle$ the map g may be factored as a product of two involutions. But even if these could be taken to be continuous on each orbit, there would typically be many choices of the factorization on each orbit, and g would only be reversible in G if these choices could be made in such a way that the factors on the orbits patched together continuously. We shall see in Chapter 8 that even for $X = \mathbb{R}$ it is impossible to factor each homeomorphism as the product of two continuous involutions.

We finish here with the observation that some permutations can *only* be reversed by involutions.

Proposition 3.5 *Suppose that g is a permutation of X, and $\langle g \rangle$ acts transitively on X. Then any permutation of X that reverses g is necessarily an involution.*

Proof Choose any element x_0 of X, and define $x_n = g^n(x_0)$ for each integer n. Since $\langle g \rangle$ has only a single orbit in X, we know that $X = \{x_n : n \in \mathbb{Z}\}$.

Suppose that h reverses g; that is, $hgh^{-1} = g^{-1}$. Define the integer m to be such that $h(x_0) = x_m$. Then, for each integer n,

$$h(x_n) = hg^n(x_0) = g^{-n}h(x_0) = g^{-n}g^m(x_0) = x_{m-n}.$$

Therefore

$$h^2(x_n) = x_{m-(m-n)} = x_n,$$

so $h^2 = \mathbb{1}$. □

In case X is finite with, say, n elements, g may be represented in n different ways as a cycle of the form $(a_1, \ldots, a_n)$, and to each such representation one obtains a corresponding reverser h of g that interchanges a_j and a_{n-j+1} for $j = 1, \ldots, n$.

3.4 Alternating groups

In this section, X is a finite set. A *transposition* is a member of $S(X)$ that interchanges two elements of X and fixes all other elements of X. Each permutation g in $S(X)$ can be expressed as a product of transpositions. It is easy to show that no matter how you represent g as a product of transpositions, the parity of the number of transpositions used is the same. We say that g is *even* if the parity is even, and otherwise g is said to be *odd*. We define the *sign* or *signature* $\mathrm{sgn}(g)$ of g to be

$$\mathrm{sgn}(g) = \begin{cases} 1 & \text{if } g \text{ is even,} \\ -1 & \text{if } g \text{ is odd.} \end{cases}$$

The *alternating group* $A(X)$ consists of the even permutations in $S(X)$. We also use the standard notation

$$A_n = A(\{1,\ldots,n\}).$$

(When X has n elements, then the groups $A(X)$ and A_n are isomorphic, but we shall find it convenient to consider permutations on various sets X.) The group $A(X)$ is a subgroup of $S(X)$ of index 2.

We let

$$\psi(r) = (-1)^{\left(\frac{r-1}{2}\right)},$$

whenever r is a positive odd integer. That is,

$$\psi(r) = \begin{cases} 1 & \text{if } r \equiv 1 \pmod 4, \\ -1 & \text{if } r \equiv 3 \pmod 4. \end{cases}$$

Lemma 3.6 *Suppose that g is a permutation of X of odd order r, and $\langle g \rangle$ acts transitively on X. If h is a permutation that reverses g, then $\mathrm{sgn}(h) = \psi(r)$.*

Proof Proposition 3.5 asserts that h is an involution, and since X has odd order, h must have a fixed point x_0. In proving Proposition 3.5, we showed that h is given by $h(x_n) = x_{-n}$, where $x_n = g^n(x_0)$. Therefore h can be expressed as a product of $(r-1)/2$ transpositions, so $\mathrm{sgn}(h) = \psi(r)$. □

Lemma 3.7 *Suppose that g is a permutation of X of even order, and $\langle g \rangle$ acts transitively on X. Then there are both odd and even permutations that reverse g.*

Proof To prove this, it helps to think of X as the vertices of a regular r-gon centred on the origin, and g as a rotation about the origin by $2\pi/r$. This rotation is reversed by both a reflection in a line that passes through two opposite vertices of the r-gon and it is also reversed by a reflection in a line that passes through two midpoints of opposite sides of the r-gon. One of these two reversing permutations is odd and the other is even. □

Recall that $\mathrm{o}(g)$ denotes the order of the element g. If g can be written in disjoint-cycle notation as just a single cycle, then $\mathrm{o}(g)$ is the length of that cycle.

Proposition 3.8 *Let X be a finite set, and let $g \in A(X)$. Let $g = g_1 g_2 \cdots g_k$ represent g as a product of pairwise-disjoint cycles, including all fixed points as singleton cycles. For each positive integer r, let m_r denote the number of cycles $g_1, g_2, \ldots, g_k$ of length r. Then the following three conditions are equivalent:*

(i) *g is strongly reversible in $A(X)$*

(ii) *g is reversible in $A(X)$*

(iii) (a) *there exists an even integer r with $m_r > 0$,* **or**

     (b) *there exists an odd integer r with $m_r > 1$,* **or**

     (c) $\displaystyle\prod_{o(g_i)\text{ odd}} \psi(o(g_i)) = 1.$

Thus g is not reversible if and only if the g_i have distinct odd orders, an odd number of which are congruent to 3 modulo 4.

Here are three corollaries of Proposition 3.8, each of which is easily verified by a systematic check of cases.

Corollary 3.9 *If an element g of $A(X)$ fixes at least two points, then it is strongly reversible.* □

Corollary 3.10 *Every element of A_n is reversible if and only if*

$$n = 1, 2, 5, 6, 10, 14.$$ □

Corollary 3.11 *Every reversible element of A_n is an involution if and only if $n = 1, 2, 3, 4$.* □

Proof of Proposition 3.8.   That (i) implies (ii) is true for all groups, not just $A(X)$.

Next we show that (ii) implies (iii). Suppose that g is reversible, and yet conditions (iii)(a) and (iii)(b) do not hold. Thus the sets X_i underlying the cycles g_i are of pairwise-distinct odd cardinalities. Choose a permutation h in $A(X)$ that reverses g. This element permutes the sets X_i, but as they are of distinct orders we see that h fixes each set X_i. It follows that, when restricted to X_i, h is a permutation of X_i that reverses g_i, so it is an involution, by Proposition 3.5, and has sign $\psi(o(g_i))$, by Lemma 3.6. Therefore

$$\prod_{o(g_i)\text{ odd}} \psi(o(g_i)) = \operatorname{sgn}(h) = 1,$$

which is condition (iii)(c).

It remains to prove that (iii) implies (i). As before, let X_i be the underlying set of the cycle g_i. For $i = 1, \ldots, r$, we may choose, by Proposition 3.4, an involution τ_i in $S(X_i)$ that reverses g_i in $S(X_i)$. If the order of g_i is even, then

Lemma 3.7 tells us that this involution may be chosen of either sign, but if the order of g_i is odd, then Lemma 3.6 tells us that the sign of the involution must be $\psi(o(g_i))$.

Let us extend the action of each involution τ_i from X_i to X by declaring that τ_i fixes each point outside X_i. Let $h = \tau_1 \cdots \tau_k$. Then h is an involution in $S(X)$ that reverses g. We will show that we can modify h (if necessary) so that it belongs to $A(X)$; thus proving (i).

Suppose first that one of the maps g_i has even order (that is, (iii)(a) holds). Then we have freedom to choose an involution τ_i with sign 1 or -1. Therefore we can also choose h with sign 1.

Suppose next that two of the maps g_i have the same odd order, say r (that is, (iii)(b) holds). We can assume that these two maps are g_1 and g_2. In disjoint-cycle notation, let

$$g_1 = (a_1, \ldots, a_r) \quad \text{and} \quad g_2 = (b_1, \ldots, b_r).$$

We define an involution τ in $S(X_1 \cup X_2)$ by

$$\tau(a_i) = b_{r-i+1}$$

for $i = 1, \ldots, r$. If we extend the action of τ to X in the usual way, then we see that the involution $h' = \tau\tau_3 \ldots \tau_k$ also reverse g. Furthermore, since

$$\operatorname{sgn}(\tau_1)\operatorname{sgn}(\tau_2) = \psi(r)^2 = 1,$$

whereas $\operatorname{sgn}(\tau) = -1$, we see that one of h or h' belongs to $A(X)$.

Suppose finally that none of the maps g_i have even order, and no two of them have the same order, but

$$\prod_{o(g_i)\ \text{odd}} \psi(o(g_i)) = 1,$$

(as in (iii)(c)). Since, by Lemma 3.6, this product represents the sign of h, we see immediately that $h \in A(X)$. $\qquad \square$

Proposition 3.8 tells us that, in general, there are elements of A_n that are not reversible. Every element of A_n is reversed by an automorphism, however, because A_n is a normal subgroup of S_n, and Proposition 3.4 implies that all elements of S_n are reversible.

3.5 Group characters

We review the elements of the theory of complex representations of finite groups. For unproved results stated in this section, refer to a text such as [55, 86, 138, 140].

Let G be a finite group, and let $\mathbb{C}G$ be the associated group algebra over the complex numbers. Let $\mathrm{GL}(V)$ be the general linear group of V; that is, the group of invertible linear maps from V to itself. Let $\mathrm{End}(V)$ be the set of endomorphisms of V; that is, the set of linear maps from V to itself.

The group representations (group homomorphisms) $\phi : G \to \mathrm{GL}(V)$ (where V is some finite-dimensional complex vector space) are in one-to-one correspondence with the algebra representations (unital algebra homomorphisms) $\tilde{\phi} : \mathbb{C}G \to \mathrm{End}(V)$. The correspondence is given by the obvious formula

$$\tilde{\phi}\left(\sum_{g \in G} \alpha_g g\right)(v) = \sum_{g \in G} \alpha_g \phi(g)(v),$$

where α_g are complex coefficients and $v \in V$. One normally drops the tilde, and uses the same symbol ϕ for $\tilde{\phi}$.

The group representation is irreducible if and only if the corresponding algebra representation is irreducible.

To each representation ϕ we associate a function χ called a *character* given by $\chi(g) = \mathrm{trace}(\phi(g))$. The value $\chi(g)$ depends only on the conjugacy class of g, and if C is a conjugacy class, we denote the common value by $\chi(C)$.

Each element g of G has finite order, so the eigenvalues $\lambda_1, \ldots, \lambda_n$ of $\phi(g)$ are roots of unity. The eigenvalues of $\phi(g^{-1})$ are $\lambda_1^{-1}, \ldots, \lambda_n^{-1}$, hence

$$\chi(g^{-1}) = \lambda_1^{-1} + \cdots + \lambda_n^{-1} = \overline{\lambda_1 + \cdots + \lambda_n} = \overline{\chi(g)}.$$

For each conjugacy class C in G, let

$$C^+ = \sum_{g \in C} g,$$

be the corresponding *class sum*.

Regarding G as a basis for $\mathbb{C}G$, we extend each character χ to a linear map (also denoted χ) from $\mathbb{C}G$ to $\mathbb{C}$, and we note that

$$\chi(C^+) = |C|\chi(C),$$

where $|C|$ denotes the number of elements in the class C.

Proposition 3.12 *The centre of $\mathbb{C}G$ is the linear span of the class sums.*

Proof   Each of the class sums lies in the centre of $\mathbb{C}G$, and hence so does the linear span of the class sums. Conversely, suppose that $\sum_{g \in G} \alpha_g g$ lies in the centre of $\mathbb{C}G$. Conjugating by elements of G we see that $\alpha_g = \alpha_{hgh^{-1}}$ for all g and h in G. It follows that $\sum_{g \in G} \alpha_g g$ is a linear combination of class sums. $\square$

Let $M(\mathbb{C}, n)$ denote the set of $n \times n$ matrices with complex coefficients. Wedderburn established the following fundamental result.

Theorem 3.13   *The group algebra $\mathbb{C}G$ is a direct sum of full matrix algebras* $\text{End}(V_i) \cong M(\mathbb{C}, d_i)$. $\square$

The terms of this direct sum are in one-to-one correspondence with the (distinct, nonisomorphic) irreducible representations. In fact, the representations are just the projections onto the factors of the direct sum, and are obtained by multiplication by corresponding idempotents in $\mathbb{C}G$. Let us fix representations $\phi_1, \ldots, \phi_k$, one for each equivalence class of irreducible representations. Let $\chi_1, \ldots, \chi_k$ be the corresponding characters, and $d_i = \chi_i(\mathbb{1})$ the corresponding degrees.

Now, the centre of $M(\mathbb{C}, d)$ is the set of scalar multiples of the identity, so the dimension of the centre of $\mathbb{C}G$ is the same as the number of irreducible representations, and we have the following corollary.

Corollary 3.14   *The number of conjugacy classes of G is the same as the number of irreducible representations.* $\square$

Let the conjugacy classes of G be $C_1, \ldots, C_k$.

The *character table* of G is the square matrix with rows indexed by the characters χ_i of irreducible representations, columns indexed by the conjugacy classes C_j, and entries $\chi_i(C_j)$.

Theorem 3.15   *The character table is an invertible matrix.* $\square$

The proof of this depends on the Schur orthogonality relations for characters, and an analysis of the decomposition of the regular representation of G (the representation induced by the permutation action of G on itself). We do not propose to discuss these details here, but we list the main identities and other properties exhibited by the characters and degrees of the irreducible representations.

In the next theorem we use the commutator subgroup $[G, G]$ of a group G, which you may recall from Chapter 2 is the subgroup of G generated by elements $aba^{-1}b^{-1}$, where $a, b \in G$. Also, we use the Dirac delta

$$\delta_{ij} = \begin{cases} 1 & \text{if } i = j, \\ 0 & \text{if } i \neq j, \end{cases}$$

where $i, j \in \{1, \ldots, k\}$.

Theorem 3.16 (Orthogonality relations)   *With the above notation,*

(i) $\displaystyle\sum_{g \in G} \chi_i(g)\overline{\chi_j(g)} = \delta_{ij}|G|$

(ii) $\displaystyle\sum_{i=1}^{k} \chi_i(C_j)\overline{\chi_i(C_r)} = \frac{|G|\delta_{jr}}{|C_j|}$

(iii) $\displaystyle\sum_{i=1}^{k} d_i^2 = |G|$

(iv) d_i *divides* $|G|$, *and indeed divides* $|G/A|$ *whenever A is a normal abelian subgroup of G*

(v) *the number of integers i with $d_i = 1$ is the order of $G/[G,G]$.*   □

These properties are useful in computing character tables. The character tables of many finite groups have been published [246], and can be accessed using software such as Gap.

As well as having the class sums as a basis, the centre of $\mathbb{C}G$ is also spanned by the idempotents

$$e_i = \frac{d_i}{|G|} \sum_{g \in G} \chi_i(g^{-1})g = \frac{d_i}{|G|} \sum_{j=1}^{k} \overline{\chi_i(C_j)}C_j^+$$

associated to the irreducible characters. The idempotents satisfy

$$e_i e_j = \delta_{ij} e_i$$

for $i, j \in \{1, \ldots, k\}$.

Since both the class sums and the idempotents are a basis for the centre of $\mathbb{C}G$, we can write an element a of the centre in two ways:

$$a = \sum_{j=1}^{k} \alpha_j C_j^+ = \sum_{i=1}^{k} \beta_i e_i,$$

for uniquely-determined complex numbers α_j and β_i. We denote α_j by $(a|C_j^+)$.

Lemma 3.17   *Let a belong to the centre of $\mathbb{C}G$, and $a = \sum_{i=1}^{k} \beta_i e_i$. Then*

$$\beta_i = \frac{\chi_i(a)}{d_i}.$$

Proof   From the first orthogonality relation (i) we obtain $\chi_i(e_j) = \delta_{ij}d_j$. Thus

$$\chi_i(a) = \sum_{j=1}^{k} \beta_j \chi_i(e_j) = \beta_i d_i,$$

and the result is immediate.   □

3.6 Characters and reversible elements

Clearly, there is a terminating algorithm that identifies all the reversible elements of a given finite group: just check all cases of the equation $g^h = g^{-1}$ for validity. This involves up to $|G|^{|G|}$ checks, which may be formidable, in practice. It is desirable to have a faster method.

It is straightforward to characterise reversibility in terms of characters.

Theorem 3.18 *An element g of a finite group G is reversible if and only if $\chi(g)$ is real for each irreducible character χ of G.*

This theorem is the reason that finite group theorists describe the reversible elements of a finite group as *real* elements. They describe strongly-reversible elements of a finite group as *strongly-real* elements.

Proof Suppose first that g is reversible. Let χ be an irreducible character of G. Then χ is constant on any conjugacy class of G, so

$$\chi(g) = \chi(g^{-1}) = \overline{\chi(g)}.$$

Therefore $\chi(g)$ is real.

Conversely, suppose that $\chi_i(g)$ is real for each of the irreducible characters $\chi_1, \ldots, \chi_k$ of G. Then

$$\chi_i(g^{-1}) = \overline{\chi_i(g)} = \chi_i(g)$$

for $i = 1, \ldots, k$. Since the character table is invertible, it follows that g and g^{-1} lie in the same conjugacy class. $\square$

Theorem 3.18 shows that the reversible classes of a finite group are immediately identifiable from its character table.

A character χ of a finite group G is said to be *real* if $\chi(g)$ is real for each element g of G. We have seen that the number of complex irreducible characters of G is equal to the number of conjugacy classes of G, and now we see that the number of real irreducible characters of G is equal to the number of real (reversible) conjugacy classes of G.

Theorem 3.19 *The number of real irreducible characters of G is equal to the number of reversible conjugacy classes of G.*

Proof Let X denote the character table of G. The rows of X correspond to irreducible characters of G and the columns correspond to conjugacy classes of G. If χ is an irreducible character of G, then so is $\overline{\chi}$. It follows that the nonreal rows of X occur in complex conjugate pairs, which implies that there is a permutation matrix P such that

$$PX = \overline{X}.$$

Next, recall that $\chi(g^{-1}) = \overline{\chi(g)}$ for each irreducible character χ and element g of G. It follows that the nonreal columns of X occur in complex conjugate pairs, which implies that there is a permutation matrix Q such that

$$XQ = \overline{X}.$$

Then

$$\text{trace}(P) = \text{trace}(\overline{X}X^{-1}) = \text{trace}(XQX^{-1}) = \text{trace}(Q).$$

But $\text{trace}(P)$ is equal to the number of real irreducible characters of G and $\text{trace}(Q)$ is equal to the number of reversible conjugacy classes of G, so the result follows. $\square$

Corollary 3.20   *A finite group G has a nontrivial real character if and only if G has even order.*

Proof   By Theorem 3.19, it suffices to prove that G contains a nontrivial reversible element if and only if G has even order, which is the first assertion of Corollary 3.2. $\square$

3.7 Characters and strongly-reversible elements

It is more difficult to obtain the strongly-reversible conjugacy classes of a finite group than it is to obtain the reversible conjugacy classes. First, we have the following reformulation of this more difficult problem.

Proposition 3.21   *Let g belong to a conjugacy class C of a finite group G. Then g is strongly reversible in G if and only if there exist conjugacy classes D and E of involutions such that*

$$(D^+E^+|C^+) \neq 0.$$ $\square$

(By a conjugacy class of involutions we mean one that contains an involution; or, equivalently, a conjugacy class all of whose members are involutions.)

Proof   Suppose first that g is strongly reversible. Then there are involutions τ_1 and τ_2 with $g = \tau_1\tau_2$. Let D and E be the conjugacy classes of τ_1 and τ_2, respectively. We can write the product D^+E^+ as a sum $\sum_{h \in G} \lambda_h h$, and the coefficient λ_g of this sum is nonzero. It follows that when D^+E^+ is written as a sum $\sum_j \mu_j C_j^+$, the coefficient of C^+ is also nonzero.

Suppose now that D and E are conjugacy classes of involutions, and

$$(D^+E^+|C^+) \neq 0.$$

Then g occurs as the product of two elements τ_1 in D and τ_2 in E.   ☐

For this result to be useful, we must have a way to compute $(D^+E^+|C^+)$.

Proposition 3.22   *Let C_1, C_2, and C_3 be conjugacy classes of a finite group G. Then*

$$(C_1^+C_2^+|C_3^+) = \frac{|C_1||C_2|}{|G|} \sum_{i=1}^{k} \frac{\chi_i(C_1)\chi_i(C_2)\overline{\chi_i(C_3)}}{d_i}.$$

Proof   Let $\omega_{ij} = \chi_i(C_j^+)/d_i$. Using Lemma 3.17,

$$C_1^+C_2^+ = \left(\sum_{i=1}^{k} \omega_{i1}e_i \right) \left(\sum_{j=1}^{k} \omega_{j2}e_j \right)$$

$$= \sum_{i=1}^{k} \omega_{i1}\omega_{i2}e_i$$

$$= \sum_{i=1}^{k} \omega_{i1}\omega_{i2} \sum_{j=1}^{k} \frac{d_i}{|G|}\overline{\chi_i(C_j)}C_j^+$$

$$= \sum_{j=1}^{k} \left\{ \sum_{i=1}^{k} \frac{\chi_i(C_1^+)\chi_i(C_2^+)\overline{\chi_i(C_j)}}{d_i|G|} \right\} C_j^+.$$

Thus

$$(C_1^+C_2^+|C_3^+) = \sum_{i=1}^{k} \frac{\chi_i(C_1^+)\chi_i(C_2^+)\overline{\chi_i(C_3)}}{d_i|G|}.$$

Therefore

$$(C_1^+C_2^+|C_3^+) = \frac{|C_1||C_2|}{|G|} \sum_{i=1}^{k} \frac{\chi_i(C_1)\chi_i(C_2)\overline{\chi_i(C_3)}}{d_i}.$$

☐

Corollary 3.23   *Let g belong to a finite group G. Then g is strongly reversible if and only if there exist involutions g_1 and g_2 in G such that*

$$\sum_{i=1}^{k} \frac{\chi_i(g_1)\chi_i(g_2)\overline{\chi_i(g)}}{d_i} \neq 0.$$

☐

3.8 Examples

3.8.1 Alternating group

As we know, the alternating group A_4 has $I = I^2 = R \neq G$. The character table is

	C_1	C_2	C_3	C_4
χ_1	1	1	1	1
χ_2	1	1	ω	ω^2
χ_3	1	1	ω^2	ω
χ_4	3	-1	0	0

where $\omega = \exp(2\pi i/3)$. The involutive classes are $C_1 = [1]$ and $C_2 = [(12)(34)]$. The fact that C_3 and C_4 are not reversible is reflected in the occurrence of non-real entries in their columns.

3.8.2 Symmetric group

The character table of each symmetric group S_n has only real integral entries. For S_4, the table is

	C_1	C_2	C_3	C_4	C_5
χ_1	1	1	1	1	1
χ_2	1	-1	1	1	-1
χ_3	2	0	2	-1	0
χ_4	3	1	-1	0	-1
χ_4	3	-1	-1	0	1

The involutive classes are $C_1 = [1]$, $C_2 = [(12)]$, and $C_3 = [(12)(34)]$. The fact that $I^2 = G$ translates into the statement that at least one of

$$(C_2^+ C_3^+ | C_4^+), (C_2^+ C_2^+ | C_4^+), (C_3^+ C_3^+ | C_4^+),$$

and at least one of

$$(C_2^+ C_3^+ | C_5^+), (C_2^+ C_2^+ | C_5^+), (C_3^+ C_3^+ | C_5^+),$$

are nonzero. In fact, we calculate

$$(C_2^+ C_3^+ | C_4^+) = \frac{|C_2||C_3|}{|G|} \sum_{i=1}^{5} \frac{\chi_i(C_2)\chi_i(C_3)\overline{\chi_i(C_4)}}{d_i}$$

$$= \frac{\binom{4}{2} \cdot \frac{1}{2}\binom{4}{2}}{4!} \left(\frac{1}{1} + \frac{-1}{1} + \frac{0}{2} + \frac{0}{3} + \frac{0}{3} \right) = 0,$$

but we find

$$(C_2^+ C_2^+ | C_4^+) = \frac{6 \cdot 6}{24} \left(\frac{1}{1} + \frac{1}{1} + \frac{0}{2} + \frac{0}{3} + \frac{0}{3} \right) = 3 \neq 0.$$

Similarly,

$$(C_2^+ C_3^+ | C_5^+) = \frac{6 \cdot 3}{24} \left(\frac{1}{1} + \frac{1}{1} + \frac{0}{2} + \frac{1}{3} + \frac{1}{3} \right) = 2 \neq 0.$$

3.8.3 Quaternion group

The character table of the quaternion group

$$Q_8 = \{\pm 1, \pm i, \pm j, \pm k\},$$

which has $R = G \neq I^2$, is

	C_1	C_2	C_3	C_4	C_5
χ_1	1	1	1	1	1
χ_2	1	1	-1	1	-1
χ_3	1	1	-1	-1	1
χ_4	1	1	1	-1	-1
χ_4	2	-2	0	0	0

The only involutive classes are those contained in the centre of Q_8, namely $C_1 = [1]$ and $C_2 = [-1]$. The fact that $C_3 = [i] = \{\pm i\}$ is not strongly reversible is reflected in the fact that

$$(C_2^+ C_2^+ | C_3^+) = \frac{1}{8} \left(\frac{1}{1} + \frac{-1}{1} + \frac{-1}{1} + \frac{1}{1} + \frac{0}{2} \right) = 0.$$

Similarly,

$$(C_2^+ C_2^+ | C_4^+) = (C_2^+ C_2^+ | C_5^+) = 0,$$

where $C_4 = [j]$ and $C_5 = [k]$.

3.8.4 Dihedral group

The dihedral group

$$D_4 = \langle a, b : a^2 = b^2 = (ab)^4 = \mathbb{1} \rangle$$

has the *same* character table as Q_8, but has four involutive classes $C_1 = [1]$, $C_2 = [(ab)^2] = \{(ab)^2\}$, $C_3 = [a] = \{a, bab\}$, and $C_4 = [b] = \{b, aba\}$. In this case, the fact that $I^2 = G$ is reflected in the value

$$(C_3^+ C_4^+ | C_5^+) = \frac{2 \cdot 2}{8} \left(\frac{1}{1} + \frac{1}{1} + \frac{1}{1} + \frac{1}{1} + \frac{0}{2} \right) = 2 \neq 0.$$

·3.9 Free groups

Free groups are far from finite, but it is convenient to include the simple story of reversibility in free groups here.

Each element of the free group $F(A)$ on the alphabet A has a unique representation as a so-called *reduced word* over A; that is, as a word in the symbols x and x^{-1}, where x ranges over A, in which no x is immediately preceded by or followed by x^{-1}. Conjugating with the last symbol in a word w moves that symbol to the front; that is, it "cycles" the word. A word is said to be *cyclically reduced* if both it and the words obtained by cycling it in this way are reduced. Each conjugacy class C of $F(A)$ contains a cyclically-reduced word r, and all other cyclically-reduced words belonging to C are obtained from r by repeated cycling.

Proposition 3.24 *Only the identity element of a free group is reversible.*

Proof Let w be a reversible element of a free group. Choose a cyclically-reduced word w_1 that is conjugate to w. Then w_1 is reversible. The word obtained by reversing the order of w_1, and replacing each symbol by its inverse, is also cyclically reduced, and represents w_1^{-1}. Since w_1 is conjugate to w_1^{-1}, this new word must be a cyclic permutation of w_1, so there exist reduced words w_2 and w_3 such that

$$w_1 = w_2 w_3 \quad \text{and} \quad w_1^{-1} = w_3 w_2.$$

Since w_1 is cyclically reduced, there is no cancellation in $w_2 w_3$ or $w_3 w_2$, so it follows from the equation $w_2 w_3 w_3 w_2 = 1$ that $w_3^2 = w_2^2 = 1$. But then $w_2 = w_3 = 1$, since G is free, and hence $w_1 = 1$. Therefore w is also equal to the identity element. $\qquad\square$

Notes

Sources

Proposition 3.1 and related results can be found in [159]. Goodson considers the dihedral group and other finite groups in [110]. Most of the material on group characters can be found in [55, 86, 138, 140]. In particular, Theorem 3.13 is [55, Theorems 5.2.2 and 5.2.4], Theorem 3.15 is [140, Theorem 14.23], and Theorem 3.19 is [140, Theorem 23.1].

The authors thank Claas Röver for remarks about free groups, and thank Derek Holt for an example of a finitely-generated group in which the reversibility problem is solvable but the conjugacy problem is unsolvable (to come, later in these Notes).

Finite simple groups

The abelian finite simple groups are the abelian groups of prime order. By Corollary 3.2, the cyclic group C_2 is the only abelian group of prime order that contains a nontrivial reversible element. According to the famous Feit–Thompson theorem [87], all groups of odd order are solvable. As nonabelian simple groups are not solvable, we see that each nonabelian finite simple group G has even order. Then Corollary 3.2 tells us that G contains an involution, so $I^\infty(G)$, which is a nontrivial normal subgroup of G, must equal G. Therefore G has plenty of reversible elements.

As stated earlier, Tiep and Zalesski [231] have classified all the finite simple groups in which all elements are reversible. In [100, 101], Gill and Singh classified the reversible and strongly-reversible conjugacy classes of the groups $SL(n,q)$, $GL(n,q)$, $PSL(n,q)$, and $PGL(n,q)$. Singh and Thakur [216, 217] have various results on reversibility in other finite simple groups, including good progress on the exceptional group G_2. See also [235] for work on reversibility in general linear and symplectic groups over finite fields.

We consider reversibility in the classical groups in more depth in Chapter 4, although we work over the real or complex fields, rather than finite fields.

Recently there has been rapid progress on word problems and growth in finite simple groups, and some of this work relates to reversibility. In [212] Shalev proved that every nontrivial word has width three in sufficiently large finite simple groups. Briefly, a word w is a function of the form $w(x_1, \ldots, x_n) = x_{i_1}^{k_1} \ldots x_{i_m}^{k_m}$, where k_i are integers and the indices $i_j \in \{1, \ldots, n\}$. Let us assume that the word w is reduced (no two consecutive powers of x_j in the product can be amalgamated). Given a group G we can consider w to be a map from $G \times \cdots \times G$ to G, and we denote the image of $G \times \cdots \times G$ by $w(G)$. Let $w(G)^3 =$

$\{abc : a,b,c, \in w(G)\}$. Shalev proved that, given a nontrivial word w, we have $w(G)^3 = G$ for sufficiently large finite simple groups G. More recently still, Liebeck, O'Brien, Shalev, and Tiep proved the Øre conjecture [168], that each element of a nonabelian finite simple group is a commutator. In the previous chapter, we gave the identity

$$[a,b] = (aba^{-1})^2 a^2 (a^{-1}b^{-1})^2$$

for expressing a commutator as a product of three squares. It follows that each element of a finite simple group can be expressed as a product of three squares.

Another famous conjecture, the Thompson conjecture, remains outstanding. It says that in any finite simple group G there exists a conjugacy class C such that $C^2 = G$. The Thompson conjecture is stronger than the Øre conjecture. To see why this is so, notice that if $C^2 = G$ then $\mathbb{1} \in C^2$, and hence $C = C^{-1}$ (that is, C is a reversible conjugacy class). Given g in G we can choose two conjugate elements x and yxy^{-1} in C such that $g = x(yxy^{-1})$. Since x is reversible there is an element z of G such that $zxz^{-1} = x^{-1}$. Therefore $g = xux^{-1}u^{-1}$, where $u = yz^{-1}$. If true the Thompson conjecture also implies that $R^2(G) = G$. Tiep and Zalesski [231] have already shown that, for some finite simple groups G, $R(G) \neq G$. An interesting discussion of the Thompson conjecture, including various related conjectures, can be found in [212].

Counting involutions

Many results about finite groups are proven by counting involutions. Among these is a theorem of Brauer and Fowler [138, Chapter 4], which says that, for any positive integer n, there are at most finitely many simple groups containing an involution with centraliser of order n. Brauer and Fowler also proved a related result that any group G of even order greater than 2 has a proper subgroup H that satisfies $|H|^3 > |G|$.

Another result about counting involutions is the *Thompson order formula* [13, 45.6]. For a finite group G with exactly two conjugacy classes of nontrivial involutions, this formula says that

$$|G| = |C_a(G)|n_b + |C_b(G)|n_a,$$

where a and b are two nonconjugate (and nontrivial) involutions, $C_a(G)$ and $C_b(G)$ are their centralisers, and n_g denotes the number of pairs (u,v) for which u is conjugate to a, v is conjugate to b, and g lies in the cyclic group generated by uv. See [14] for more on counting involutions.

Group presentations

Free groups have the simplest presentations, but, as we saw in Proposition 3.24, only trivial reversibility properties. More complicated group presentations can be constructed with more interesting reversibility properties. Consider, for example,

$$G = \langle a_1, a_2, \ldots \,|\, a_1^2 = a_2^2 = \cdots = \mathbb{1} \rangle.$$

Since $a_1 a_2 \cdots a_n \in I^n \setminus I^{n-1}$ for $n = 2, 3, \ldots$ we see that G satisfies

$$I^1 \subsetneq I^2 \subsetneq \cdots \subsetneq I^\infty.$$

In Chapter 7 we meet a group with the simple presentation $\langle a, b \,|\, a^2 = b^3 = \mathbb{1} \rangle$, namely the modular group, and we find that reversibility in this group is far from trivial.

The reversibility problem in free groups

Let F be free group of rank at least two. In [183] Miller constructed a finitely-generated subgroup L of $F \times F$ that has unsolvable conjugacy problem. The only reversible element in $F \times F$ is, by Propositions 2.16 and 3.24, the identity element. Therefore L is a finitely-generated subgroup of $F \times F$ in which the reversibility problem is solvable, but the conjugacy problem is unsolvable.

Coxeter groups

A *Coxeter group* is a group that admits a presentation

$$\langle a_1, a_2, \ldots, a_n \,|\, (a_i a_j)^{m_{ij}} = \mathbb{1} \rangle,$$

where m_{ij} are positive integers such that $m_{ii} = 1$, $m_{ij} \geqslant 2$ for $i \neq j$, and $m_{ij} = m_{ji}$. The condition $m_{ii} = 1$ ensures that each generator a_i is an involution.

Springer proved in 1974 that every element in a *finite* Coxeter group is strongly reversible. (See [218, Theorem 8.7]. The result had previously been proved for the subclass of finite Weyl groups by Carter, in [50, page 45].) We are unaware of any work on reversibility in infinite Coxeter groups.

Open problems

The Kourovka notebook

There are many questions on reversibility in finite and infinite groups in the Kourovka notebook [178]. Two are listed below.

11.67 Does there exist a torsion-free group with exactly three classes of conjugate elements such that no nontrivial conjugacy class contains a pair of inverse elements? (Posed by B. Neumann.)

In other words, is there an infinite group G with just two nontrivial conjugacy classes and $R(G) = \{\mathbb{1}\}$?

16.76 Which groups G, obtained as maximal unipotent subgroups of a group of Lie type over a field of characteristic 2, have $I^2(G) = G$? (Posed by Ya.N. Nuzhin.) He uses the term *strictly real* to refer to a group with $I^2(G) = G$.

There is a large swathe of finite group theory about the least number of involutions, or the least number of reversibles, needed to generate a given group, and there are many problems of this type in the Notebook.

The finite simple groups

According to the classification theorem for finite simple groups, a nonabelian finite simple group is either alternating, a group of Lie type, or one of the sporadic groups. The real and strongly-real elements of the alternating groups were determined in Section 3.4, and there are only finitely many sporadic groups (and their character tables are known). Gill and Singh classified the real and strongly-real classes of the projective special linear groups, but the problem of classifying real and strongly-real classes in most of the remaining finite groups of Lie type remains. See [22, 81, 93, 101, 100, 121, 122, 153, 154, 202, 216, 217, 231, 233, 235] for recent related results.

The Thompson conjecture

The Thompson conjecture, described above, is a significant unsolved problem in the theory of finite simple groups.

Coxeter groups

We mentioned already that every element of a finite Coxeter group is strongly reversible. Is the same true of all Coxeter groups? If not then which elements of an infinite Coxeter group are strongly reversible?

4

The classical groups

4.1 The classical groups

Hermann Weyl [243] coined the phrase 'classical groups' to describe certain subgroups of general linear groups, and other groups derived from these subgroups, that preserve particular sesquilinear forms. We discuss the collection of all classical groups in this section, but only study reversibility in a selection of them.

Let V denote a finite-dimensional vector space over a commutative field F. The *general linear group* of V is the group of all invertible linear transformations of V. It is denoted $GL(V)$. Let σ denote a field automorphism of F that is an involution. A *sesquilinear form* on V, relative to σ, is a map B from $V \times V$ to F such that for all vectors u, v, and w in V, and scalars λ and μ in F,

(i) $B(\lambda u + \mu v, w) = \lambda B(u, w) + \mu B(v, w)$
(ii) $B(w, \lambda u + \mu v) = \sigma(\lambda) B(w, u) + \sigma(\mu) B(w, v)$.

If σ is the identity map, then B is said to be *bilinear*. The sesquilinear form B is *nondegenerate* if each of the conditions $B(u, v) = 0$ for all vectors v in V, or $B(v, u) = 0$ for all vectors v in V, imply that $u = 0$. A nondegenerate sesquilinear form B is *reflexive* if the equation $B(u, v) = 0$ implies that $B(v, u) = 0$.

A reflexive, nondegenerate sesquilinear form V is said to be

○ *alternating*, if B is bilinear, and $B(v, v) = 0$ for each vector v in V
○ *Hermitian*, if σ is of order 2, and is such that $B(u, v) = \sigma(B(v, u))$ for each pair of vectors u and v
○ *symmetric*, if B is bilinear, and $B(u, v) = B(v, u)$ for each pair of vectors u and v.

A famous theorem of Birkhoff and von Neumann [225, Theorem 7.1] says that each reflexive, nondegenerate sesquilinear form is either alternating, Hermitian, or symmetric. We define the *isometry group* of B, Isom(B), to be the subgroup of GL(V) consisting of those maps that preserve B. That is

$$\text{Isom}(B) = \{g \in \text{GL}(V) : B(g(u),g(v)) = B(u,v) \text{ for each pair } u \text{ and } v \text{ in } V\}.$$

The group Isom(B) is called *symplectic* if B is alternating, *unitary* if B is Hermitian, and *orthogonal* if B is symmetric. The collection of groups known as the classical groups includes the general linear, symplectic, unitary, and orthogonal groups, as well as certain subgroups and quotient groups of these parent groups.

There is a vast literature on questions of reversibility in the classical groups (refer to the Notes for references). In this chapter we study the general linear, orthogonal, and unitary groups, as well as certain subgroups of these groups. We consider the general linear group over a general commutative field F, but for the other groups we use either the real or complex fields. In the Hermitian case, the field automorphism is complex conjugation. We consider our groups to be matrix groups, for simplicity and clarity. We denote the n-by-n identity matrix by $\mathbb{1}_n$. Then the three main classes of groups we consider are

- ○ $\text{GL}(n,F) = \{n \times n \text{ invertible matrices over } F\}$
- ○ $\text{O}(n,\mathbb{R}) = \{A \in \text{GL}(n,\mathbb{R}) : A^t A = \mathbb{1}_n\}$
- ○ $\text{U}(n,\mathbb{C}) = \{A \in \text{GL}(n,\mathbb{C}) : A^* A = \mathbb{1}_n\}$

where A^t denotes the transpose of the matrix A (not a group theoretic conjugate of A), $\overline{A}$ denotes the complex conjugate of A, and $A^* = \overline{A}^t$. (Explicitly, A^t and $\overline{A}$ are n-by-n matrices with coefficients $(A^t)_{ij} = A_{ji}$ and $(\overline{A})_{ij} = \overline{A_{ij}}$.) The symplectic group is discussed in the Notes at the end of the chapter.

4.2 The general linear group

4.2.1 The general linear group over the complex numbers

Reversibility in GL(n,F) is simpler when $F = \mathbb{C}$ because of the Jordan normal form, so we begin by studying GL($n,\mathbb{C}$), and then move on to GL(n,F). We briefly recall the theory of the Jordan normal form. A *Jordan block* is an m-by-

m matrix

$$J(m,\lambda) = \begin{pmatrix} \lambda & 1 & & & \\ & \lambda & 1 & & \\ & & \ddots & \ddots & \\ & & & \lambda & 1 \\ & & & & \lambda \end{pmatrix},$$

in which zeros occupy empty spaces, and λ is a complex number. The *Jordan normal form* of an n-by-n complex matrix A is a matrix conjugate to A of the form

$$\begin{pmatrix} J_1 & & & \\ & J_2 & & \\ & & \ddots & \\ & & & J_k \end{pmatrix},$$

where $J_1, J_2, \ldots, J_k$ are Jordan blocks. The next theorem can be found in [62, Theorem 25.16].

Theorem 4.1   *Each member of* $\mathrm{GL}(n, \mathbb{C})$ *has a Jordan normal form, and this Jordan normal form is unique up to permutations of the Jordan blocks.*   □

The inverse of $J(m, \lambda)$, where $\lambda \neq 0$, is an m-by-m upper-triangular matrix with entries $1/\lambda$ on the diagonal. Since the geometric multiplicity of $J(m, \lambda)$ is 1 we can deduce that the Jordan normal form of $J(m, \lambda)^{-1}$ is a single block $J(m, \lambda^{-1})$. More generally, if $A \in \mathrm{GL}(n, \mathbb{C})$ is in Jordan normal form, with blocks $J(n_1, \lambda_1), \ldots, J(n_k, \lambda_k)$ (where $\lambda_i \neq 0$ for each i), then the Jordan normal form of A^{-1} has blocks $J(n_1, \lambda_1^{-1}), \ldots, J(n_k, \lambda_k^{-1})$.

Theorem 4.2   *An element A of* $\mathrm{GL}(n, \mathbb{C})$ *with Jordan normal form consisting of blocks $J(n_1, \lambda_1), \ldots, J(n_k, \lambda_k)$, where $\lambda_i \neq 0$, is reversible if and only if the blocks can be partitioned into pairs $\{J(m, \lambda), J(m, \lambda^{-1})\}$ or singletons $\{J(m, \mu)\}$, where $\mu = \pm 1$.*

Proof   By Theorem 4.1, A is conjugate to A^{-1} if and only if

$$\{J(n_1, \lambda_1), \ldots, J(n_k, \lambda_k)\} = \{J(n_1, \lambda_1^{-1}), \ldots, J(n_k, \lambda_k^{-1})\},$$

and the result follows immediately.   □

Recall that an element g of a group G is *reversed by an automorphism* if there is a group automorphism ϕ of G with $\phi(g) = g^{-1}$.

Theorem 4.3   *Every element of* $\mathrm{GL}(n, \mathbb{C})$ *is reversed by an automorphism.*

Proof Let α be the automorphism of $GL(n, \mathbb{C})$ given by $\alpha(A) = (A^{-1})^t$. Any invertible complex matrix is conjugate to its transpose; in particular, A^{-1} and $\alpha(A)$ are conjugate. Therefore A is reversed by an automorphism.   □

We finish with a useful proposition for simultaneously conjugating two involutive matrices in $GL(2, \mathbb{C})$ to a normal form. We let $\mathrm{diag}(\mu_1, \ldots, \mu_n)$ denote a diagonal matrix with entries $\mu_1, \ldots, \mu_n$ down the leading diagonal, in that order.

Proposition 4.4 *Given two involutions A and B in $GL(2, \mathbb{C})$ that do not share a common eigenvector, there is a matrix P in $GL(2, \mathbb{C})$, and a nonzero complex number $\lambda \neq \pm 1$, such that*

$$PAP^{-1} = \begin{pmatrix} 0 & \lambda \\ \lambda^{-1} & 0 \end{pmatrix}, \qquad PBP^{-1} = \begin{pmatrix} 0 & \lambda^{-1} \\ \lambda & 0 \end{pmatrix}.$$

Proof Let $C = AB$, and suppose by conjugation that C is in Jordan normal form. From Theorem 4.2, C is equal to either $J(2, 1)$, $J(2, -1)$, or $\mathrm{diag}(\mu, \mu^{-1})$, for $\mu \neq 0$. In the first case, the only eigenspace of C is a one-dimensional vector space V_2 spanned by the standard basis vector e_2. From the equation $ACA = C^{-1}$ we see that $CA(V_2) = A(V_2)$. Hence $A(V_2) = V_2$, and likewise $B(V_2) = V_2$. This is a contradiction, because A and B have no common eigenvectors. In a similar way we see that $C \neq J(2, -1)$. Therefore $C = \mathrm{diag}(\mu, \mu^{-1})$. This time C has two one-dimensional eigenspaces V_1 and V_2, where V_1 is spanned by the standard basis vector e_1. From the equations $ACA^{-1} = C^{-1}$ and $BCB^{-1} = C^{-1}$ we deduce that A and B each interchange V_1 and V_2. After conjugating A and B by a suitable matrix $\mathrm{diag}(v, v^{-1})$, for some $v \neq 0$, we obtain the desired result.   □

4.2.2 The rational canonical form

For the remainder of this section we consider reversibility in the general linear group $GL(n, F)$, where F is a commutative field.

Within each conjugacy class in $GL(n, F)$, we identify an element that takes a particularly simple form, known as the *rational canonical form*. Information on the rational canonical form can be found, for example, in [62].

Let m be a monic polynomial. There are monic, irreducible, coprime polynomials $p_1, \ldots, p_k$, and positive integers $s_1, \ldots, s_k$, such that

$$m(x) = p_1(x)^{s_1} \cdots p_k(x)^{s_k}.$$

For each integer s_i we choose an increasing sequence of positive integers $s_{i1} \leqslant s_{i2} \leqslant \cdots \leqslant s_{ir_i}$, such that $s_{ir_i} = s_i$. This collection of polynomials and integers defines a unique matrix, in the following fashion. Given a monic polynomial $q(x) = x^n + a_{n-1}x^{n-1} + \cdots + a_0$, the matrix

$$\begin{pmatrix} & & & & -a_0 \\ 1 & & & & -a_1 \\ & 1 & & & -a_2 \\ & & \ddots & & \vdots \\ & & & 1 & -a_{n-2} \\ & & & 1 & -a_{n-1} \end{pmatrix} \qquad (4.1)$$

is known as the *companion matrix* associated with q. Returning to our polynomial m, we let A_{ij} denote the companion matrix associated with $p_i^{s_{ij}}$. Now define

$$\cdot \; B_i = \begin{pmatrix} A_{i1} & & & \\ & A_{i2} & & \\ & & \ddots & \\ & & & A_{ir_i} \end{pmatrix}$$

and consider the matrix

$$\begin{pmatrix} B_1 & & & \\ & B_2 & & \\ & & \ddots & \\ & & & B_k \end{pmatrix}. \qquad (4.2)$$

Such a matrix is said to be in *rational canonical form*.

The *minimum polynomial s* of an invertible matrix A is the monic polynomial of smallest degree such that $s(A) = 0$. The minimum polynomial is an invariant of conjugacy, but alone it is insufficient to classify conjugacy in $GL(n, F)$. The minimum polynomial of the matrix from (4.1) is q, and the minimum polynomial of (4.2) is m. The next theorem can be found in, for example, [62, Section 25].

Theorem 4.5 *Let m be the minimum polynomial of an element A of $GL(n, F)$, and suppose that $p_1, \ldots, p_k$ are monic, irreducible, coprime polynomials such that $m = p_1^{s_1} \cdots p_k^{s_k}$ for positive integers $s_1, \ldots, s_k$. Then there are positive integers $s_{i1} \leqslant \cdots \leqslant s_{ir_i} = s_i$ for each i such that A is conjugate to the matrix of* (4.2); *that is, A can be put into rational canonical form.* $\square$

The rational canonical form of a matrix is unique up to permutations of the blocks $B_1, B_2, \ldots, B_k$.

4.2.3 Reversible and strongly-reversible elements

A monic polynomial p with roots $\alpha_1, \ldots, \alpha_n$, all nonzero, is called the *reciprocal* of the monic polynomial with roots $\alpha_1^{-1}, \ldots, \alpha_n^{-1}$. Equivalently, p is the reciprocal of q if

$$p(t) = q(0)^{-1} t^n q(t^{-1}). \tag{4.3}$$

A monic polynomial with nonzero roots that is the reciprocal of itself is described as *self reciprocal*. If $p(t) = a_n t^n + a_{n-1} t^{n-1} + \cdots + a_0$ is monic and self reciprocal then we see from (4.3) that either $a_i = a_{n-i}$ for each i, or $a_i = -a_{n-i}$ for each i. It is straightforward to check that the companion matrix associated to a self-reciprocal polynomial is reversed by the n-by-n involution

$$J_n = \begin{pmatrix} & & & & 1 \\ & & & 1 & \\ & & \cdot^{\cdot^{\cdot}} & & \\ & 1 & & & \\ 1 & & & & \end{pmatrix}. \tag{4.4}$$

The next lemma shows that, in particular, reversible matrices have self-reciprocal minimum polynomials.

Lemma 4.6 *Let A and B be elements of $\mathrm{GL}(n, F)$ (possibly equal) with minimum polynomials m_A and m_B. If A is conjugate to B^{-1} then m_A and m_B are reciprocal.*

Proof The lemma is true because the minimum polynomial is preserved under conjugation, and, if k is the degree of m_A, then

$$m_{A^{-1}}(t) = m_A(0)^{-1} t^k m_A(t^{-1}). \qquad \square$$

In the following theorem, first proven by Wonenburger in [247] (see also [70, 134]), we classify the reversible and strongly-reversible elements of $\mathrm{GL}(n, F)$ that are in rational canonical form. Some preliminary notation is needed in order to state the theorem concisely.

Let A be an invertible matrix in rational canonical form. Let $A_1, \ldots, A_r$ be the companion matrix blocks of A, in order, down the leading diagonal. Let m_i be the minimum polynomial of A_i. Let n_i be the size of A_i, let $N_1 = 0$ and $N_i = n_1 + \cdots + n_{i-1}$ for $i > 1$, and let V_i denote the subspace spanned by the standard basis vectors $e_{N_i+1}, \ldots, e_{N_i+n_i}$. Each subspace V_i is fixed, as a set, by A, and there is no proper subspace of V_i that is fixed by A.

Theorem 4.7   *For an element A of* $GL(n,F)$ *in rational canonical form, the following are equivalent:*

(i) *A is strongly reversible*

(ii) *A is reversible*

(iii) *the polynomials* $m_1, \ldots, m_r$ *can be partitioned into pairs of reciprocal polynomials, and singleton sets of self-reciprocal polynomials.*

Proof   That (i) implies (ii) is true in all groups. Next we show that (ii) implies (iii). Suppose there is an element P of $GL(n,F)$ such that $PAP^{-1} = A^{-1}$. Notice that

$$AP(V_i) = APA(V_i) = P(V_i).$$

Since V_j has no proper A invariant subspaces, the subspace $P(V_i)$ can intersect V_j either in 0 or in V_j. Therefore $P(V_i) = V_k$ for some k. In other words, P acts on the collection of subspaces $V_1, \ldots, V_r$. Observe, using Lemma 4.6, that if P maps V_i to V_j then m_i and m_j are reciprocal. Using this observation we can partition $\{m_1, \ldots, m_r\}$ as required.

It remains to prove that (iii) implies (i). We construct an involution S such that $SAS^{-1} = A^{-1}$ using the blocks J_n of (4.4). Recall that r denotes the number of blocks down the diagonal in A. We think of S as an r-by-r array of blocks; each block is either occupied entirely by 0 entries, or else it is a matrix J_n. The (i,j)th block has size n_j-by-n_i, corresponding to the dimensions of V_j and V_i. Now let us specify the blocks of S. If i is an index for which m_i is self reciprocal, then we occupy the (i,i)th block with a matrix S_{n_i} If i and j are indices for which m_i and m_j are reciprocal, then we occupy the (i,j)th and (j,i)th blocks with matrices S_{n_i}. All other blocks contain 0 entries. The resulting matrix S has the required properties.   □

Corollary 4.8   *An element of* $GL(n,F)$ *is reversible if and only if it is strongly reversible.*   □

4.2.4   Products of involutions

For square matrices A and B we have the equation $\det(AB) = \det(A)\det(B)$, and from this equation we deduce, first, that involutions have determinant either 1 or -1, and second, that $I^\infty(GL(n,F))$ is contained within the subgroup H of $GL(n,F)$ consisting of matrices with determinant either 1 or -1. The next theorem, which was first proven by Gustafson, Halmos, and Radjavi in [126], shows that $H = I^4(GL(n,F))$.

Theorem 4.9 *Each member of* $GL(n,F)$ *with determinant* 1 *or* -1 *can be expressed as a composite of four involutions.*

The proof of Theorem 4.9 follows the next two propositions. A *weighted permutation matrix* (over F) is a square matrix that has a single nonzero entry in each row and each column.

Proposition 4.10 *Each weighted permutation matrix of determinant either* 1 *or* -1 *is expressible as a composite of three involutions.*

Proof Let L_i denote the line in F^n spanned by the standard basis vector e_i. Each weighted permutation matrix permutes the lines $L_1, L_2, \ldots, L_n$, and thus gives rise to a unique element of S_n. Conversely, each element of S_n corresponds to a unique permutation matrix (with all weights equal to 1). Now let A be a weighted permutation matrix, and let α denote the corresponding element of S_n. We can decompose α as $\beta\gamma$, where β is an involution and γ has no fixed points. For instance, if, in cycle notation,

$$\alpha = (a_1, \ldots, a_p)(b_1, \ldots, b_q)(c_1, \ldots, c_r)(d_1, \ldots, d_s)$$

then we can choose

$$\beta = (a_p, b_1)(b_q, c_1)(c_r, d_1)$$

and

$$\gamma = (a_1, \ldots, a_p, b_2, \ldots, b_{q-1}, c_2, \ldots, c_{r-1}, d_2, \ldots, d_s, d_1, c_1, b_1).$$

Let B denote the involutive permutation matrix corresponding to β, and let C denote the weighted permutation matrix BA. Then γ is the unique fixed-point-free element of S_n corresponding to C, and, after conjugation of A, B, and C we may assume that $\gamma = (1, 2, \ldots, n)$. Note also that $\det(C) = \det(BA) = \pm 1$. Hence, for example when $n = 6$,

$$C = \begin{pmatrix} & & & & & u_1 \\ u_2 & & & & & \\ & u_3 & & & & \\ & & u_4 & & & \\ & & & u_5 & & \\ & & & & u_6 & \end{pmatrix},$$

where $u_1 u_2 \cdots u_6 = \pm 1$. Then $C = DE$, where D and E are involutions

$$
D = \begin{pmatrix}
& & & & & u_1 \\
& & & & u_1 u_2 u_6 & \\
& & & u_1 u_2 u_3 u_5 u_6 & & \\
& & \dfrac{1}{u_1 u_2 u_3 u_5 u_6} & & & \\
& \dfrac{1}{u_1 u_2 u_6} & & & & \\
\dfrac{1}{u_1} & & & & &
\end{pmatrix}
$$

and

$$
E = \begin{pmatrix}
& & & & u_1 u_6 \\
& & & u_1 u_2 u_5 u_6 & \\
& & u_1 u_2 u_3 u_4 u_5 u_6 & & \\
& \dfrac{1}{u_1 u_2 u_5 u_6} & & & \\
\dfrac{1}{u_1 u_6} & & & & \\
& & & & 1
\end{pmatrix}.
$$

A similar formula holds in n-dimensions. Thus $A = BDE$, a product of three involutions, as required. $\qquad\square$

Proposition 4.11 *Each invertible matrix A in rational canonical form can be decomposed as $A = JW$ where J is an involution and W is a weighted permutation matrix.*

Proof The companion matrix (4.1) can be expressed as a product of an involution followed by a weighted permutation matrix as follows:

$$
\begin{pmatrix}
-1 & & & & \\
-\dfrac{a_1}{a_0} & 1 & & & \\
-\dfrac{a_2}{a_0} & & 1 & & \\
& & & \ddots & \\
& & & & 1 \\
-\dfrac{a_{n-1}}{a_0} & & & & 1
\end{pmatrix}
\begin{pmatrix}
& & & & a_0 \\
1 & & & & \\
& 1 & & & \\
& & \ddots & & \\
& & & 1 & \\
& & & & 1
\end{pmatrix}. \qquad (4.5)
$$

Also, trivially, an invertible diagonal matrix D can be expressed as a product of an involution (the identity) followed by a weighted permutation matrix (D itself). We can suppose that A is in rational canonical form, so that it consists of blocks of the form (4.1) down the diagonal (each of size at least 2) followed by a diagonal matrix. Working block by block we can construct the required matrices J and W. $\qquad\square$

Proof of Theorem 4.9 Let A be an element of $\mathrm{GL}(n, F)$ with determinant ± 1.

By Proposition 4.11 there is an involution J_1 and a weighted permutation matrix W such that $A = J_1 W$. Observe that $\det(W) = \det(J_1 A) = \pm 1$. Hence, by Proposition 4.10, there are involutions J_2, J_3, and J_4 such that $W = J_2 J_3 J_4$. Therefore $A = J_1 J_2 J_3 J_4$. $\qquad\qquad\qquad\qquad\qquad\qquad\qquad\qquad\qquad\qquad\square$

There are elements of $GL(n, F)$ with determinant 1 that are *not* expressible as a composite of three involutions. For example, let $n = 3$ and $F = \mathbb{C}$, and let

$$A = \begin{pmatrix} \omega & & \\ & \omega & \\ & & \omega \end{pmatrix}, \qquad (4.6)$$

where ω is a nonreal third root of unity. Then $A \notin I^3(G)$, by Proposition 2.15. The problem of describing I^3 is open for $GL(n, F)$, for most fields.

4.3 The orthogonal group

As well as considering the standard orthogonal group $O(n, \mathbb{R})$ we also consider reversibility in the *special orthogonal group*, $SO(n, \mathbb{R})$, which consists of orthogonal matrices with determinant 1. Our results on this group are needed in later chapters.

4.3.1 Conjugacy

The results of this section rest on the following well known principal axis theorem, which is [62, Theorem 30.5]. Let

$$R_\theta = \begin{pmatrix} \cos\theta & -\sin\theta \\ \sin\theta & \cos\theta \end{pmatrix}$$

for a real number θ, and let $\mathbb{1}_k$ denote the k-by-k identity matrix.

Theorem 4.12 *In either of the groups $O(n, \mathbb{R})$ or $SO(n, \mathbb{R})$, each matrix is conjugate to another matrix of the form*

$$\begin{pmatrix} R_{\theta_1} & & & & \\ & \ddots & & & \\ & & R_{\theta_r} & & \\ & & & -\mathbb{1}_s & \\ & & & & \mathbb{1}_t \end{pmatrix}, \qquad (4.7)$$

where $\theta_1, \ldots, \theta_r \in (-\pi, \pi) \setminus \{0\}$.

An orthogonal matrix is in *standard form* if it is of the same form as the matrix of (4.7). For an orthogonal matrix A, let r_A, s_A, and t_A denote the integers r, s, and t of a matrix in standard form that is conjugate to A. There is no ambiguity in this definition as s_A and t_A are the dimensions of the eigenspaces of -1 and 1, and $r_A = (n - s_A - t_A)/2$. The involutions in $O(n, \mathbb{R})$ are those orthogonal maps A with $r_A = 0$.

4.3.2 Reversible elements of the orthogonal group

Coxeter proved in [61] that $O(n, \mathbb{R})$ is bireflectional; that is, every element is strongly reversible.

Theorem 4.13 *Each member of $O(n, \mathbb{R})$ is strongly reversible.*

Proof It suffices to prove the theorem for an orthogonal map A in standard form. Define

$$
B = \begin{pmatrix} K & & & & & \\ & \ddots & & & & \\ & & K & & & \\ & & & \varepsilon_1 & & \\ & & & & \ddots & \\ & & & & & \varepsilon_{s_A + t_A} \end{pmatrix}, \tag{4.8}
$$

where there are r_A blocks

$$
K = \begin{pmatrix} -1 & 0 \\ 0 & 1 \end{pmatrix},
$$

and the ε_i are either 1 or -1. Then B is both orthogonal and an involution, and $BAB = A^{-1}$, no matter how the ε_i are chosen from $\{-1, 1\}$. $\qquad\square$

In fact, each member of the orthogonal group over any field is strongly reversible; see [81, 121, 247].

4.3.3 Reversible elements of the special orthogonal group

Proposition 4.14 *Let A be an element of $SO(n, \mathbb{R})$ with $s_A = t_A = 0$, and let S be an element of $O(n, \mathbb{R})$ such that $SAS^{-1} = A^{-1}$. Then*

$$
\det(S) = \begin{cases} 1 & \text{if } n \equiv 0 \pmod 4, \\ -1 & \text{if } n \equiv 2 \pmod 4. \end{cases}
$$

Proof Assume, by conjugation, that A is in standard form (4.7). For $i = 1, \ldots, r_A$, let V_i be the subspace spanned by the standard basis vectors e_{2i-1} and e_{2i}. Then $AS(V_i) = S(V_i)$. Since neither -1 nor 1 are eigenvalues of A, we see that A does not fix any one-dimensional space. On the other hand, for $i, j \in \{1, \ldots, r_A\}$, A fixes $S(V_i) \cap V_j$, which implies that either $S(V_i) = V_j$ or else $S(V_i) \cap V_j = \{0\}$. Therefore S permutes the collection $\{V_1, \ldots, V_{r_A}\}$. Let σ denote the induced permutation of $\{1, \ldots, r_A\}$. Since A fixes each of the subspaces V_i, the equation $SAS^{-1} = A^{-1}$ shows us that σ is an involution.

The matrix S is an r_A-by-r_A array of two-by-two blocks such that all but one two-by-two block in each row and each column is the 0 block. Let X_i denote the nonzero orthogonal two-by-two block in the $(\sigma(i), i)$th position. Then $X_i R_{\theta_i} X_i^{-1} = R_{-\theta_{\sigma(i)}}$. If $\det(X_i) = 1$ then X_i and R_{θ_i} commute (they are both two-dimensional rotations), so $\theta_i = -\theta_{\sigma(i)}$. If $\det(X_i) = -1$ then $X_i R_{\theta_i} X_i^{-1} = R_{-\theta_i}$ (X_i is a two-dimensional reflection), so $\theta_i = \theta_{\sigma(i)}$. Since $\theta_i = -\theta_{\sigma(i)}$ for an even number of values i from $\{1, \ldots, r_A\}$, we deduce that there are an even number of blocks X_i with determinant 1. Therefore the number of blocks X_i with determinant -1 has the same parity as r_A. Now, we can bring the blocks X_i to the leading diagonal of S by interchanging an even number of columns of S (since moving each block requires us to interchange two columns). It follows that the determinant of S is equal to the product of the determinants of the blocks X_i. Hence $\det(S) = (-1)^{r_A}$. $\square$

In contrast to Proposition 4.14, if $s_A + t_A \geqslant 1$ (for example, if n is odd) then we can choose whether $\det(S) = 1$ or $\det(S) = -1$ by using the matrix B from (4.8) and adjusting the entry ε_1.

Theorem 4.15 *For a matrix A in $\mathrm{SO}(n, \mathbb{R})$, the following are equivalent:*

 (i) *A is strongly reversible in $\mathrm{SO}(n, \mathbb{R})$*
 (ii) *A is reversible in $\mathrm{SO}(n, \mathbb{R})$*
 (iii) *$n \not\equiv 2 \pmod 4$ or $s_A + t_A \geqslant 1$.*

Proof That (i) implies (ii) is true in all groups. That (ii) implies (iii) follows from Proposition 4.14 and the remarks following that proposition. Assume condition (iii). This condition ensures that the involution B from (4.8) can be chosen to have positive determinant. Hence, by conjugating A to standard form, we see that A is strongly reversible in $\mathrm{SO}(n, \mathbb{R})$, which is condition (i). $\square$

Since $\mathrm{SO}(n, \mathbb{R})$ is a normal subgroup of the entirely reversible group $\mathrm{O}(n, \mathbb{R})$, we see that each element of $\mathrm{SO}(n, \mathbb{R})$ is reversed by an automorphism.

The only involutions in $\mathrm{SO}(2, \mathbb{R})$ are $\pm \mathbb{1}_2$, and these two matrices lie in the centre of the group. Therefore no other elements of $\mathrm{SO}(2, \mathbb{R})$ can be expressed as a product of involutions. For all other special orthogonal groups we have the following theorem.

Theorem 4.16   *Each member of* $\mathrm{SO}(n, \mathbb{R})$, $n \geqslant 3$, *can be expressed as the composite of three involutions in* $\mathrm{SO}(n, \mathbb{R})$. *Also, each member of* $\mathrm{SO}(n, \mathbb{R})$ *can be expressed as the composite of three involutions in* $\mathrm{O}(n, \mathbb{R})$, *two of which lie in* $\mathrm{O}(n, \mathbb{R}) \setminus \mathrm{SO}(n, \mathbb{R})$.

Proof   Consider an element A of $\mathrm{SO}(n, \mathbb{R})$ which, by conjugation, we may assume is in standard form. In Theorem 4.13 we saw that there is an orthogonal involution B given by (4.8) and another orthogonal involution

$$
C = \begin{pmatrix}
KR_{\theta_1} & & & & & \\
& \ddots & & & & \\
& & KR_{\theta_{r_A}} & & & \\
& & & \varepsilon_1 & & \\
& & & & \ddots & \\
& & & & & \varepsilon_{s_A + t_A}
\end{pmatrix},
$$

such that $A = \mathbb{1}_n BC$.

The determinants of B and C have the same sign. By modifying the factors B, C, and $\mathbb{1}_n$, we can simultaneously switch the signs of the determinants of both B and C. To achieve this, if $s_A + t_A > 0$ (for example, if n is odd) then we switch the sign of the $(2r_A + 1)$th diagonal entries in both B and C. If $s_A + t_A = 0$ then we adjust a pair of two-by-two blocks in each of $\mathbb{1}_n$, B, and C using the identity below:

$$
\begin{pmatrix} K & \\ & K \end{pmatrix} \begin{pmatrix} KR_{\theta_1} & \\ & KR_{\theta_2} \end{pmatrix} = \begin{pmatrix} K & \\ & K \end{pmatrix} \begin{pmatrix} KR_{\theta_1} & \\ & \mathbb{1}_2 \end{pmatrix} \begin{pmatrix} \mathbb{1}_2 & \\ & KR_{\theta_2} \end{pmatrix}.
$$

Both parts of Theorem 4.16 have thereby been accounted for.   □

4.4   The unitary group

As well as considering the standard unitary group $\mathrm{U}(n, \mathbb{C})$ we also consider reversibility in the *special unitary group* $\mathrm{SU}(n, \mathbb{C})$ which consists of unitary matrices with determinant 1. Our results on this group are needed in later chapters. For reversibility in unitary groups over fields other than $\mathbb{C}$, see [44, 72, 77].

4.4.1 Conjugacy

We make use of the following principal axis theorem, which can be found in [62].

Theorem 4.17 *Each conjugacy class in the two groups* $U(n, \mathbb{C})$ *and* $SU(n, \mathbb{C})$ *contains an element of the form*

$$
\begin{pmatrix}
z_1 & & & \\
& z_2 & & \\
& & \ddots & \\
& & & z_n
\end{pmatrix},
\tag{4.9}
$$

where $|z_i| = 1$ *for each index i.*   □

Corollary 4.18 *Every element of* $U(n, \mathbb{C})$ *and* $SU(n, \mathbb{C})$ *is reversed by an automorphism.*

Proof This follows immediately from Theorem 4.17, because the automorphism $A \mapsto \bar{A}$ reverses the matrix (4.9)   □

We need the next elementary lemma.

Lemma 4.19 *Each permutation matrix is unitary.*   □

It follows from Lemma 4.19 that the order of the entries z_i in (4.9) can be permuted without leaving the conjugacy class. (This holds in both groups $U(n, \mathbb{C})$ and $SU(n, \mathbb{C})$; we can force a permutation matrix to have determinant 1 by, if necessary, replacing a matrix entry 1 with -1.)

4.4.2 Reversible elements of the unitary group

Recall that a monic complex polynomial with roots $\alpha_1, \ldots, \alpha_n$, all nonzero, is *self reciprocal* if it is equal to the monic polynomial with roots $\alpha_1^{-1}, \ldots, \alpha_n^{-1}$. The next theorem is proven in [76].

Theorem 4.20 *For a unitary matrix A, the following are equivalent:*

(i) *A is strongly reversible*
(ii) *A is reversible*
(iii) *the characteristic polynomial of A is self reciprocal.*

Proof That (i) implies (ii) is true in all groups, and that (ii) implies (iii) is true because A is diagonalisable, and the conjugate matrices A and A^{-1} have the same characteristic polynomial. It remains only to prove that (iii) implies (i). By Theorem 4.17 we can assume that A is diagonal (with eigenvalues along the diagonal). Lemma 4.19 says that permutation matrices are unitary, and after conjugating A by a permutation matrix, we can assume that

$$A = \operatorname{diag}(\lambda_1, \lambda_1^{-1}, \ldots, \lambda_r, \lambda_r^{-1}, 1, \ldots, 1, -1, \ldots, -1).$$

Let

$$J = \begin{pmatrix} 0 & 1 \\ 1 & 0 \end{pmatrix},$$

and let

$$P = \begin{pmatrix} J & & & \\ & J & & \\ & & \ddots & \\ & & & \mathbb{1}_{n-2r} \end{pmatrix}.$$

Then P is an involutive permutation matrix, and $PAP^{-1} = A^{-1}$. Thus A is strongly reversible.   □

We finish by studying the collections $I^n(\mathrm{U}(n, \mathbb{C}))$. Only unitary matrices of determinant either 1 or -1 can be expressed as products of involutions.

Theorem 4.21 *Each element of* $\mathrm{U}(n, \mathbb{C})$ *with determinant either* 1 *or* -1 *can be expressed as a product of four involutions.*

Proof It suffices to prove the theorem for an element A of the form (4.9), where $\det(A)$ is either 1 or -1. Let

$$B = \operatorname{diag}(z_1, \overline{z_1}, z_1 z_2 z_3, \overline{z_1 z_2 z_3}, z_1 z_2 z_3 z_4 z_5, \ldots),$$
$$C = \operatorname{diag}(1, z_1 z_2, \overline{z_1 z_2}, z_1 z_2 z_3 z_4, \overline{z_1 z_2 z_3 z_4}, \ldots).$$

Each of these matrices is strongly reversible in $\mathrm{U}(n, \mathbb{C})$, by Theorem 4.20, and $A = BC$.   □

The example at the end of Section 4.2.4 shows that the four involutions from Theorem 4.21 cannot be reduced to three.

4.4.3 Reversible elements of the special unitary group

Theorem 4.22 *An element* A *of* $\mathrm{SU}(n, \mathbb{C})$ *is reversible if and only if its characteristic polynomial is self reciprocal.*

Proof If A is reversible, then A and A^{-1} have the same characteristic polynomial. Since A is diagonalisable, it follows immediately that its characteristic polynomial is self reciprocal. Conversely, suppose that the characteristic polynomial of A is self reciprocal. By conjugation, we can assume that A has the form

$$\begin{pmatrix} e^{i\theta_1} & & & & & & & \\ & e^{-i\theta_1} & & & & & & \\ & & \ddots & & & & & \\ & & & e^{i\theta_r} & & & & \\ & & & & e^{-i\theta_r} & & & \\ & & & & & \mathbb{1}_s & \\ & & & & & & -\mathbb{1}_t \end{pmatrix},$$

where $\theta_1, \ldots, \theta_r \in (0, \pi)$. This matrix is reversed by

$$\begin{pmatrix} 0 & 1 & & & & & \\ -1 & 0 & & & & & \\ & & \ddots & & & & \\ & & & 0 & 1 & & \\ & & & -1 & 0 & & \\ & & & & & \mathbb{1}_{s+t} \end{pmatrix}. \qquad (4.10)$$

$\square$

The reversing matrix (4.10) is not in general an involution. It can be made into an involution by replacing each block

$$\begin{pmatrix} 0 & 1 \\ -1 & 0 \end{pmatrix} \quad \text{with} \quad \begin{pmatrix} 0 & 1 \\ 1 & 0 \end{pmatrix};$$

however, if the number of such blocks is odd, the resulting involution has determinant -1. If $s + t > 0$ then we can replace one of the entries 1 on the leading diagonal by -1 to bring the determinant back to -1. Some reversible elements are not strongly reversible though. A complete classification was provided by Gongopadhyay and Parker [106, Proposition 3.3].

Theorem 4.23 *A reversible element A of $\mathrm{SU}(n, \mathbb{C})$ is strongly reversible unless $n \equiv 2 \pmod 4$ and neither -1 nor 1 are eigenvalues of A.* $\qquad \square$

We turn now to products of involutions in $\mathrm{SU}(n, \mathbb{C})$. The only involutions in $\mathrm{SU}(2, \mathbb{C})$ are $\pm \mathbb{1}_2$. For higher values of n, we have the following theorem.

Theorem 4.24 *For $n > 2$, each element of $\mathrm{SU}(n, \mathbb{C})$ can be expressed as a product of six involutions.*

Proof Let $A \in \mathrm{SU}(n, \mathbb{C})$. By Theorem 4.17 we can assume that

$$A = \mathrm{diag}(z_1, z_2, \ldots, z_n),$$

where $|z_i| = 1$ for each n. Define

$$J_1 = \begin{pmatrix} 0 & 1 & & \\ 1 & 0 & & \\ & & -1 & \\ & & & \mathbb{1}_{n-3} \end{pmatrix}, \quad J_2 = \begin{pmatrix} 0 & \overline{z_1} & & \\ z_1 & 0 & & \\ & & -1 & \\ & & & \mathbb{1}_{n-3} \end{pmatrix}.$$

Then J_1 and J_2 are both involutions in $\mathrm{SU}(n, \mathbb{C})$, and

$$J_2 J_1 A = \mathrm{diag}(1, z_1 z_2, z_3, \ldots, z_n).$$

Now let

$$B = \mathrm{diag}(1, 1, z_1 z_2 z_3, \overline{z_1 z_2 z_3}, z_1 z_2 z_3 z_4 z_5, \ldots),$$
$$C = \mathrm{diag}(1, z_1 z_2, \overline{z_1 z_2}, z_1 z_2 z_3 z_4, \overline{z_1 z_2 z_3 z_4}, \ldots),$$

so that $J_2 J_1 A = BC$. Since both B and C contain entries 1 in their main diagonals, we see from Theorem 4.22 (and the comments following Theorem 4.22), that there are involutions J_3, J_4, J_5, and J_6 in $\mathrm{SU}(n, \mathbb{C})$ such that $B = J_3 J_4$ and $C = J_5 J_6$. Therefore $A = J_1 J_2 J_3 J_4 J_5 J_6$. $\square$

We do not know whether four or five involutions are sufficient in Theorem 4.24.

4.5 Summary

We finish by summarising most of the key results of this chapter for the groups $\mathrm{GL}(n, \mathbb{C})$, $\mathrm{SL}(n, \mathbb{C})$, $\mathrm{O}(n, \mathbb{R})$, $\mathrm{SO}(n, \mathbb{R})$, $\mathrm{U}(n, \mathbb{C})$, and $\mathrm{SU}(n, \mathbb{C})$. The special linear group $\mathrm{SL}(n, \mathbb{C})$ consists of those members of $\mathrm{GL}(n, \mathbb{C})$ that have determinant 1, and it has not so far been considered. We discuss reversibility in this group briefly in the Notes, and include it here for completeness.

4.5.1 Reversible and strongly reversible

The class of reversible elements coincides with the class of strongly-reversible elements in $\mathrm{GL}(n, \mathbb{C})$, $\mathrm{O}(n, \mathbb{R})$, $\mathrm{SO}(n, \mathbb{R})$, and $\mathrm{U}(n, \mathbb{C})$, but not, in general, in $\mathrm{SL}(n, \mathbb{C})$ and $\mathrm{SU}(n, \mathbb{C})$. Of the six classes of groups, $\mathrm{O}(n, \mathbb{R})$ is the only class in which every element is reversible, and in fact strongly reversible.

Every element from each of the groups is reversed by an automorphism.

4.5.2 Products of involutions

The following table describes, for each class of groups, the least known integer m such that any element of a group G from the class can be expressed as a product of m involutions in G. The tick or cross indicates whether it is proven that m is the least possible such integer.

Group	m	least
$\mathrm{GL}(n,\mathbb{C})$	4	✓
$\mathrm{SL}(n,\mathbb{C})$	4	✓
$\mathrm{O}(n,\mathbb{R})$	2	✓
$\mathrm{SO}(n,\mathbb{R})$	3	✓
$\mathrm{U}(n,\mathbb{C})$	4	✓
$\mathrm{SU}(n,\mathbb{C})$	6	×

Notes

Sources

Most of the material of this chapter, and much more, is known by experts such as Djoković, Ellers, Knüppel, Nielsen, and Radjavi, among others. A sample of the vast literature is [27, 44, 70, 71, 75, 78, 79, 80, 81, 82, 88, 109, 121, 125, 126, 134, 149, 150, 151, 152, 153, 167, 169, 170, 200, 201, 216, 217, 238, 247]. The rational canonical form is described in [62]. Theorem 4.7 is proven in [70, 134, 247], and Theorem 4.9 is proven in [126]. Reversibility in the orthogonal group is dealt with in [75, 78, 81, 149, 150, 151, 153]. Unitary groups are considered in [44, 77].

Symplectic groups

One collection of classical groups that we have not considered are the *symplectic groups*. The 2*n*-by-2*n real symplectic group* is the group

$$\mathrm{Symp}(2n,\mathbb{R}) = \{A \in \mathrm{GL}(2n,\mathbb{R}) : AJA^t = J\},$$

where J is the 2*n*-by-2*n matrix

$$\begin{pmatrix} 0 & \mathbb{1}_n \\ -\mathbb{1}_n & 0 \end{pmatrix}.$$

Ellers [75] proved that, unlike $\mathrm{O}(n,\mathbb{R})$, there are elements of $\mathrm{Symp}(2n,\mathbb{R})$ that are not strongly reversible. On the other hand, there are symplectic groups over certain fields of characteristic 2 in which every element *is* strongly reversible [78, 81]. Wonenburger [247] proved that each element of $\mathrm{Symp}(2n, \mathbb{R})$ can be expressed as a product of two *skew*-symplectic involutions. In the next chapter we investigate a subgroup of $\mathrm{Symp}(2n, \mathbb{R})$, the *compact* symplectic group.

Special linear group

Knüppel and Nielsen [152] obtained a result similar to Theorem 4.9 for the special linear group $\mathrm{SL}(n, F)$ over any field F: they proved that each element of $\mathrm{SL}(n, F)$ can be expressed as a product of four involutions.

If F is algebraically closed then an element A of $\mathrm{SL}(n, F)$ is reversible in $\mathrm{SL}(n, F)$ if and only if it is reversible in $\mathrm{GL}(n, F)$. To see this, suppose that $BAB^{-1} = A^{-1}$, where $B \in \mathrm{GL}(n, F)$. Either $B \in \mathrm{SL}(n, F)$ or $B \notin \mathrm{SL}(n, F)$. In the latter case, $(BD)A(BD)^{-1} = A^{-1}$, where $D = \lambda \mathbb{1}_n$, and λ is any nth root of $1/\det(B)$ (so that $\det(D) = 1/\det(B)$). The matrix D lies in the centre of the group. Since $BD \in \mathrm{SL}(n, F)$ we conclude that A is reversible in $\mathrm{SL}(n, F)$.

Unlike $\mathrm{GL}(n, F)$, however, there are reversible elements in $\mathrm{SL}(n, F)$ that are not strongly reversible. For instance, the matrix

$$\begin{pmatrix} 2 & 0 \\ 0 & \frac{1}{2} \end{pmatrix}$$

in $\mathrm{SL}(2, \mathbb{C})$ is only reversed by involutions of the form

$$\begin{pmatrix} 0 & \lambda \\ \frac{1}{\lambda} & 0 \end{pmatrix},$$

and such involutions have determinant -1. One can argue similarly in $\mathrm{SL}(n, \mathbb{C})$ for certain larger values of n.

Discrete groups

The are numerous interesting subgroups of $\mathrm{SL}(2, \mathbb{R})$. Particularly useful are the discrete subgroups. We shall study $\mathrm{SL}(2, \mathbb{Z})$ in Chapter 7.

Finite simple groups of type G_2

Singh and Thakur [216, 217] considered the relationship between reversibility and strong reversibility in more general algebraic groups than those considered here. They obtain results on the finite simple groups of type G_2.

Automorphism groups

The automorphism groups of the classical groups were studied in depth by Dieudonné [67]. The approach of Dieudonné (building on techniques of Cartan, Mackey, Schreier, and van der Waerden) involves examining the manner in which automorphisms of a group permute involutions.

Commutators in the general linear group

Thompson [228, 229, 230] proved that, given a field F and an integer $n > 2$, each element of $SL(n, F)$ can be expressed as a commutator $XYX^{-1}Y^{-1}$, where $X, Y \in GL(n, F)$. See also [35] for a unified approach to certain matrix factorisation results.

Infinite dimensions

When one considers groups of linear transformations of infinite-dimensional spaces, one normally adds topology to the story, and considers continuous linear maps of Hilbert spaces, Banach spaces, or more general topological vector spaces.

For instance, Gustafson, Halmos, and Radjavi [126] showed that that for infinite-dimensional Hilbert spaces H, we have $I^4 \neq I^7 = GL(H)$. We do not know whether this holds if 7 is reduced to 6 or 5.

Hladnik, Omladič and Radjavi [133] proved that every invertible bounded linear operator on a complex infinite-dimensional Hilbert space can be expressed as a composition of five nth roots of the identity, for each integer $n > 2$.

Open problems

General linear group

In [126], Gustafson, Halmos, and Radjavi ask for a characterisation of $I^3(GL(n, F))$. There has been work on this question; see, for example, [170].

Affine group

Let $AF(n, \mathbb{R})$ denote the *affine group* that acts on $\mathbb{R}^n$. Each member g of $AF(n, \mathbb{R})$ can be expressed uniquely in the form $g(x) = Ax + b$, where $A \in GL(n, \mathbb{R})$ and $b \in \mathbb{R}^n$. By conjugating by a translation we may assume that b lies in the eigenspace E of A corresponding to the eigenvalue 1 (this eigenspace may be zero), in which case $g^{-1}(x) = A^{-1}x - b$. Given the depth of understanding of reversibility in $\mathrm{GL}(n, F)$, a description of reversibility in $\mathrm{AF}(n, \mathbb{R})$ seems within reach (even with more general fields than $\mathbb{R}$).

Symplectic groups

Classify reversibility comprehensively in the symplectic groups. Questions of reversibility in the symplectic group have been studied (as indicated above, see [75, 81, 78]) but complete answers are not known. Of particular interest here, of course, are the symplectic finite simple groups.

Unitary groups

What is the least integer m for which $I^m(\mathrm{SU}(n, \mathbb{C})) = \mathrm{SU}(n, \mathbb{C})$? We proved in Theorem 4.24 that $m \leqslant 6$, and certainly $m \geqslant 3$ by Theorem 4.22. Can we characterise the sets I^3, I^4, and I^5?

Groups associated to Banach Algebras

Let A be a Banach algebra (a complete normed complex algebra), which has an identity $\mathbb{1}$, with $\|\mathbb{1}\| = 1$. Little is known in general about reversibility in the group A^{-1} of invertible elements of A (when A is not commutative). One could also study reversibility in distinguished subgroups of A, such as

$$\mathrm{Iso}(A) = \{x \in A : \|x\| = \|x^{-1}\| = 1\}.$$

This coincides with the subgroup (often denoted $\mathrm{U}(A)$ [5]) of unitary elements, in case A is a C^* algebra. Another distinguished subgroup of A^{-1} is the normal subgroup

$$\{x \in A : \|a - 1\| < 1\}^{\infty},$$

which lies in the group $(\exp A)^{\infty}$.

Also of interest is the group $\mathrm{Aut}(A)$, and the subgroup of isometric automorphism of A. For example, suppose that X is a locally-compact Hausdorff space and A is the Banach algebra of continuous complex functions on X that vanish at ∞. Then $\mathrm{Aut}(A)$ is isomorphic to $\mathrm{Homeo}(X)$. The case $X = \mathbb{R}$ is dealt with in Chapter 8 and the case $X = \mathbb{S}^1$ is dealt with in Chapter 9. For the disk algebra, the automorphism group is isomorphic to $\mathrm{PSL}(2, \mathbb{R})$, which is dealt with in Chapter 6. As a final example, the algebra of all formal power series in n indeterminates, with complex coefficients, has a natural Frechet algebra structure, and embeds in some Banach algebras [5]. Its automorphism group is isomorphic to the group of formally-invertible formal germs, considered in Chapter 10.

5
Compact groups

5.1 Reversibility in compact groups

A *compact group* is a topological group that is compact. We assume that all our compact groups are Hausdorff, although that is not usually part of the definition. It is a well known maxim that theorems on finite groups have counterparts in compact groups. We aim to mimic some of the earlier theory of reversibility in finite groups for compact groups. We assume a little knowledge of the representation theory of compact groups, which can be found, for example, in [209, 215].

Given a compact group G, let $C(G)$ denote the Banach space of complex-valued continuous functions on G, equipped with the supremum norm. An *invariant function* on G is a complex-valued function that is constant on conjugacy classes of G. Let $IF(G)$ denote the closed subspace of $C(G)$ consisting of continuous invariant functions. Let $X(G)$ denote the linear span of the characters of finite-dimensional irreducible unitary representations of G. The next lemma follows from the Peter–Weyl theorem [215, Theorem VII.10.1].

Lemma 5.1 *The space $X(G)$ is uniformly dense in $IF(G)$.* □

Corollary 5.2 *Given nonconjugate elements a and b of a compact group G, there is a character χ on G such that $\chi(a) \neq \chi(b)$.*

Proof The conjugacy class C_a of a is compact, because it is the continuous image of G under the map $g \mapsto gag^{-1}$. Likewise the conjugacy class C_b of b is compact. Now, G is compact and Hausdorff, and hence normal, which implies that, by Urysohn's Lemma, there is a continuous function $\phi : G \to [0,1]$ that maps C_a to 1 and C_b to 0. Let λ denote Haar measure on G. Then we can define an element ψ of $IF(G)$ by

$$\psi(g) = \int_G \phi(xgx^{-1}) \, d\lambda(x).$$

The function ψ satisfies $\psi(a) = 1$ and $\psi(b) = 0$. Using Lemma 5.1 we deduce the existence of a character χ such that $\chi(a) \neq \chi(b)$.   □

Theorem 5.3 *An element g of a compact group G is reversible if and only if each irreducible character on G is real when evaluated at g.*

Proof Suppose that g is reversible, and that χ is an irreducible character on G. We can assume that χ is the character of a unitary representation. Thus $\chi(g)$ is a sum of roots of unity (the eigenvalues of the representation of g). Therefore

$$\chi(g) = \chi(g^{-1}) = \overline{\chi(g)}.$$

Conversely, if g and g^{-1} are not conjugate then Corollary 5.2 shows that there is an irreducible character χ such that $\chi(g) \neq \chi(g^{-1})$. This implies that $\chi(g) \neq \overline{\chi(g)}$, so $\chi(g)$ is not real.   □

5.2 Compact Lie groups

We move on to consider *compact Lie groups*. Again, the reader is referred to [209, 215] for the theory of compact Lie groups. The simple Lie groups were classified by Cartan. The simple, simply-connected compact Lie groups are $SU(n, \mathbb{C})$ $(n \geqslant 2)$, $Sp(n, \mathbb{C})$ $(n \geqslant 2)$, $Spin(n, \mathbb{C})$ $(n \geqslant 5)$, E_6, E_7, E_8, F_4, and G_2. There is a small amount of repetition in this list because $Spin(5, \mathbb{C}) = Sp(2, \mathbb{C})$ and $Spin(6, \mathbb{C}) = SU(4, \mathbb{C})$. We have already met $SU(n, \mathbb{C})$ in Section 4.4.3; we briefly return to this group in the next section, before studying $Sp(n, \mathbb{C})$ and then $Spin(n, \mathbb{C})$. We do not consider the five exceptional Lie groups.

Theorem 5.4 *Each compact, connected Lie group G satisfies*

$$G = K/H, \qquad K = G_1 \times G_2 \times \cdots \times G_m,$$

where each G_i is either a simple, simply-connected compact Lie group, or else a copy of $SO(2, \mathbb{R})$, and H is a discrete subgroup of $Z(K)$, the centre of K.   □

From the theorem we see that, given an understanding of reversibility in the simple, simply-connected compact Lie groups, much can be said about reversibility in general compact, connected Lie groups. In the remaining sections of this chapter, we study the three infinite families of simple, simply-connected compact Lie groups.

5.3   The special unitary group

We have already discussed the special unitary group in Section 4.4.3, and we found, in Theorem 4.22, that an element A of $\mathrm{SU}(n, \mathbb{C})$ whose nonreal eigenvalues occur in conjugate pairs is reversible. According to Theorem 5.3, each irreducible character on $\mathrm{SU}(n, \mathbb{C})$ is real when evaluated at the reversible matrix A. In this section we reaffirm this observation.

Let μ denote a sequence of integers $0 = \mu_n \leqslant \mu_{n-1} \leqslant \ldots \leqslant \mu_1$. Given a matrix A in $\mathrm{SU}(n, \mathbb{C})$ with eigenvalues $z_1, \ldots, z_n$ we define $X_\mu(A)$ to be the matrix with (i, j)th coefficient z_i^{n-j}, and we define $Y_\mu(A)$ to be the matrix with (i, j)th coefficient $z_i^{\mu_j + n - j}$. Finally, we define $\chi_\mu(A) = \det(Y_\mu(A)) / \det(X_\mu(A))$. The map χ_μ is an irreducible character of $\mathrm{SU}(n, \mathbb{C})$, and all irreducible characters take this form (see [215, Theorem IX.9.1]).

Suppose that the nonreal eigenvalues of an element A of $\mathrm{SU}(n, \mathbb{C})$ occur in conjugate pairs. Then the rows of each of the matrices $X_\mu(A)$ and $Y_\mu(A)$ can be partitioned into complex conjugate pairs, and singleton real rows.

Lemma 5.5   *Let B be a square matrix whose rows can be partitioned into m pairs of complex conjugate rows, and a number of singleton real rows. Then $\det(B)$ is real if m is even, and $\det(B)$ is purely imaginary if m is odd.*

Proof   Interchanging two rows of a matrix reverses the sign of the determinant. In the matrix B, interchange each pair of complex conjugate rows. We have transformed B to $\overline{B}$. Thus

$$\det(B) = (-1)^m \det(\overline{B}) = (-1)^m \overline{\det(B)},$$

from which the result follows.   □

From Lemma 5.5 we see that $\det(X_\mu(A))$ and $\det(Y_\mu(A))$ are either both real or both purely imaginary. Hence $\chi_\mu(A)$ is real, as required.

5.4   Compact symplectic groups

Recall that the symplectic group over a field F is the group

$$\mathrm{Symp}(2n, F) = \{A \in \mathrm{GL}(n, F) \,|\, AJA^t = J\},$$

where

$$J = \begin{pmatrix} 0 & \mathbb{1}_n \\ -\mathbb{1}_n & 0 \end{pmatrix},$$

$\mathbb{1}_n$ is the n-by-n identity matrix, and A^t denotes the transpose matrix of A. For each integer $n \geqslant 2$, the group

$$\mathrm{Sp}(n,\mathbb{C}) = \mathrm{Symp}(2n,\mathbb{C}) \cap \mathrm{U}(2n,\mathbb{C})$$

is the *compact symplectic group*, and it is these groups that we study in this section. Unlike $\mathrm{Symp}(2n,\mathbb{C})$, the group $\mathrm{Sp}(n,\mathbb{C})$ is a simple, simply-connected compact Lie group. It can also be described as the group of n-by-n unitary quaternionic matrices.

The next lemma is relatively straightforward to prove; see [215, Proposition VII.5.4] for help.

Lemma 5.6 *Each element of* $\mathrm{Sp}(n,\mathbb{C})$ *is conjugate to a matrix of the form* $\mathrm{diag}(e^{i\theta_1},\dots,e^{i\theta_n},e^{-i\theta_1},\dots,e^{-i\theta_n})$. $\qquad\qquad\square$

Using this normal form we see immediately that all members of $\mathrm{Sp}(n,\mathbb{C})$ are reversible.

Theorem 5.7 *Each element of* $\mathrm{Sp}(n,\mathbb{C})$ *is reversible.*

Proof It is sufficient to prove the theorem for matrices in the normal form of Lemma 5.6. Let $A = \mathrm{diag}(e^{i\theta_1},\dots,e^{i\theta_n},e^{-i\theta_1},\dots,e^{-i\theta_n})$. Then $\overline{A} = A^{-1}$. Hence, using the definition of $\mathrm{Sp}(n,\mathbb{C})$,

$$J^{-1}AJ = (A^t)^{-1} = \overline{A} = A^{-1}. \qquad\qquad\square$$

This theorem can also be proven by observing that all the irreducible characters on $\mathrm{Sp}(n,\mathbb{C})$ are real valued, and then following a similar procedure to that given at the end of Section 5.3. This time, we may assume that our element A of $\mathrm{Sp}(n,\mathbb{C})$ has eigenvalues $z_1,\dots,z_n,\overline{z_1},\dots,\overline{z_n}$. Given a sequence μ of the form $0 = \mu_n \leqslant \mu_{n-1} \leqslant \dots \leqslant \mu_1$ we define $X_\mu(A)$ to be the matrix with (i,j)th coefficient $z_i^{n+1-j} - z_i^{-(n+1-j)}$, and we define $Y_\mu(A)$ to be the matrix with (i,j)th coefficient $z_i^{\mu_j+n+1-j} - z_i^{-(\mu_j+n+1-j)}$. Finally, we define $\chi_\mu(A) = \det(Y_\mu(A))/\det(X_\mu(A))$. The map χ_μ is an irreducible character of $\mathrm{Sp}(n,\mathbb{C})$, and all irreducible characters take this form (see [215, Theorem IX.9.3]). Using Lemma 5.5 we can, as before, check that $\chi_\mu(A)$ is real.

The involutions in $\mathrm{Sp}(n,\mathbb{C})$ are the elements with only real eigenvalues. We make particular use of two classes of involutions. First, if U is an n-by-n unitary matrix involution then

$$X = \begin{pmatrix} U & \\ & \overline{U} \end{pmatrix} \qquad\qquad (5.1)$$

is an involution in $\mathrm{Sp}(n,\mathbb{C})$. Next, suppose that n is even. Define, for a complex number z of unit modulus,

$$V_z = \begin{pmatrix} & & & & z \\ & & & 1 & \\ & & \cdot\cdot\cdot & & \\ & -1 & & & \\ -z & & & & \end{pmatrix}.$$

The matrix V is antisymmetric; that is, $V = -V^t$. The $2n$-by-$2n$ matrix

$$Y_z = \begin{pmatrix} & V_z \\ -\overline{V_z} & \end{pmatrix} \tag{5.2}$$

is an involution in $\mathrm{Sp}(n,\mathbb{C})$. Now suppose that n is odd. Define, for a complex number z of unit modulus,

$$W_z = \begin{pmatrix} & & & & & & z \\ & & & & & 1 & \\ & & & & \cdot\cdot\cdot & & \\ & & & 1 & & & \\ & & 0 & & & & \\ & -1 & & & & & \\ & \cdot\cdot\cdot & & & & & \\ -1 & & & & & & \\ -z & & & & & & \end{pmatrix}.$$

Again, W_z is antisymmetric. Let E be the n-by-n matrix with a 1 in the single middlemost position in the matrix, and 0s elsewhere. The $2n$-by-$2n$ matrix

$$Y_z = \begin{pmatrix} E & W_z \\ -\overline{W_z} & E \end{pmatrix} \tag{5.3}$$

is an involution in $\mathrm{Sp}(n,\mathbb{C})$. Notice that, whether n is odd or even,

$$Y_z Y_1 = \mathrm{diag}(z, 1, \ldots, 1, z; \bar{z}, 1, \ldots, 1, \bar{z}) \tag{5.4}$$

(the semicolon separates the first n terms from the second n terms).

Now we consider strongly-reversible elements in $\mathrm{Sp}(n,\mathbb{C})$. We prove that, even though every element in $\mathrm{Sp}(n,\mathbb{C})$ is reversible, most elements are *not* strongly reversible.

Theorem 5.8 *Choose nonreal complex numbers of unit modulus $z_1, \ldots, z_n$ such that, for $i \neq j$, $z_i \neq z_j$ and $z_i \neq \overline{z_j}$. Then $\mathrm{diag}(z_1, \ldots, z_n, \overline{z_1}, \ldots, \overline{z_n})$ is not strongly reversible in $\mathrm{Sp}(n,\mathbb{C})$.*

Proof Let $D = \mathrm{diag}(z_1, \ldots, z_n, \overline{z_1}, \ldots, \overline{z_n})$. Suppose there is an involution P in $\mathrm{GL}(n, \mathbb{C})$ such that $PDP = D^{-1}$. For each pair (i,j) we have $P_{ij}D_{jj} = D_{ii}^{-1}P_{ij}$. Hence $P_{ij} \neq 0$ if and only if $|i-j| = n$. Since also $J_{ij} \neq 0$ if and only if $|i-j| = n$ we see that

$$(J^t P J)_{ij} = \sum_{k,l} J_{ki} P_{kl} J_{lj} = -P_{ji}.$$

That is, $J^t P J = -P^t$. On the other hand, if $P \in \mathrm{Symp}(2n, \mathbb{C})$ then $J^t P J = P^t$, which implies that $P = 0$. Thus $P \notin \mathrm{Symp}(2n, \mathbb{C})$, and hence $P \notin \mathrm{Sp}(n, \mathbb{C})$. $\square$

Our final result is on products of involutions in $\mathrm{Sp}(n, \mathbb{C})$.

Theorem 5.9 *Each element of* $\mathrm{Sp}(n, \mathbb{C})$ *can be expressed as a product of six involutions.*

Proof Choose an element A of $\mathrm{Sp}(n, \mathbb{C})$; by conjugation we can assume that $A = \mathrm{diag}(z_1, \ldots, z_n, \overline{z_1}, \ldots, \overline{z_n})$. Let z be one of the square-roots of $1/(z_1 z_2 \cdots z_n)$. Define involution Y_z and Y_1 as in (5.2) or (5.3) (depending on whether n is even or odd). Using (5.4) we see that the matrix $Y_z Y_1 A$ is a diagonal matrix of the form

$$\begin{pmatrix} A_0 & \\ & \overline{A_0} \end{pmatrix},$$

where $A_0 = \mathrm{diag}(\zeta_1, \ldots, \zeta_n)$, and $\zeta_1 \zeta_2 \cdots \zeta_n = 1$.

Next, just as in Theorem 4.21 we define n-by-n matrices

$$B_0 = \mathrm{diag}(\zeta_1, \overline{\zeta_2}, \zeta_1 \zeta_2 \zeta_3, \overline{\zeta_1 \zeta_2 \zeta_3}, \zeta_1 \zeta_2 \zeta_3 \zeta_4 \zeta_5, \ldots),$$
$$C_0 = \mathrm{diag}(1, \zeta, \overline{\zeta_1 \zeta_2}, \zeta_1 \zeta_2 \zeta_3 \zeta, \overline{\zeta_1 \zeta_2 \zeta_3 \zeta_4}, \ldots).$$

Each of these matrices is strongly reversible in $\mathrm{U}(n, \mathbb{C})$, by Theorem 4.20. In other words, we can find involutions U_1, U_2, U_3, and U_4 in $\mathrm{U}(n, \mathbb{C})$ such that $B_0 = U_1 U_2$ and $C_0 = U_3 U_4$. Since $A_0 = B_0 C_0$ we see that $A_0 = U_1 U_2 U_3 U_4$. Furthermore, by (5.1) each of the matrices

$$X_i = \begin{pmatrix} U_i & \\ & \overline{U_i} \end{pmatrix}$$

is an involution in $\mathrm{Sp}(n, \mathbb{C})$. Since $Y_z Y_1 A = X_1 X_2 X_3 X_4$ it follows that $A = Y_1 Y_z X_1 X_2 X_3 X_4$, and we have thereby expressed A as a product of six involutions in $\mathrm{Sp}(n, \mathbb{C})$. $\square$

We do not know whether the number six in Theorem 5.9 can be lowered.

5.5   The spinor groups

We provide a brief introduction to spinor groups; the reader is referred to [23, Chapter 5] to fill in details.

For each integer $n = 0, 1, 2, \ldots$, we will define the *Clifford algebra* C_n to be a real algebra of dimension 2^n. This algebra contains elements $e_1, \ldots, e_n$ that satisfy $e_i^2 = -1$ for each integer i, and $e_i e_j = -e_j e_i$ whenever $i \neq j$. A basis for C_n is comprised of all elements of the form $e_{i_1} \cdots e_{i_k}$, where $0 \leqslant k \leqslant n$ and $1 \leqslant i_1 < \cdots < i_k \leqslant n$. If $k = 0$ then $e_{i_1} \cdots e_{i_k}$ is considered to be equal to the identity element 1. This basis uniquely specifies C_n.

Let us consider some examples of Clifford algebras C_n for low values of n. The Clifford algebra C_0 is one dimensional, with basis element 1, so it is isomorphic to $\mathbb{R}$. The Clifford algebra C_1 has basis elements 1 and e_1, where $e_1^2 = -1$, so it is isomorphic to the complex numbers $\mathbb{C}$. The Clifford algebra C_2 has basis elements 1, e_1, e_2, and $e_1 e_2$. If we label the latter three basis elements by i, j, and k, respectively, then we can check that $i^2 = j^2 = k^2 = -1$, $ij = -ji$, $jk = -kj$, and $ki = -ik$ (and all other relations between i, j, and k are consequences of these). It follows that C_2 is isomorphic to the quaternions. Only C_0 and C_1 are fields, and only C_2 is a skew-field. The remaining Clifford algebras are not division algebras.

There is an automorphism of C_n, denoted $x \mapsto x^\dagger$, that is defined by the property that $e_i \mapsto -e_i$ for $i = 1, \ldots, n$. There is an antiautomorphism of C_n, denoted $x \mapsto x'$, that is defined by the property that $e_{i_1} \cdots e_{i_k} \mapsto e_{i_k} \cdots e_{i_1}$ for $i_1, \ldots, i_k \in \{1, \ldots, n\}$. There is another antiautomorphism of C_n, denoted $x \mapsto \bar{x}$, that is the composite, in either order, of the maps $x \mapsto x^\dagger$ and $x \mapsto x'$. On C_1 and C_2, the antiautomorphism $x \mapsto \bar{x}$ is the usual conjugation.

As a vector space, C_n inherits the Euclidean norm from $\mathbb{R}^{2^n}$ by identifying the basis vectors $e_{i_1} \cdots e_{i_k}$ with the standard basis of $\mathbb{R}^{2^n}$. We denote this norm by $|x|$, where $x \in C_n$.

The collection of elements of the form $x_1 e_1 + \cdots + x_n e_n$ is a subspace of C_n, which we identify with $\mathbb{R}^n$ (let e_i correspond to the ith standard basis vector). Notice that, for elements x and y of $\mathbb{R}^n$, $|x|^2 = x\bar{x} = \bar{x}x$ and $|xy| = |x||y|$. The first of these identities shows that if x is a nonzero element of $\mathbb{R}^n$, then it has a multiplicative inverse, namely $\bar{x}/|x|^2$. The *Clifford group* Γ_n is the multiplicative group generated by $\mathbb{R}^n \setminus \{0\}$. As usual, we denote by $\mathbb{S}^{n-1}$ the elements in $\mathbb{R}^n$ of modulus 1.

The next lemma can be proven by expressing each element of Γ_n as a product of elements of $\mathbb{R}^n$.

Lemma 5.10   *For elements x and y of Γ_n,*

(i) $|x|^2 = x\bar{x} = \bar{x}x$
(ii) $|xy| = |x||y|$. □

For $n \geqslant 2$, the *pinor group* $\mathrm{Pin}(n, \mathbb{C})$ consists of those elements of Γ_n with norm 1. The *spinor group* is the subgroup $\mathrm{Spin}(n, \mathbb{C})$ of the pinor group defined by

$$\mathrm{Spin}(n, \mathbb{C}) = \{x \in \mathrm{Pin}(n, \mathbb{C}) : x = x^\dagger\}.$$

It can also be defined as the subgroup of $\mathrm{Pin}(n, \mathbb{C})$ consisting of products of even numbers of elements of $\mathbb{R}^n$.

Given an element u of $\mathrm{Pin}(n, \mathbb{C})$, one can check that if $x \in \mathbb{R}^n$ then $u^\dagger x \bar{u}$ also belongs to $\mathbb{R}^n$. (To see this, first check the statement for elements u of $\mathbb{S}^{n-1}$.) Therefore we can define a map $\rho_u : \mathbb{R}^n \to \mathbb{R}^n$ by $\rho_u(x) = u^\dagger x \bar{u}$. This map is linear. It is orthogonal because $|\rho_u(x)| = |x|$. Thus we have a homomorphism ρ from $\mathrm{Pin}(n, \mathbb{C})$ to $\mathrm{O}(n, \mathbb{R})$ given by $u \mapsto \rho_u$.

The kernel of the homomorphism ρ is $\{-1, 1\}$ [23, Proposition 5.25]. Let us now show that ρ is surjective. If $u \in \mathbb{S}^{n-1}$, then the map ρ_u is a reflection in the hyperplane perpendicular to u. This is because $\rho_u(u) = -u$, and if v is an element of $\mathbb{S}^{n-1}$ orthogonal to u (so that $u \cdot v = 0$ or, equivalently, $uv = -vu$) then $\rho_u(v) = v$. Since $\mathrm{O}(n, \mathbb{R})$ is generated by such reflections, we see that ρ is indeed surjective.

An element of $\mathrm{Pin}(n, \mathbb{C})$ lies in $\mathrm{Spin}(n, \mathbb{C})$ if and only if it can be expressed as a product of an even number of elements from $\mathbb{S}^{n-1}$. Given an element u of $\mathbb{S}^{n-1}$, the determinant of ρ_u is -1, so we conclude that ρ maps $\mathrm{Spin}(n, \mathbb{C})$ onto $\mathrm{SO}(n, \mathbb{R})$ with kernel $\{-1, 1\}$. Thus $\mathrm{Spin}(n, \mathbb{C})$ is a two-fold covering group of $\mathrm{SO}(n, \mathbb{R})$. It is both compact and path connected [23, Theorem 5.28]. If $n \geqslant 3$ then it is simply connected, so it is the universal cover of $\mathrm{SO}(n, \mathbb{R})$, and it is often defined as such.

If we lift the normal form for elements of $\mathrm{SO}(n, \mathbb{R})$ given by equation (4.7) to $\mathrm{Spin}(n, \mathbb{C})$ using the just defined covering, then we obtain the following lemma.

Lemma 5.11 *Each element of* $\mathrm{Spin}(n, \mathbb{C})$ *is conjugate to an element of the form*

$$(\lambda_1 + \mu_1 e_1 e_2) \cdots (\lambda_m + \mu_m e_{2m-1} e_{2m}), \tag{5.5}$$

where m is the integer part of $n/2$, and λ_i and μ_i are complex numbers such that $\lambda_i^2 + \mu_i^2 = 1$. □

We say that an element of $\mathrm{Spin}(n, \mathbb{C})$ of the form (5.5) has *normal form*. Note that the inverse of an element of normal form is obtained by replacing each μ_i

with $-\mu_i$. If x is the element (5.5), and θ_i are chosen such that $\lambda_i = \cos\theta_i$ and $\mu_i = \sin\theta_i$, then ρ_x is the n-by-n matrix with two-by-two blocks

$$\begin{pmatrix} \cos\theta_i & -\sin\theta_i \\ \sin\theta_i & \cos\theta_i \end{pmatrix}$$

down the diagonal, followed, if n is odd, by a single 1-by-1 block containing the number 1.

We can now study reversibility questions in $\mathrm{Spin}(n,\mathbb{C})$.

Lemma 5.12   *Let x be the element $(\lambda_1 + \mu_1 e_1 e_2) \cdots (\lambda_m + \mu_m e_{2m-1} e_{2m})$ of $\mathrm{Spin}(n,\mathbb{C})$ described in (5.5). Suppose either that one of the μ_i vanishes, or that $n \equiv 0,1,3 \pmod 4$. Then x is reversible in $\mathrm{Spin}(n,\mathbb{C})$.*

Proof   Choose integers $1 \leqslant i_1 < i_2 < \cdots < i_k$ such that k is even, and such that for each pair $(2j-1, 2j)$ for which $\mu_j \neq 0$, precisely one of the indices $2j-1$ and $2j$ occurs in the set $\{i_1, \ldots, i_k\}$. It is possible to do this if at least one of the μ_i vanishes, or if $n \equiv 0,1,3 \pmod 4$. Define $g = e_{i_1} \cdots e_{i_k}$; then $gxg^{-1} = x^{-1}$. $\square$

Theorem 5.13   *An element of $\mathrm{Spin}(n,\mathbb{C})$ is reversible if and only if either $n \equiv 0,1,3 \pmod 4$ or one of the numbers μ_i in the normal form of x vanishes.*

Proof   If x is a reversible member of $\mathrm{Spin}(n,\mathbb{C})$ then ρ_x is reversible in the special orthogonal group $\mathrm{SO}(n,\mathbb{R})$. It follows from Theorem 4.15 that either $n \not\equiv 2 \pmod 4$, or one of the μ_i vanishes. Conversely, suppose that either $n \equiv 0,1,3 \pmod 4$ or one of the μ_i vanishes; then by Lemma 5.12 we may assume that our element x is in normal form, and we can apply Lemma 5.12 to deduce that x is reversible. $\square$

We turn now to strongly-reversible elements of $\mathrm{Spin}(n,\mathbb{C})$. The reversing element in Lemma 5.12 took the form $e_{i_1} \cdots e_{i_k}$. This is an involution in the spinor group $\mathrm{Spin}(n,\mathbb{C})$ if and only if $k \equiv 0 \pmod 4$. The relationship between reversibility and strong reversibility depends on the value of n, and we examine various cases rather than supplying a full explanation.

For certain values of n, strong reversibility and reversibility coincide. Part of the next proposition was proven in [234, Theorem 2.2].

Proposition 5.14   *Suppose that $n \equiv 0,1,7 \pmod 8$. Then every element of $\mathrm{Spin}(n,\mathbb{C})$ is strongly reversible.*

Proof   Consider an element x of $\mathrm{Spin}(n,\mathbb{C})$, which we can assume is in the normal form of (5.5) (with $m \equiv 0 \pmod 4$). If $n \equiv 0,1 \pmod 8$ then define $g = e_1 e_3 \cdots e_{2m-1}$. If $n \equiv 7 \pmod 8$ then define $g = e_1 e_3 \cdots e_{2m-1} e_{2m+1}$. Then g is an involution and $gxg = x^{-1}$. $\square$

Furthermore, there are large classes of strongly-reversible elements in every group $\mathrm{Spin}(n, \mathbb{C})$.

Proposition 5.15 *An element* $x = (\lambda_1 + \mu_1 e_1 e_2) \cdots (\lambda_m + \mu_m e_{2m-1} e_{2m})$ *of* $\mathrm{Spin}(n, \mathbb{C})$, $n \geqslant 4$, *for which two of the* μ_i *vanish is strongly reversible.*

Proof Without loss of generality we can assume that $\mu_{m-1} = \mu_m = 0$. Adjoin either zero, one, two, or three of the elements $\{e_{2m-3}, e_{2m-2}, e_{2m-1}, e_{2m}\}$ to the end of the product $e_1 e_3 \cdots e_{2m-5}$ so that the number of terms of the resulting product g is congruent to 0 modulo 4. Then g is an involution such that $gxg = x^{-1}$. $\square$

For some values of n there are reversible elements that are not strongly reversible; for example, in $\mathrm{Spin}(4, \mathbb{C})$ the only involutions are $\{\pm 1, \pm e_1 e_2 e_3 e_4\}$, and these involutions lie in the centre of $\mathrm{Spin}(4, \mathbb{C})$. (Whereas, in contrast, every element of $\mathrm{Spin}(4, \mathbb{C})$ is reversible.)

It remains only to consider products of involutions.

Theorem 5.16 *Every element of* $\mathrm{Spin}(n, \mathbb{C})$, $n \geqslant 5$, *can be expressed as a product of four involutions.*

Proof Choose an element $x = (\lambda_1 + \mu_1 e_1 e_2) \cdots (\lambda_m + \mu_m e_{2m-1} e_{2m})$ (in normal form). Observe that $x = uv$ where $u = (\lambda_1 + \mu_1 e_1 e_2)(\lambda_2 + \mu_2 e_3 e_4)$ and $v = (\lambda_3 + \mu_3 e_5 e_6) \cdots (\lambda_m + \mu_m e_{2m-1} e_{2m})$. The element u is strongly reversible by the involution $e_1 e_2 e_3 e_4$, and v is strongly reversible by Proposition 5.15. Hence x is a product of four involutions. $\square$

We do not know whether every element of $\mathrm{Spin}(n, \mathbb{C})$, $n \geqslant 5$, can be expressed as a product of three involutions.

Notes

Sources

Texts by Sepanski [209] and Simon [215] describe the representation theory of compact groups. In particular, Theorem 5.4 is [215, Theorem VII.8.1] and Lemma 5.6 is proven on [215, page 143]. Clifford algebras and matrix groups are discussed in [23].

Elements reversed by an automorphism in spinor groups

Every element of $\mathrm{Spin}(n, \mathbb{C})$ is reversed by an automorphism; just embed in a larger spinor group. Feit and Zuckerman [88] found small extensions of spinor and compact symplectic groups in which all elements become reversible.

Open problems

The exceptional Lie groups

Classify reversibility in the exceptional Lie groups E_6, E_7, E_8, F_4, and G_2. Singh and Thakur [216, 217] have worked on G_2.

Strongly-reversible elements of compact symplectic groups

We are unaware of a complete characterisation of the strongly-reversible elements of $\mathrm{Sp}(n, \mathbb{C})$.

Products of involutions in compact symplectic groups

What is the least integer m such that $I^m(\mathrm{Sp}(n, \mathbb{C})) = \mathrm{Sp}(n, \mathbb{C})$? We proved in Theorem 5.8 that $m \geqslant 3$, and it follows from Theorem 5.9 that $m \leqslant 6$. We know little about the sets I^3, I^4, or I^5.

Strongly-reversible elements of spinor groups

Propositions 5.14 and 5.15 both discuss the problem of characterising the strongly-reversible elements of spinor groups, but we do not have a complete characterisation of such elements.

Products of involutions in spinor groups

We prove in Theorem 5.16 that, for $n \geqslant 5$, each element of $\mathrm{Spin}(n, \mathbb{C})$ can be expressed as a product of four involutions. Will three involutions suffice? If not then is it possible to characterise the set $I^3(\mathrm{Spin}(n, \mathbb{C}))$ neatly?

6

Isometry groups

6.1 Isometries of spherical, Euclidean, and hyperbolic space

For each positive integer n, there are three simply-connected, complete Riemannian n-manifolds with constant curvature, namely n-dimensional spherical space $\mathbb{S}^n$, n-dimensional Euclidean space $\mathbb{R}^n$, and n-dimensional hyperbolic space $\mathbb{H}^n$. Note that $\mathbb{R}^1$ and $\mathbb{H}^1$ are isometric, but otherwise there are no repetitions in this list. We denote the isometry groups of these manifolds by $\mathrm{Isom}(\mathbb{S}^n)$, $\mathrm{Isom}(\mathbb{R}^n)$, and $\mathrm{Isom}(\mathbb{H}^n)$. These groups are each generated by reflections. We denote the three subgroups of these three isometry groups, comprised of orientation-preserving isometries, by $\mathrm{Isom}^+(\mathbb{S}^n)$, $\mathrm{Isom}^+(\mathbb{R}^n)$, and $\mathrm{Isom}^+(\mathbb{H}^n)$. A map in $\mathrm{Isom}(\mathbb{S}^n)$ lies in $\mathrm{Isom}^+(\mathbb{S}^n)$ if and only if it can be expressed as a composite of an even number of reflections. Similar comments apply to the groups $\mathrm{Isom}^+(\mathbb{R}^n)$ and $\mathrm{Isom}^+(\mathbb{H}^n)$.

We studied the orthogonal group $\mathrm{O}(n, \mathbb{R})$ and the special orthogonal group $\mathrm{SO}(n, \mathbb{R})$ in Chapter 4; these two groups are $\mathrm{Isom}(\mathbb{S}^{n-1})$ and $\mathrm{Isom}^+(\mathbb{S}^{n-1})$, respectively. In this chapter we consider reversibility in the remaining four isometry groups $\mathrm{Isom}(\mathbb{R}^n)$, $\mathrm{Isom}(\mathbb{H}^n)$, $\mathrm{Isom}^+(\mathbb{R}^n)$, and $\mathrm{Isom}^+(\mathbb{H}^n)$.

6.2 Hyperbolic geometry in two and three dimensions

In Chapter 1 we briefly discussed reversibility in the Euclidean isometry groups $\mathrm{Isom}^+(\mathbb{R}^2)$ and $\mathrm{Isom}^+(\mathbb{R}^3)$. We found that, in two dimensions, the only elements that are strongly reversible, other than involutions, are translations. In three dimensions we found that all isometries are strongly reversible. Before we tackle higher-dimensional isometry groups we first, in this section, consider isometry groups of two- and three-dimensional hyperbolic space. We use different methods to handle each of $\mathbb{H}^2$ and $\mathbb{H}^3$, and later on $\mathbb{H}^n$ (for $n > 3$)

is handled in yet another way. The full isometry groups of $\mathbb{H}^2$ and $\mathbb{H}^3$ both consist entirely of strongly-reversible elements; it is more interesting to focus on the orientation-preserving subgroups.

Let us first consider reversibility in the group of orientation-preserving isometries of $\mathbb{H}^2$. We use the unit disc model of hyperbolic space, denoted $\mathbb{D}$. Each orientation-preserving hyperbolic isometry of $\mathbb{D}$ can be represented as $\tau_1 \tau_2$, where τ_1 is a reflection in a hyperbolic line ℓ_1 and τ_2 is a reflection in a different hyperbolic line ℓ_2 – see Figure 6.1. The map f is *elliptic* if ℓ_1 and ℓ_2 intersect in $\mathbb{D}$, *parabolic* if ℓ_1 and ℓ_2 intersect only on the ideal boundary $\partial \mathbb{D}$, or *loxodromic* if ℓ_1 and ℓ_2 are disjoint in $\overline{\mathbb{D}}$. (See Section 6.4 for more on this trichotomy.) The involutions are elliptic maps for which ℓ_1 and ℓ_2 intersect in an angle $\pi/2$.

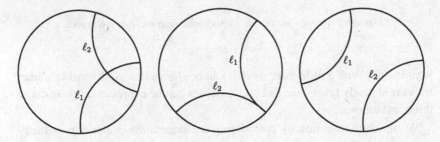

Figure 6.1 A reflection in ℓ_1 followed by a reflection in ℓ_2 gives rise to, from left to right, an elliptic, parabolic, or loxodromic map

We have just described how each orientation-preserving isometry of $\mathbb{D}$ can be written as a product of two *orientation-reversing* involutions. Now we prove that an orientation-preserving isometry f, that is not involutive, is a product of two *orientation-preserving* involutions if and only if it is loxodromic. Suppose first that f is loxodromic; say $f = \tau_1 \tau_2$ for reflections τ_1 and τ_2 in disjoint and not asymptotically-parallel lines ℓ_1 and ℓ_2. Define ℓ_3 to be the unique hyperbolic line orthogonal to both ℓ_1 and ℓ_2, and define τ_3 to be the reflection in ℓ_3. The maps $\sigma_1 = \tau_1 \tau_3$ and $\sigma_2 = \tau_3 \tau_2$ are both elliptic maps of order two, and $f = \sigma_1 \sigma_2$. This decomposition is illustrated in Figure 6.2.

Conversely, suppose now that $f = \sigma_1 \sigma_2$ for elliptic rotations σ_1 and σ_2 of order two about distinct points p_1 and p_2. Define ℓ_3 to be the unique hyperbolic line containing p_1 and p_2. Define ℓ_1 to be the unique hyperbolic line containing p_1 that is orthogonal to ℓ_3 and define ℓ_2 to be the unique hyperbolic line containing p_2 that is orthogonal to ℓ_3. Note that ℓ_1 and ℓ_2 are disjoint and do not meet on the ideal boundary. Let τ_1, τ_2, and τ_3 be the corresponding reflections in ℓ_1, ℓ_2, and ℓ_3. Then $\sigma_1 = \tau_1 \tau_3$ and $\sigma_2 = \tau_3 \tau_2$. Therefore $f = \tau_1 \tau_2$, so f is

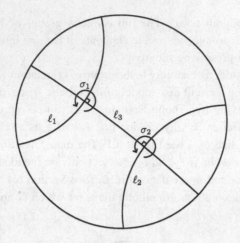

Figure 6.2 How to decompose a loxodromic map into two involutions

loxodromic. With a little more work it can be shown that all reversible isometries are strongly reversible, and all isometries can be expressed as products of three involutions.

We finish this section by considering the orientation-preserving isometry group of $\mathbb{H}^3$. This group can be realised as the collection of complex Möbius transformations; that is, maps of the form

$$f(z) = \frac{az+b}{cz+d},$$

where a, b, c, and d are complex numbers and $ad - bc \neq 0$ (see [28]). A nonidentity element f has one or two fixed points, and, after conjugating, we can assume that either it fixes ∞ only, or else fixes both 0 and ∞. In the first case $f(z) = z + v$, where $v \neq 0$, and in the second case $f(z) = \lambda z$, where $\lambda \neq 0, 1$. Now $f(z) = z + v$ is conjugate to $f^{-1}(z) = z - v$ by the involution $z \mapsto -z$, and $f(z) = \lambda z$ is conjugate to $f^{-1}(z) = \lambda^{-1} z$ by the involution $z \mapsto 1/z$. Thus, in all cases, f is strongly reversible. Therefore all elements of the group of orientation-preserving isometries of $\mathbb{H}^3$ are strongly reversible. In fact, a more careful analysis shows that the conjugating maps are all involutions. In other words, every reverser in this group is an involution. This is not so in higher dimensions.

The results of this section are considered in more detail and in greater generality in Section 6.4.2.

6.3 Euclidean isometries

We use $\mathbb{R}^n$ to denote the standard model of n-dimensional Euclidean space. The group $\mathrm{Isom}(\mathbb{R}^n)$ consists of maps of the form $x \mapsto A(x) + v$, where $A \in \mathrm{O}(n, \mathbb{R})$ and $v \in \mathbb{R}^n$. Such maps preserve orientation if and only if $A \in \mathrm{SO}(n, \mathbb{R})$.

Each member of $\mathrm{Isom}(\mathbb{R}^n)$ either has a fixed point, in which case it is conjugate to an orthogonal map, or it does not have a fixed point. For isometries from the latter category, we have the following lemma.

Lemma 6.1 *Each element g of $\mathrm{Isom}(\mathbb{R}^n)$ that does not have a fixed point is conjugate to an isometry of the form $f(x) = A(x) + v$, where A is an orthogonal map, $v \neq 0$, and $A(v) = v$. If $g \in \mathrm{Isom}^+(\mathbb{R}^n)$ then the conjugating map can be chosen from $\mathrm{Isom}^+(\mathbb{R}^n)$.*

Proof After conjugating by an orthogonal map we can assume, by Theorem 4.12, that $g(x) = A(x) + w$, where $w \in \mathbb{R}^n$,

$$
A = \begin{pmatrix} R_{\theta_1} & & & & \\ & \ddots & & & \\ & & R_{\theta_r} & & \\ & & & -\mathbb{1}_s & \\ & & & & \mathbb{1}_t \end{pmatrix}, \quad R_{\theta_i} = \begin{pmatrix} \cos\theta_i & -\sin\theta_i \\ \sin\theta_i & \cos\theta_i \end{pmatrix},
$$

and $\theta_1, \ldots, \theta_r \in (-\pi, \pi) \setminus \{0\}$. Now choose a vector c such that the first $r+s$ coordinates of $(A - \mathbb{1}_n)(c)$ are equal to those of w (that is, $[(A - \mathbb{1}_n)(c)]_i = w_i$ for $i = 1, \ldots, r+s$). Let h be the translation $h(x) = x + c$. Then

$$
hgh^{-1}(x) = A(x) + w + c - A(c) = A(x) + w - (A - \mathbb{1}_n)(c).
$$

Let $v = w - (A - \mathbb{1}_n)(c)$ and $f = hgh^{-1}$. Then $v_i = 0$ for $i = 1, \ldots, r+s$, and $v \neq 0$ else f, and therefore g, has a fixed point. This map f is of the required form. $\square$

Corollary 6.2 *Each involution in $\mathrm{Isom}(\mathbb{R}^n)$ is conjugate to an orthogonal involution.*

Proof Lemma 6.1 shows that maps without fixed points have infinite order, so they are not involutions. As we have seen, Euclidean isometries with fixed points are conjugate to orthogonal maps. $\square$

6.3.1 The full group of Euclidean isometries

We saw in Theorem 4.13 that each element of $\mathrm{Isom}(\mathbb{S}^n)$ is strongly reversible, and now we see that the same is true of $\mathrm{Isom}(\mathbb{R}^n)$.

Theorem 6.3 *Each element of* $\mathrm{Isom}(\mathbb{R}^n)$ *is strongly reversible.*

Proof Select a member f of $\mathrm{Isom}(\mathbb{R}^n)$. If f has a fixed point then it is conjugate to an orthogonal map, which implies that it is strongly reversible, by Theorem 4.13. Otherwise $f(x) = A(x) + v$, where A is an orthogonal map, $v \neq 0$, and $A(v) = v$. By Theorem 4.13, we may choose orthogonal involutions S_1 and S_2 that both fix v such that $A = S_1 S_2$. Define Euclidean isometries σ_1 and σ_2 by the formulae $\sigma_1(x) = -S_1(x) + v$ and $\sigma_2(x) = -S_2(x)$. Both maps are involutions, and they satisfy $f = \sigma_1 \sigma_2$. $\qquad\square$

6.3.2 Orientation-preserving Euclidean isometries

We begin by considering isometries with fixed points and isometries without fixed points separately.

Proposition 6.4 *Let A belong to $\mathrm{SO}(n,\mathbb{R})$. The following are equivalent:*

 (i) *A is reversible in $\mathrm{SO}(n,\mathbb{R})$*
 (ii) *A is strongly reversible in $\mathrm{Isom}^+(\mathbb{R}^n)$*
(iii) *A is reversible in $\mathrm{Isom}^+(\mathbb{R}^n)$.*

Proof The implications (i) implies (ii) and (ii) implies (iii) are straightforward. Assume (iii); that is, assume there is a Euclidean isometry g given by $g(x) = B(x) + w$, where $B \in \mathrm{SO}(n,\mathbb{R})$ and $w \in \mathbb{R}^n$, such that $gAg^{-1} = A^{-1}$. By expanding out this equation we see that $BAB^{-1} = A^{-1}$, so A is reversible in $\mathrm{SO}(n,\mathbb{R})$, which is statement (i). $\qquad\square$

Proposition 6.5 *Let f be an element of $\mathrm{Isom}^+(\mathbb{R}^n)$ such that $f(x) = A(x) + v$, where A is an orthogonal map, $v \neq 0$, and $A(v) = v$. The following are equivalent:*

 (i) *there is a map B in $\mathrm{SO}(n,\mathbb{R})$ such that $BAB^{-1} = A^{-1}$ and $B(v) = -v$*
 (ii) *f is strongly reversible in $\mathrm{Isom}^+(\mathbb{R}^n)$*
(iii) *f is reversible in $\mathrm{Isom}^+(\mathbb{R}^n)$.*

Proof The implications (i) implies (ii) and (ii) implies (iii) are straightforward. Assume (iii); that is, assume there is a Euclidean isometry g given by $g(x) = B(x) + w$, where $B \in \mathrm{SO}(n,\mathbb{R})$ and $w \in \mathbb{R}^n$, such that $gfg^{-1} = f^{-1}$. By expanding out this equation we see that

$$BAB^{-1} = A^{-1} \quad \text{and} \quad B(v) + w - A^{-1}(w) = -v.$$

Since A is orthogonal and fixes v, we can take the scalar product of both sides of the second equation with v to obtain $\langle B(v), v \rangle = -\langle v, v \rangle$. Thus, because B is orthogonal, $B(v) = -v$, so statement (i) is satisfied. $\qquad\square$

Theorem 6.6   *A map in* Isom$^+(\mathbb{R}^n)$ *is strongly reversible if and only if it is reversible.*

Proof   This is immediate from the equivalence of (ii) and (iii) in Propositions 6.4 and 6.5.   □

We turn now to classifying the reversible isometries of Isom$^+(\mathbb{R}^n)$.

Lemma 6.7   *Let f be an element of* Isom$^+(\mathbb{R}^n)$.

(i) *If $f \in \mathrm{SO}(n,\mathbb{R})$, then there is an f-invariant line in $\mathbb{R}^n$ if and only if there is an f-invariant line through the origin.*

(ii) *If $f(x) = A(x) + v$, where $A \in \mathrm{SO}(n,\mathbb{R})$, $v \neq 0$, and $A(v) = v$, then there is an f-invariant two-dimensional plane in $\mathbb{R}^n$ if and only if there is an A-invariant line through the origin that is orthogonal to v.*

Proof   To prove (i), suppose that f fixes a line ℓ that does not pass through the origin. Then f fixes the unique point w on ℓ that is closest to 0. Therefore the line spanned by w is fixed by f.

To prove (ii), suppose that f fixes a two-dimensional plane Π that does not pass through the origin. The line spanned by v does not intersect Π because otherwise f would fix the intersection point, which is impossible. Therefore v is parallel to Π. It follows that A fixes Π, which implies that the unique point w on Π that is closest to 0 is fixed by A. This point w satisfies $\langle w, v \rangle = 0$, and the line spanned by w is fixed by A.   □

Theorem 6.8   *Given an element f of* Isom$^+(\mathbb{R}^n)$, *either*

(i) $n \equiv 0,3 \pmod 4$, *in which case f is strongly reversible*

(ii) $n \equiv 1 \pmod 4$, *in which case f is strongly reversible if and only if either f has a fixed point or there is a two-dimensional f-invariant plane*

(iii) $n \equiv 2 \pmod 4$, *in which case f is strongly reversible if and only if either f is fixed-point free or there is an f-invariant line.*

Proof   If f has a fixed point then we may assume, by conjugation, that f is an orthogonal map. In this case, by Proposition 6.4, f is strongly reversible in Isom$^+(\mathbb{R}^n)$ if and only if it is reversible in $\mathrm{SO}(n,\mathbb{R})$. By Theorem 4.15, f is reversible in $\mathrm{SO}(n,\mathbb{R})$ unless $n - 1 \equiv 1 \pmod 4$ and there is no f-invariant line through the origin. By Lemma 6.7, there is an f-invariant line through the origin if and only if there is an f-invariant line. Theorem 6.8 has now been established for maps with fixed points.

Now suppose that $f(x) = A(x) + v$, where A is an orthogonal map, $v \neq 0$, and $A(v) = v$. By Proposition 6.5, f is strongly reversible if and only if there is a map B in $\mathrm{SO}(n,\mathbb{R})$ such that $BAB^{-1} = A^{-1}$ and $B(v) = -v$. Let A_0 and B_0 be the restrictions of A and B to the orthogonal complement of v. Since B preserves orientation if and only if B_0 reverses orientation, we see from Proposition 4.14 (and the remarks following Proposition 4.14) that such a map B exists unless $n - 1 \equiv 0 \pmod 4$ and $s_{A_0} + t_{A_0} = 0$. Since $s_{A_0} + t_{A_0} \geqslant 1$ if and only if there is an A_0-invariant line through the origin, we can apply Lemma 6.7 to deduce Theorem 6.8 for maps without fixed points.   □

Theorem 6.9 *For $n \geqslant 3$, each map in $\mathrm{Isom}^+(\mathbb{R}^n)$ can be written as a composite of three involutions.*

Theorem 6.9 fails when n is 1 or 2. The group $\mathrm{Isom}^+(\mathbb{R})$ consists only of translations $x \mapsto x + v$, where $v \in \mathbb{R}$. It is an abelian group and the only involution is the identity. In $\mathrm{Isom}^+(\mathbb{R}^2)$, the only maps of order two are of the form $x \mapsto -x + v$, for $v \in \mathbb{R}^2$. The collection of strongly-reversible maps is equal to the collection of involutions and translations, and this collection is a group. The remaining maps in $\mathrm{Isom}^+(\mathbb{R}^2)$ are rotations by angles that are not integer multiples of π, and these maps are not expressible as composites of involutions.

Proof of Theorem 6.9 Let f be an element of $\mathrm{Isom}^+(\mathbb{R}^n)$. If f has a fixed point then it is conjugate to an orthogonal map and we can apply Theorem 4.16 for spherical isometries. Otherwise, $f(x) = A(x) + v$, where A is an orthogonal map, $v \neq 0$, and $A(v) = v$. By Theorem 6.8, the map f is strongly reversible if n is even, so we assume that n is odd. By Theorem 4.16 we can find orthogonal involutions S_1, S_2, and S_3, two from $\mathrm{O}(n,\mathbb{R}) \setminus \mathrm{SO}(n,\mathbb{R})$ and one from $\mathrm{SO}(n,\mathbb{R})$, that each fix v and satisfy $A = S_1 S_2 S_3$. We can assume that $S_1, S_2 \in \mathrm{O}(n,\mathbb{R}) \setminus \mathrm{SO}(n,\mathbb{R})$ by composing suitably with the map $x \mapsto -x$. Define $\sigma_1(x) = -S_1(x) + v$, $\sigma_2(x) = -S_2(x)$, and $\sigma_3(x) = S_3(x)$. These maps are involutions in $\mathrm{Isom}^+(\mathbb{R}^n)$ that satisfy $f = \sigma_1 \sigma_2 \sigma_3$.   □

6.4 Hyperbolic isometries

The *Möbius group* is the group of bijections of $\mathbb{R}^n_\infty$, the one-point compactification of $\mathbb{R}^n$, generated by reflections in $(n-1)$-dimensional planes and spheres. Let $\mathbb{H}^{n+1}$ denote the upper half-space model of $(n+1)$-dimensional hyperbolic space. We identify $\mathbb{R}^n_\infty$ with the ideal boundary of $\mathbb{H}^{n+1}$ in the usual way, and identify the Möbius group with $\mathrm{Isom}(\mathbb{H}^{n+1})$. The subgroup $\mathrm{Isom}^+(\mathbb{H}^{n+1})$ consists of orientation-preserving hyperbolic isometries. This is made up of those members of $\mathrm{Isom}(\mathbb{H}^{n+1})$ that can be expressed as a composite of an even number of reflections in n-dimensional hyperbolic planes. Note that $\mathrm{Isom}(\mathbb{H}^{n+1})$ contains $\mathrm{Isom}(\mathbb{R}^n)$.

Alternatively, the unit ball $\mathbb{B}^{n+1} = \{x \in \mathbb{R}^n : |x| < 1\}$ is a model of $(n + 1)$-dimensional hyperbolic space, with corresponding isometry group $\mathrm{Isom}(\mathbb{B}^{n+1})$. This is the group of Möbius transformations acting on $\mathbb{R}^{n+1}_\infty$ that fix $\mathbb{B}^{n+1}$ as a set. See [28, 203] for more on hyperbolic geometry. One-dimensional hyperbolic space is isometric to one-dimensional Euclidean space, so we work with $\mathrm{Isom}(\mathbb{H}^{n+1})$ or $\mathrm{Isom}(\mathbb{B}^{n+1})$ with $n \geqslant 1$, henceforth.

A nonidentity Möbius transformation is said to be *elliptic* if it has a fixed point in $\mathbb{H}^{n+1}$, and if it does not have a fixed point in $\mathbb{H}^{n+1}$, then it is *parabolic* if it has a unique fixed point in $\mathbb{R}^n_\infty$, and *loxodromic* otherwise [203, page 142]. It is natural to use the ball model of hyperbolic space to deal with elliptic maps, because, after conjugating, the fixed point of an elliptic map can be chosen to be the origin of the unit ball, in which case the map is orthogonal. For parabolic and loxodromic maps we use the upper half-space model of hyperbolic space. Parabolic maps are conjugate in $\mathrm{Isom}(\mathbb{H}^{n+1})$ to Euclidean isometries, and loxodromic maps are conjugate to maps of the form $x \mapsto \lambda A(x)$, where A is an orthogonal map and $\lambda > 1$. These comments on conjugacy apply even if we restrict to orientation-preserving Möbius transformations [28, Theorem 3.5.1]. Since parabolic and loxodromic maps are of infinite order, the only Möbius transformation involutions are those elliptic maps that, in the ball model, are conjugate to orthogonal involutions.

Let γ denote the inversion $x \mapsto x/|x|^2$, which is an orientation-reversing involution. For $\lambda > 0$ we let λ also denote the map $x \mapsto \lambda x$. The proof of the next lemma is straightforward and omitted.

Lemma 6.10 *Let A be an orthogonal map and let $\lambda > 0$. Then*

(i) $\gamma A = A\gamma$

(ii) $\gamma \lambda = \lambda^{-1}\gamma$. $\qquad\qquad\qquad\qquad\qquad\qquad\qquad\qquad\qquad\qquad\qquad\qquad$ $\square$

6.4.1 The full group of hyperbolic isometries

Theorem 6.11 *Each element of $\mathrm{Isom}(\mathbb{H}^{n+1})$ is strongly reversible.*

Proof Elliptic and parabolic maps are conjugate to Euclidean isometries, therefore both types of map are strongly reversible, by Theorem 6.3. Suppose then that f is a loxodromic member of $\mathrm{Isom}(\mathbb{H}^{n+1})$. By conjugation we may assume that $f(x) = \lambda A(x)$, where $\lambda > 1$ and $A \in \mathrm{O}(n, \mathbb{R})$. Choose orthogonal involutions S and T such that $A = ST$, and define involutions $\sigma = \lambda\gamma S$ and $\tau = \gamma T$. Then $\sigma, \tau \in \mathrm{Isom}(\mathbb{H}^{n+1})$ and $g = \sigma\tau$. $\qquad\qquad\qquad\qquad$ $\square$

6.4.2 Orientation-preserving hyperbolic isometries

To determine whether a map in $\mathrm{Isom}^+(\mathbb{H}^{n+1})$ is strongly reversible, it is useful to consider elliptic, parabolic, and loxodromic maps separately.

Proposition 6.12 *Let A be an orthogonal map in $\mathrm{Isom}^+(\mathbb{B}^{n+1})$. The following are equivalent:*

(i) *A is reversible in $\mathrm{SO}(n+1,\mathbb{R})$*
(ii) *A is strongly reversible in $\mathrm{Isom}^+(\mathbb{B}^{n+1})$*
(iii) *A is reversible in $\mathrm{Isom}^+(\mathbb{B}^{n+1})$.*

Proof The implications (i) implies (ii) and (ii) implies (iii) are straightforward. Suppose that (i) is false. Then Theorem 4.15 tells us $s_A + t_A = 0$, so A does not have 1 as an eigenvalue, which implies that the only fixed point of A is 0. If g is an element of $\mathrm{Isom}(\mathbb{B}^{n+1})$ and $gAg^{-1} = A^{-1}$, then $Ag(0) = 0$. Therefore $g(0) = 0$. This implies that g is an element of $\mathrm{O}(n+1,\mathbb{R}) \setminus \mathrm{SO}(n+1,\mathbb{R})$. Hence (iii) is false. □

Proposition 6.13 *Let f be a member of $\mathrm{Isom}^+(\mathbb{R}^n)$ without a fixed point. The following are equivalent:*

(i) *f is reversible in $\mathrm{Isom}^+(\mathbb{R}^n)$*
(ii) *f is strongly reversible in $\mathrm{Isom}^+(\mathbb{H}^{n+1})$*
(iii) *f is reversible in $\mathrm{Isom}^+(\mathbb{H}^{n+1})$.*

Proof The implications (i) implies (ii) and (ii) implies (iii) are straightforward. Assume (iii); that is, assume there is an orientation-preserving Möbius transformation g such that $gfg^{-1} = f^{-1}$. As usual, there is an orthogonal map A that fixes a nonzero vector v such that $f(x) = A(x) + v$. Since $fg(\infty) = g(\infty)$, and ∞ is the only fixed point of f, we see that $g(\infty) = \infty$. Therefore there is an orthogonal map W, a positive number λ, and an element w of $\mathbb{R}^n$ such that $g(x) = \lambda B(x) + w$. By expanding out the equation $gfg^{-1} = f^{-1}$ we see that $w - A^{-1}(w) + \lambda B(v) = -v$. Take the scalar product of each side of this equation with v to see that $\lambda \langle B(v), v \rangle = -\langle v, v \rangle$. We deduce that $\lambda = 1$. Hence $g \in \mathrm{Isom}^+(\mathbb{R}^n)$, so statement (i) has been verified. □

Proposition 6.14 *Suppose that an element f of $\mathrm{Isom}^+(\mathbb{H}^{n+1})$ is given by $f(x) = \lambda A(x)$, where $A \in \mathrm{SO}(n,\mathbb{R})$ and $\lambda > 1$. The following are equivalent:*

(i) *there exists an element B of $\mathrm{O}(n,\mathbb{R}) \setminus \mathrm{SO}(n,\mathbb{R})$ such that $BAB^{-1} = A^{-1}$*
(ii) *f is strongly reversible in $\mathrm{Isom}^+(\mathbb{H}^{n+1})$*
(iii) *f is reversible in $\mathrm{Isom}^+(\mathbb{H}^{n+1})$.*

Proof   That (i) implies (ii) is true because B can be chosen to be an involution, and the map $g = B\gamma$, where $\gamma(x) = x/|x|^2$, satisfies $gfg^{-1} = f^{-1}$. The implication (ii) implies (iii) is straightforward. Assume (iii); that is, assume there is an element g of $\mathrm{Isom}^+(\mathbb{H}^{n+1})$ such that $gfg^{-1} = f^{-1}$. The map g must either fix each of the two fixed points of f, 0 and ∞, or else interchange them. If g fixes each of 0 and ∞, then g is itself an orthogonal map followed by a dilation, in which case gfg^{-1} has the same dilation coefficient as f, namely λ. However, the dilation coefficient of f^{-1} is λ^{-1}, so we conclude that in fact $g(0) = \infty$ and $g(\infty) = 0$. In this case we can express g in the form $g(x) = \mu B\gamma(x)$, where $\mu > 1$ and B is an element of $\mathrm{O}(n,\mathbb{R}) \setminus \mathrm{SO}(n,\mathbb{R})$. Statement (i) now follows after expanding out the equation $gfg^{-1} = f^{-1}$, using Lemma 6.10. $\square$

Theorem 6.15   *Each map in* $\mathrm{Isom}^+(\mathbb{H}^n)$ *is strongly reversible if and only if it is reversible.*

Proof   This is immediate, as (ii) and (iii) are equivalent in Propositions 6.12, 6.13, and 6.14. $\square$

We turn now to classifying the reversible isometries of $\mathrm{Isom}^+(\mathbb{H}^n)$.

Lemma 6.16   *Let f be an n-dimensional Möbius transformation.*

 (i) *If, when considered as an element of* $\mathrm{Isom}^+(\mathbb{B}^{n+1})$, f *is orthogonal, then there is an f-invariant hyperbolic line if and only if there is an f-invariant Euclidean line through the origin.*

 (ii) *If, when considered as an element of* $\mathrm{Isom}^+(\mathbb{H}^{n+1})$, f *is a Euclidean isometry without a fixed point, then there is an f-invariant three-dimensional hyperbolic plane if and only if there is an f-invariant two-dimensional Euclidean plane.*

 (iii) *If, when considered as an element of* $\mathrm{Isom}^+(\mathbb{H}^{n+1})$, $f(x) = \lambda A(x)$, *for $A \in \mathrm{SO}(n,\mathbb{R})$ and $\lambda > 0$, then there is an f-invariant two-dimensional hyperbolic plane if and only if there is an A-invariant Euclidean line through the origin.*

Proof   To prove (i), first notice that a hyperbolic line through the origin is a portion of a Euclidean line through the origin. Suppose then that f fixes a hyperbolic line ℓ in $\mathbb{B}^{n+1}$ that does not pass through the origin. Let p be the point on ℓ nearest the origin in hyperbolic distance. Then f fixes p, so f fixes the Euclidean line through 0 and p.

To prove (ii), suppose that f fixes a three-dimensional hyperbolic plane Π. Then f fixes the two-dimensional boundary of Π, which must be a Euclidean plane (rather than a Euclidean sphere), since f does not fix any compact sets in $\mathbb{R}^n$.

To prove (iii), suppose that f fixes a two-dimensional hyperbolic plane Π. Then f fixes the one-dimensional boundary of Π, which must be a Euclidean line through 0 and ∞, since these points are the attracting and repelling fixed points of f.   □

Theorem 6.17 *Given an element f of $\mathrm{Isom}^+(\mathbb{H}^n)$, either*

(i) $n \equiv 0,3 \pmod 4$, *in which case f is strongly reversible*

(ii) $n \equiv 1 \pmod 4$, *in which case f is strongly reversible if and only if either f is elliptic, parabolic, or there is an f-invariant two-dimensional plane*

(iii) $n \equiv 2 \pmod 4$, *in which case f is strongly reversible if and only if either f is loxodromic, or f is elliptic and there is an f-invariant line, or f is parabolic and there is an f-invariant three-dimensional plane.*

Proof   Only the identity in $\mathrm{Isom}^+(\mathbb{H}^1)$ is reversible, so the theorem is correct for $n = 1$. Suppose that $n \geqslant 2$. If f is elliptic then, by conjugation, we consider it to be an orthogonal map in $\mathrm{Isom}^+(\mathbb{B}^n)$. Using Theorem 4.15 and Proposition 6.12 we see that f is strongly reversible unless $n - 1 \equiv 1 \pmod 4$ and there is no f-invariant line through the origin. Using Lemma 6.16(i), we deduce Theorem 6.17 for elliptic maps.

If f is a parabolic map then, by conjugation, we consider it to be a fixed-point-free element of $\mathrm{Isom}^+(\mathbb{R}^{n-1})$ within $\mathrm{Isom}^+(\mathbb{H}^n)$. Using Theorem 6.8 and Proposition 6.13 we see that f is strongly reversible unless $n - 1 \equiv 1 \pmod 4$ and there is no f-invariant two-dimensional Euclidean plane in $\mathbb{R}^{n-1}$. Using Lemma 6.16(ii), we deduce Theorem 6.17 for parabolic maps.

If f is loxodromic then, by conjugation, we may assume that f is an element of $\mathrm{Isom}^+(\mathbb{H}^n)$ of the form $f(x) = \lambda A(x)$, where $A \in \mathrm{SO}(n-1,\mathbb{R})$ and $\lambda > 1$. Using Proposition 4.14 and Proposition 6.14 we see that f is strongly reversible unless $n - 1 \equiv 0 \pmod 4$ and there is no A-invariant Euclidean line through the origin in $\mathbb{R}^{n-1}$. Using Lemma 6.16(iii), we deduce Theorem 6.17 for loxodromic maps.   □

Finally, we consider products of involutions in $\mathrm{Isom}^+(\mathbb{H}^n)$.

Theorem 6.18   *For $n \geqslant 2$, each member of $\mathrm{Isom}^+(\mathbb{H}^n)$ can be written as a composite of three involutions.*

Proof   Choose an element f of $\mathrm{Isom}^+(\mathbb{H}^{n+1})$, where $n \geqslant 1$. If f is elliptic or parabolic, then f can be expressed as a composite of three involutions, by Theorem 6.9. If f is loxodromic then, by conjugation, we can assume that $f(x) = \lambda A(x)$, where A is orthogonal and $\lambda > 1$. By Theorem 4.16 there are involutions $S_1, S_2 \in \mathrm{O}(n,\mathbb{R}) \setminus \mathrm{SO}(n,\mathbb{R})$ and $S_3 \in \mathrm{SO}(n,\mathbb{R})$ such that $A = S_1 S_2 S_3$.

Define involutions $\sigma_1 = \lambda \gamma S_1$, $\sigma_2 = \gamma S_2$, and $\sigma_3 = S_3$ in $\mathrm{Isom}^+(\mathbb{H}^{n+1})$. Then $f = \sigma_1 \sigma_2 \sigma_3$. $\qquad\qquad\qquad\qquad\qquad\qquad\qquad\qquad\qquad\qquad\qquad\qquad$ $\square$

Notes

Sources

All the results of this chapter can be found in [213]. (See also [105].) Most prior results on composites of involutions in isometry groups of spherical, Euclidean, and hyperbolic space involve reflections. Many such results can be found, for example, in the work of Coxeter (see [59] or [60, page 99]). Since we allow ourselves the freedom of working with more general involutions than just reflections, and since there are no reflections in the orientation-preserving isometry groups, our results tend to differ from the classic results on composites of reflections.

– Lorentz space

The results on hyperbolic space have been proven by Gongopadhyay using the Lorentz model of hyperbolic space [105]. Theorem 6.18 was proven in a different context by Knüppel and Thomsen [153, Theorem 8.8]. Knüppel and Thomsen were working with orthogonal groups with respect to general symmetric bilinear forms (not necessarily Euclidean inner products) and they were examining composites of involutions in commutator subgroups of orthogonal groups. If you equip $\mathbb{R}^{N+1}$ with the Lorentz inner product $\langle x, y \rangle = x_1 y_1 + \cdots + x_N y_N - x_{N+1} y_{N+1}$, then the commutator subgroup of the corresponding orthogonal group is isomorphic to $\mathrm{Isom}^+(\mathbb{H}^N)$.

Quaternionic Möbius transformations

Theorems 6.11, 6.15, 6.18, and 6.17 have all been proven for $\mathbb{H}^5$ by the authors and Lavička [166] using quaternionic Möbius transformations.

Complex hyperbolic space

Gongopadhyay and Parker [106] have tackled the reversibility problem in complex hyperbolic space.

Open problems

Metric Spaces

Every metric space has an associated isometry group, and it may be that the reversibility problem for isometry groups relates to the geometry of the metric space. A good collection of metric spaces to start such an investigation are graphs equipped with the graph metric.

Discrete groups

The discrete subgroups of $\mathrm{Isom}(\mathbb{H}^n)$ have a rich geometry associated to them, which may provide insight into the reversibility problem. Only discrete groups that contain elliptic elements will contain reversible elements. One such group is the modular group, studied in the next chapter.

7

Groups of integer matrices

The group $GL(2, \mathbb{Z})$ consists of those two-by-two matrices with integer entries and determinant ± 1. Its index-two subgroup $SL(2, \mathbb{Z})$ consists of those matrices with determinant 1. Let us denote the identity matrix by $\mathbb{1}$. The centre of both groups is $\{\pm \mathbb{1}\}$ (isomorphic to the cyclic group C_2), and the respective quotients are $PGL(2, \mathbb{Z})$ and the *modular group* $PSL(2, \mathbb{Z})$.

$$
\begin{array}{ccccc}
\{\pm \mathbb{1}\} & \rightarrow & SL(2, \mathbb{Z}) & \rightarrow & PSL(2, \mathbb{Z}) \\
& & \downarrow & & \downarrow \\
\{\pm \mathbb{1}\} & \rightarrow & GL(2, \mathbb{Z}) & \rightarrow & PGL(2, \mathbb{Z})
\end{array}
$$

These groups arise in many applications. The group $GL(2, \mathbb{Z})$ is isomorphic to the group of conformal automorphisms of a torus. It is isomorphic to the extended mapping class group of the torus, and $SL(2, \mathbb{Z})$ is the oriented mapping class group. The group $GL(2, \mathbb{Z})$ is also the mapping class group of the unique genus-three nonorientable compact surface. Furthermore, it is the group of two-dimensional lattice isomorphisms, and it has been used to construct interesting foliations of three-manifolds. As we shall see shortly, it has an intimate relation to binary integral quadratic forms.

The group $PGL(2, \mathbb{Z})$ may be regarded as a group of maps of the four-punctured sphere (because you obtain a sphere with four distinguished points when you identify pairs $\pm z$ in the complex plane before taking the quotient by $\mathbb{Z}^2$). It may also be viewed as a group acting on the punctured torus, and this illuminates an aspect of diophantine approximation theory [210]. For some other uses, see [207].

The group $PSL(2, \mathbb{Z})$ is called the modular group because, when realised as the group of Möbius tranformations with integral coefficients that preserve the upper half-plane, its orbits on the half-plane classify (that is, provide 'moduli' for) the various possible conformal structures on the torus. As a group, it is isomorphic to the free product $C_2 * C_3$. In fact, it may be described as

$$\langle s, t : s^2 = t^3 = \mathbb{1} \rangle,$$

where

$$s(z) = -\frac{1}{z}, \qquad t(z) = -\frac{1}{z+1}.$$

The elements of finite order in these groups other than $\pm \mathbb{1}$ are called the *elliptic* elements of the group. This language is motivated by the fact that the corresponding elements of $\mathrm{PSL}(2, \mathbb{Z})$ are elliptic when regarded as Möbius transformations (see Section 6.4). The *parabolic* elements are those represented by matrices other than $\pm \mathbb{1}$ that have just one eigenvalue (necessarily 1 or -1). These correspond to parabolic Möbius transformations, which are conjugate to translations of the upper half-plane. The remaining elements are known as *hyperbolic* elements. These correspond to loxodromic Möbius transformations that fix the upper half-plane. The use of the word 'hyperbolic' here is different from the use of the same word in the phrase 'hyperbolic geometry'. The terminology clash is unfortunate, but standard.

This chapter is longer than the others because we discuss the significance of the groups of integer matrices to quadratic forms. The chapter is structured as follows. We begin in Section 7.1 with some basic facts about conjugacy in $\mathrm{GL}(2, \mathbb{Z})$, and show a first connection with the theory of binary integral quadratic forms. In Sections 7.2 to 7.4 we review some relevant parts of this theory. In Section 7.5 we deal with the conjugacy problem for elliptic elements of $\mathrm{GL}(2, \mathbb{Z})$ and the other groups. Section 7.6 deals with centralisers: In Section 7.7 we begin the study of the reversible elements, and resolve some of the standard questions. For hyperbolic elements, the results leave something to be desired, in that they do not provide an explicit algorithm to determine reversibility. In Section 7.8 we connect this problem to the representation of ± 1 by a second integral quadratic form, which is of indefinite type. In Section 7.9 we review algorithms for such forms, completing the algorithmic identification of the reversibles. Finally, in Sections 7.10 and 7.11 we review alternative methods for handling reversibility in these groups.

7.1 Conjugacy to rational canonical form

Let us begin with basic theory about conjugacy. Throughout, we adopt the convention that, given an element g of $\mathrm{GL}(2, \mathbb{Z})$, we denote the corresponding equivalence class in $\mathrm{PGL}(2, \mathbb{Z})$ also by g, rather than using the more cumbersome notation $\{\pm g\}$.

Given an element g of $GL(2,\mathbb{Z})$, the determinant Δ and trace τ are integers, and $\Delta = \pm 1$. The characteristic polynomial χ is given by

$$\chi(\lambda) = \lambda^2 - \tau\lambda + \Delta,$$

so, by the Cayley–Hamilton theorem,

$$g^2 = \tau g - \Delta \mathbb{1}. \qquad (7.1)$$

Evidently, the eigenvalues of g are algebraic integers. If not rational integers, then they are two distinct elements of an irrational quadratic extension K of $\mathbb{Q}$, so belong to the *maximal order O* of K, the ring of integers of K. To be explicit, observe that the discriminant D of χ is given by

$$D = \tau^2 - 4\Delta,$$

and we let r and k be integers, where k is square free, such that $D = r^2 k$. Then $K = \mathbb{Q}(\sqrt{k})$, and O is either $\mathbb{Z}[\sqrt{k}]$ or $\mathbb{Z}\left[\frac{1}{2}(1+\sqrt{k})\right]$ (see [51, 54]).

The determinant and trace are invariant under conjugacy in $GL(2,\mathbb{Z})$, and Δ remains a well-defined invariant in $PGL(2,\mathbb{Z})$. The trace τ is not an invariant in $PGL(2,\mathbb{Z})$, however, the modulus of trace, $|\tau|$, is an invariant.

Each element g of $GL(2,\mathbb{Z})$ is rationally conjugate (that is, conjugate in the group $GL(2,\mathbb{Q})$) to its rational canonical form (see Chapter 4). In case the characteristic polynomial is irreducible, this means that g is rationally conjugate to

$$\begin{pmatrix} 0 & -\Delta \\ 1 & \tau \end{pmatrix}.$$

It may or may not be integrally conjugate to it.

Proposition 7.1 *Suppose that an element g of $GL(2,\mathbb{Z})$, given by*

$$g = \begin{pmatrix} a & b \\ c & d \end{pmatrix},$$

has an irreducible characteristic polynomial. Then g is conjugate in $GL(2,\mathbb{Z})$ to its rational canonical form if and only if there exist integers x and y such that

$$cx^2 + (d-a)xy - by^2 = \pm 1.$$

Proof The matrix g is conjugate in $GL(2,\mathbb{Z})$ to its rational canonical form (which in this case is the companion matrix of its characteristic polynomial) if and only if there exists an integer vector $v = (x,y)$ in $\mathbb{Z}^2$ such that the matrix with column representation (v, gv) has determinant ± 1. Since the determinant of (v, gv) is $cx^2 + (d-a)xy - by^2$, the proposition follows. $\square$

Using the same sort of proof we obtain the following result.

Corollary 7.2 *Suppose that an element g of* $\mathrm{SL}(2,\mathbb{Z})$, *given by*

$$g = \begin{pmatrix} a & b \\ c & d \end{pmatrix},$$

has an irreducible characteristic polynomial. Then g is conjugate in $\mathrm{SL}(2,\mathbb{Z})$ *to its rational canonical form if and only if there exist integers x and y such that*

$$cx^2 + (d-a)xy - by^2 = 1. \qquad\qquad \square$$

These results relate the conjugacy problem in $\mathrm{GL}(2,\mathbb{Z})$ and $\mathrm{SL}(2,\mathbb{Z})$ to the theory of integral quadratic forms, and later on we shall see more connections of this type. We now review the theory of integral quadratic forms.

7.2 Integral quadratic forms

7.2.1 Proper and improper equivalence

We consider binary quadratic forms

$$[A,B,C](x,y) = Ax^2 + Bxy + Cy^2$$

with integral coefficients. A binary quadratic form is said to be *positive definite* if it takes only positive values, and it is said to be *positive semidefinite* if it takes only nonnegative values. The terms *negative definite* and *negative semidefinite* are defined in the obvious way. A quadratic form is said to be *indefinite* if it is neither positive semidefinite nor negative semidefinite.

The *discriminant* of $[A,B,C]$ is

$$\mathrm{discr}[A,B,C] = B^2 - 4AC.$$

It is straightforward to show that a form is positive or negative definite if and only if its discriminant is negative, and it is indefinite if and only if its discriminant is positive.

The group $\mathrm{GL}(2,\mathbb{Z})$ acts on binary quadratic forms by composition on the right[1]:

$$g([A,B,C])(x,y) = [A,B,C]((x,y)g^t),$$

---

[1]   One could, in line with normal practice among group theorists, use g^{-1} instead of g^t here, without affecting any of our subsequent statements. It is traditional among workers on forms to use the transpose.

(regarding the transpose g^t as a linear map of $\mathbb{R}^2$). That is,

$$g([A,B,C])(x,y) = [A,B,C](ax+by,cx+dy).$$

The action depends only on the class of g in $\mathrm{PGL}(2,\mathbb{Z})$. It is known as the 'symmetric square representation' of $\mathrm{PGL}(2,\mathbb{Z})$. We say that $g([A,B,C])$ is *equivalent* to $[A,B,C]$. The equivalence is *proper* if $\det(g) = 1$, and *improper* if $\det(g) = -1$. Equivalence and proper equivalence are equivalence relations. We denote equivalence by $\sim$ and proper equivalence by $\approx$. Improper equivalence is not an equivalence relation. A form may be improperly equivalent to itself. For instance, the form $x^2 + y^2$ is taken to itself by either of the involutions

$$\begin{pmatrix} 1 & 0 \\ 0 & -1 \end{pmatrix} \quad \text{or} \quad \begin{pmatrix} 0 & 1 \\ 1 & 0 \end{pmatrix}.$$

Equivalent forms *represent* the same set of integers; that is, they have the same image when restricted to $\mathbb{Z}^2$.

The most important class invariant of a form is the discriminant $\mathrm{discr}[A,B,C]$ introduced earlier. It is invariant under proper or improper equivalence, because

$$\mathrm{discr}(g[A,B,C]) = \mathrm{discr}[A,B,C] \cdot \det(g)^2$$

whenever $g \in \mathrm{GL}(2,\mathbb{Z})$. We note that many authors (following Gauss) use *one quarter of this value* as the definition of the discriminant. This is only integral when B is even, so is better suited to contexts in which only even B are of interest. Care should be taken to check the definition in use, when referring to other sources.

The greatest common divisor (gcd) of A, B, and C is also a class invariant. The form is said to be *primitive* if this gcd is 1.

The next lemma contains some elementary observations about equivalence.

Lemma 7.3

(i) $[A,B,C] \approx [C,-B,A]$

(ii) $[A,B,C] \sim [A,-B,C]$ *(improperly)*

(iii) $[A,B,C] \sim [C,B,A]$ *(improperly)*

(iv) $[A,B,C] \approx [A,B+2tA,C+tB+t^2A]$ *(where $t \in \mathbb{Z}$)*

(v)   *If C is nonzero, $B+B'$ is divisible by $2C$, and*
      $\mathrm{discr}[A,B,C] = \mathrm{discr}[C,B',C']$, *then $[A,B,C] \approx [C,B',C']$.*

Proof   The proofs of (i), (ii), and (iii) are straightforward. To prove (iv), use the matrix

$$g = \begin{pmatrix} 1 & t \\ 0 & 1 \end{pmatrix}.$$

To prove (v), use the matrix

$$g = \begin{pmatrix} 0 & -1 \\ 1 & \frac{B+B'}{2C} \end{pmatrix}. \qquad\qquad \square$$

Let us write $[A,B,C] \to [C,B',C']$ if these forms are as in (v). In Gauss' language, $[A,B,C]$ *is a neighbour by the last part* to $[C,B',C']$, and $[C,B',C']$ *is a neighbour by the first part* to $[A,B,C]$.

We say that the form $[A,B,C]$ *represents* the integer N if there are integers x and y with $Ax^2 + Bxy + Cy^2 = N$. We say that the representation is *primitive* if $\gcd(x,y) = 1$. We shall only be interested in representations of 1 and -1; these are necessarily primitive.

7.2.2 First associated form

Let g be an element of $\mathrm{GL}(2,\mathbb{Z})$, given by

$$g = \begin{pmatrix} a & b \\ c & d \end{pmatrix}.$$

We call the form $F(g) = [c, d-a, -b]$ that appeared in Proposition 7.1 the *first associated form* to g. Recall that Δ is the determinant of g, and τ is the trace of g.

Lemma 7.4 *If $g \in \mathrm{GL}(2,\mathbb{Z})$, then its first associated form has the same discriminant $\tau^2 - 4\Delta$ as its characteristic polynomial.*

Proof Since $d = \tau - a$, the form has discriminant

$$(\tau - 2a)^2 - 4c(-b) = \tau^2 - 4a\tau + 4a^2 + 4bc = \tau^2 - 4\Delta. \qquad \square$$

The first form associated to g may not be primitive, and this prompts us to define the *modified first form*, $[c/r, (d-a)/r, -b/r]$, where $r = \gcd(b,c,d-a)$, at least when $g \neq \pm\mathbb{1}$.

The map to the first associated form is almost injective.

Proposition 7.5 *Suppose that g and g' belong to $\mathrm{GL}(2,\mathbb{Z})$ and have the same first associated form and determinant. Then either $g' = g$ or $g' = -g^{-1}$.*

Proof Let

$$g = \begin{pmatrix} a & b \\ c & d \end{pmatrix} \quad \text{and} \quad g' = \begin{pmatrix} a' & b' \\ c' & d' \end{pmatrix}.$$

Since g and g' have the same first associated forms and determinants, we obtain $b' = b$ and $c' = c$, and

$$d' - a' = d - a \quad \text{and} \quad a'd' = ad.$$

By examining the intersection points of the line $y - x = p$ with the curve $xy = q$ (for constants p and q), we see that (a', d') is either (a, d) or $(-d, -a)$ In the latter case, $g' = -g^{-1}$. $\qquad\square$

Thus the map is two-to-one on $\mathrm{GL}(2, \mathbb{Z})$ except on the trace zero elements, where it is injective. We obtain an injective map if we restrict attention to those matrices with nonnegative trace.

7.2.3 Representing ± 1

The problem of deciding whether a given integer is represented by a given binary quadratic form was discussed by Gauss [58, 95], and his approach was reworked by Hermite and Minkowskii, using ideal theory of quadratic fields and geometry of numbers [51, page 298 and following]. The resulting algorithm is, in general, difficult to use, and its justification is quite deep. This is particularly the case for indefinite forms (those with positive discriminant), and in our application we are mainly interested in indefinite forms.

The basic idea in all variants of the method is to pick out a distinguished set of primitive forms, called the reduced forms (The precise definition of reduced form differs from expert to expert). One shows that there are just a finite number of reduced forms having a given discriminant, and every primitive form is properly equivalent to some reduced form. One specifies a reduction algorithm (such as Gauss reduction, Hermite reduction, or Minkowski reduction) for constructing a reduced form that is properly equivalent to a given form. The algorithm is always efficient, comparable in complexity to the Euclidean algorithm. In fact, it is quite like the Euclidean algorithm, employing some variant of parts (i) and (iv) of Lemma 7.3 applied repeatedly. To finish the algorithm, one needs an a priori upper bound on the size of the integers x and y needed to represent the integer N by a reduced primitive form of discriminant D. Given this, one can just reduce the given form, and check all pairs x and y up to that bound.

More recently, Conway [57] gave an astonishingly simple algorithm, justified using only the most elementary methods. We start with this, and then go on to discuss the use of Gauss' algorithm, which simplifies markedly when it comes to the cases $N = \pm 1$. Conway's algorithm is rather slow compared to Gauss', but it provides an easy way to understand what Gauss' does.

7.3 Conway's topograph

It would be impossible to improve upon the sparkling account given by Conway, so we just summarise the conclusions. The proofs that we omit are elementary, and straightforward. The hard part was thinking of the concept of the topograph, not justifying its properties.

Regarding $\mathbb{Z}^2$ as a $\mathbb{Z}$-module, we consider all ordered bases (e_1, e_2) for $\mathbb{Z}^2$ (here e_1 and e_2 are elements of $\mathbb{Z}^2$, which together form a basis of $\mathbb{Z}^2$). Related to an ordered base (e_1, e_2) is an unordered base $\{e_1, e_2\}$. We call the class of four unordered bases $\{\pm e_1, \pm e_2\}$ a *lax base*. We also consider ordered triples (e_1, e_2, e_3), where $e_1 + e_2 + e_3 = 0$ and (e_1, e_2) is an ordered base. These triples are called *strict superbases*. Related to a strict superbase (e_1, e_2, e_3) is a three-element set $\{e_1, e_2, e_3\}$. We call the class of eight sets $\{\pm e_1, \pm e_2, \pm e_3\}$ a *lax superbase*. Last, we call a pair $\{\pm e\}$, where e is a vector in $\mathbb{Z}^2$ that belongs to some basis (that is, $e = (a, b)$ with $\gcd(a, b) = 1$) a *lax vector*. For convenience, we sometimes represent a lax vector by e (rather than $\{\pm e\}$) and we follow similar conventions for lax bases and superbases.

One may form, in a straightforward way, a two-dimensional 'cell complex' in which the vertices are the lax superbases, the edges are the lax bases, and the faces are lax vectors (Figure 7.1). This cell complex is Conway's topograph.

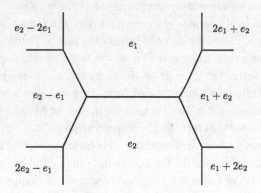

Figure 7.1 Conway's topograph

The underlying graph of Conway's topograph is a tree in which each vertex has degree three (Figure 7.2). It is sometimes called the Markoff tree.

A quadratic form $[A, B, C]$ assigns a value $[A, B, C](v)$ to each face v. We abbreviate this to $\bar{v}$. These values are the *primitive values* of the form. To check whether a given integer N is represented by the form, it suffices to check for primitive representations of N/m^2, where m^2 is the square part of N. For each

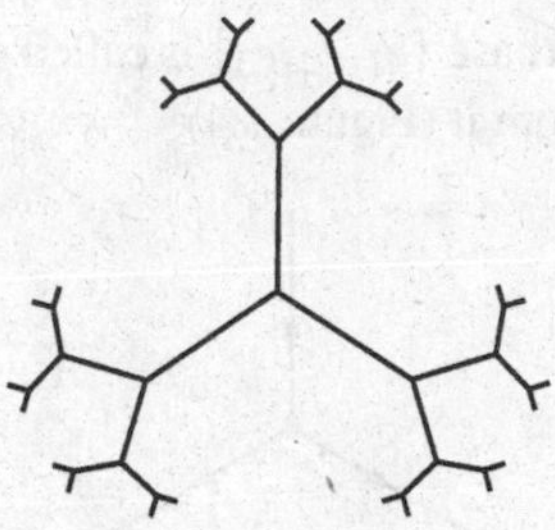

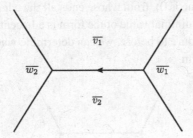

Figure 7.2 The Markoff tree

edge there are two faces v_1 and v_2 that meet along the edge, and two other faces $w_1 = v_1 + v_2$ and $w_2 = v_1 - v_2$ that meet it at its ends. One can check that

$$\overline{w_1} + \overline{w_2} = 2(\overline{v_1} + \overline{v_2}),$$

so $\overline{w_1}$, $\overline{v_1} + \overline{v_2}$, and $\overline{w_2}$ are in arithmetic progression. We assign to the edge the value of the step

$$\tfrac{1}{2}\left(\overline{w_2} - \overline{w_1}\right),$$

and we orient the edge so that this is positive (if the step is zero, then we don't orient the edge). This progression rule is illustrated in Figure 7.3.

Figure 7.3 The arithmetic-progression rule

The arithmetic-progression rule yields the climbing lemma, which says that if two adjacent faces have positive values for some form $[A, B, C]$ and the edge separating them has a nonzero value, then the face at the forward end of the edge also has a positive value

The arithmetic-progression rule allows us to label in turn as many faces as we please with their correct values for a given form, once we have the values at three faces meeting at a point. Given a form $[A, B, C]$, an obvious starting point is given by the values A, B, and $A + B + C$ at $(1, 0)$, $(0, 1)$, and $(1, 1)$, respectively.

For a given form, a superbase $\{e_1, e_2, e_3\}$ is called a *well* if the three adjacent edges are directed away from it (Figure 7.4).

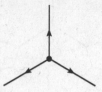

Figure 7.4 A well

It can be shown that for a positive-definite form there can be at most one well. One of the values on faces adjacent to the well is the least value represented by the form. At each vertex that is not a well there are two edges oriented out and one in, so we have an algorithm for locating the well, if it exists: just travel against the arrows. Once we find the well, we can find all values primitively represented by the form by working outwards from the well. If the value N has a primitive representation, then it will (by the climbing lemma) be found at a distance no more than N from the well.

If a positive-definite form does not have a well, then the above algorithm leads to a *double well*, which is an unoriented edge (the corresponding arithmetic progression value is 0), from whose ends all the edges are oriented away (see Figure 7.5). The minimal value of the form is adjacent to this special edge. Using a similar procedure to before, we can determine whether an integer N is represented by the form.

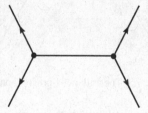

Figure 7.5 A double well

For a positive semidefinite form, not definite, one obtains a unique *lake*, which is a face (representing a vector) with value 0. All adjacent faces are assigned the same positive value, and all the adjacent edges point away from the lake (Figure 7.6). Again, one can locate this distinguished vector by tracking back from any initial superbase. Such forms are equivalent to Ax^2, for some positive integer A.

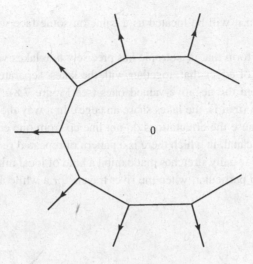

Figure 7.6 A single lake

The story for negative definite and negative semidefinite forms is similar to that for positive definite and positive semidefinite forms.

For indefinite forms, there are two cases, depending on whether or not zero has a (nontrivial) representation. Forms that represent 0 nontrivially are called *isotropic*. The other indefinite forms are called *anisotropic*.

An indefinite form that is anisotropic has a *river* of edges that separates all the positive-valued faces from the negative-valued ones (Figure 7.7).

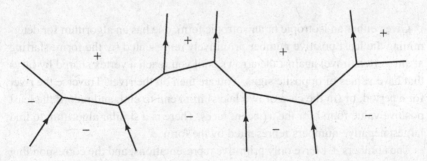

Figure 7.7 A river

This chain of edges is periodic, in the sense that after a finite number of steps along the river away from a given edge, one arrives at another edge with the same values on it and on the adjacent faces. If a number N is primitively represented, then it will be located as a value on some face within $|N|$ steps from the river.

An indefinite form that is isotropic has precisely two lakes valued 0, joined by a finite river of edges that, together with the lakes, separates the positive-valued faces from the negative-valued ones (see Figure 7.8). The river may have zero length (that is, the lakes share an edge). In a way the term 'river' is misleading, because the orientations do not line up from one end to the other. It is more like a canal, in which there is a pattern of repeated rising and falling levels. There are usually stretches that exhibit a kind of local mirror symmetry. This happens, in particular, when the river travels for a while around a single face.

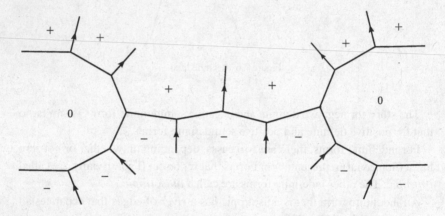

Figure 7.8 Two lakes and a river

Given either an isotropic or anisotropic form, one has an algorithm for determining the least positive number primitively represented by the form: starting at any vertex, travel against the arrows until you reach a vertex shared by faces that have values of opposite signs. You are then on the river. Traverse the river for a period, or (in the case of two lakes) from end to end, and record the least positive value found for the adjacent faces. There is a similar algorithm to find largest negative numbers represented by the form.

The numbers ± 1 have only primitive representations, and the corresponding faces can only occur on the 'riverbank', that is, they have an edge on the river. Thus we can determine whether or not ± 1 is represented by traversing one period of the river.

Incidentally, Conway also shows that the subgroup of $GL(2, \mathbb{Z})$ that leaves an indefinite form invariant is dihedral, and indicates how to calculate it as the composition of two reflections.

We remark that $GL(2,\mathbb{Z})$ acts on the topograph, and $\{\pm\mathbb{1}\}$ acts trivially, so we have another action of $PGL(2,\mathbb{Z})$, as a group of automorphisms of the Markoff tree.

7.4   Gauss' method for definite forms

A positive-definite form $[A,B,C]$ with (negative) discriminant D is said to be *Gauss reduced* if $0 \leqslant B \leqslant A \leqslant \sqrt{-D/3}$ and $A \leqslant C$. Gauss gave an algorithm for constructing, from any given positive-definite form, a reduced form that is properly equivalent to it. The algorithm is simple, and based on repeated application of Lemma 7.3(v). If $[A,B,C]$ is not yet reduced, take $A' = C$, take B' to be the least nonnegative number congruent to $-B$ modulo $2A'$, and then let $C' = (B'^2 - D)/4A'$ (which is an integer!) and replace $[A,B,C]$ by $[A',B',C']$. Continue until the form is reduced. This will happen as soon as we reach a form $[A'',B'',C'']$ in which $C'' \geqslant A''$.

In Conway's terms this procedure is the same as taking a short cut across a face of the topograph whenever possible.

Gauss showed that for a positive-definite form there are only one or two reduced forms in each proper equivalence class. There are two precisely when there is a reduced form $[A,B,C]$ that is properly equivalent to $[A,-B,C]$. He called any form $[A,B,C]$ (whether definite or not, reduced or not) *ambiguous* if $[A,B,C]$ is properly equivalent to $[A,-B,C]$. A form is ambiguous if $A = C$, by Lemma 7.3(v), but may also be ambiguous when $A \neq C$. A necessary and sufficient condition for a reduced form to be ambiguous is that it is equivalent to one with B divisible by A [95, Art. 164].

Here are two corollaries of these observations.

Proposition 7.6   *Let $D < 0$. If there is only one proper equivalence class of Gauss reduced forms with discriminant D, then any two positive-definite forms with this discriminant are properly equivalent, and hence represent the same integers.*   □

Proposition 7.7   *Let $D < 0$. If all Gauss reduced positive-definite forms with discriminant D represent 1, then all positive-definite forms with discriminant D represent 1.*   □

We defer consideration of Gauss' method for indefinite forms for the time being, and return to $GL(2,\mathbb{Z})$ for a while.

7.5 Elliptic elements of $GL(2, \mathbb{Z})$

The following proposition is well known to crystallographers. In fact, $GL(2, \mathbb{Z})$ is the group of linear maps of $\mathbb{R}^2$ that preserve the vertices of the lattice $\mathbb{Z}^2$, so the finite order elements of the group belong to finite symmetry groups of the lattice.

Proposition 7.8 *The only possible orders of finite-order elements of* $GL(2, \mathbb{Z})$ *are 1,2,3,4, and 6.*

Proof The eigenvalues of a finite-order element must have absolute value 1, and if not real must be a pair of complex conjugates whose sum is an integer. So the possible sets of eigenvalues are:

$$\{1\}, \quad \{-1\}, \quad \{\pm 1\}, \quad \{\pm i\}, \quad \left\{\tfrac{1}{2} + \pm i \tfrac{\sqrt{3}}{2}\right\}, \quad \left\{-\tfrac{1}{2} + \pm i \tfrac{\sqrt{3}}{2}\right\}.$$

The result follows, because these give elements of orders 1, 2, 2, 4, 6, and 3, respectively. $\square$

All these orders occur, as instanced by

$$\mathbb{1}, \quad -\mathbb{1}, \quad \begin{pmatrix} 0 & -1 \\ 1 & -1 \end{pmatrix}, \quad \begin{pmatrix} 0 & -1 \\ 1 & 0 \end{pmatrix}, \quad \text{and} \quad \begin{pmatrix} 0 & -1 \\ 1 & 1 \end{pmatrix},$$

which have orders 1, 2, 3, 4, and 6, respectively.

An element g of $GL(2, \mathbb{Z})$ is elliptic if and only if its trace τ and determinant Δ satisfy $0 \leqslant \tau^2/\Delta < 4$. The only possibilities are (i) $\tau = 0$, (ii) $\tau = 1$ and $\Delta = 1$, or (iii) $\tau = -1$ and $\Delta = 1$. Let us work out the possible orders of g in each of these cases. In case (i), $g^2 = \pm\mathbb{1}$ by (7.1), so g has order 1, 2, or 4. In case (ii), $g^2 = g - \mathbb{1}$, and we calculate that $g^6 = \mathbb{1}$. In case (iii), $g^2 = -g - \mathbb{1}$, and we find that $g^3 = \mathbb{1}$.

The element g is parabolic if and only if $\tau^2/\Delta = 4$ and $g \neq \pm\mathbb{1}$. The only possibilities are $\tau = \pm 2$ and $\Delta = 1$. All other values of τ and Δ correspond to elements g that are hyperbolic.

We note that if

$$g = \begin{pmatrix} a & b \\ c & d \end{pmatrix},$$

and at least one of b or c is zero, then $ad = \pm 1$, so each of a and d is ± 1. Therefore g is either elliptic (order 1,2, or 4) if $a \neq d$ or parabolic if $a = d$. Thus all hyperbolic elements of $GL(2, \mathbb{Z})$ satisfy $b \neq 0$ and $c \neq 0$.

7.5.1 Involutions

The elements $\pm\mathbb{1}$ of the centre of $GL(2, \mathbb{Z})$ are involutions.

Proposition 7.9 *Let g be an involution in $\mathrm{GL}(2,\mathbb{Z})$, not equal to ± 1. Then g is conjugate to precisely one of*

$$\begin{pmatrix} 1 & 0 \\ 0 & -1 \end{pmatrix} \quad or \quad \begin{pmatrix} 1 & 0 \\ 1 & -1 \end{pmatrix},$$

or, equivalently, to precisely one of

$$\begin{pmatrix} 1 & 0 \\ 0 & -1 \end{pmatrix} \quad or \quad \begin{pmatrix} 0 & 1 \\ 1 & 0 \end{pmatrix}$$

Proof The matrix g has characteristic polynomial $\lambda^2 - 1$, so $\Delta = -1$ and $\tau = 0$, and g has eigenvalues $+1$ and -1.

Let $g = \begin{pmatrix} a & b \\ c & d \end{pmatrix}$. Then $d = -a$ and $bc = 1 - a^2$, so the matrix

$$T_1 = \begin{pmatrix} -b & a-1 \\ a-1 & c \end{pmatrix}$$

conjugates g to $\begin{pmatrix} 1 & 0 \\ 0 & -1 \end{pmatrix}$ in $\mathrm{GL}(2,\mathbb{Q})$. Note that $\det(T_1) = 2$.

Since $a - 1$ divides bc, we may write $a - 1 = rs$, with $r|b$ and $s|c$. Then

$$T_2 = \begin{pmatrix} -b/r & r \\ s & c/s \end{pmatrix}$$

also has determinant 2 and conjugates g to $\begin{pmatrix} 1 & 0 \\ 0 & -1 \end{pmatrix}$.

If both entries in any row or column are even, we may divide them by 2 and get a matrix $T_3 \in \mathrm{GL}(2,\mathbb{Z})$ that conjugates g to $\begin{pmatrix} 1 & 0 \\ 0 & -1 \end{pmatrix}$. Otherwise, one may see that a must be even, so both r and s are odd, hence all entries of T_2 are odd. In that case, write T_2 in columns as (c_1, c_2). Then $gc_1 = c_1$, $gc_2 = -c_2$, so

$$T_3 = \left(\frac{c_1 + c_2}{2}, \frac{c_1 - c_2}{2} \right)$$

conjugates g to $\begin{pmatrix} 0 & 1 \\ 1 & 0 \end{pmatrix}$, and has integral entries and determinant 1. This proves the first assertion. For the second assertion, observe that

$$\begin{pmatrix} 1 & -1 \\ 0 & 1 \end{pmatrix} \begin{pmatrix} 1 & 0 \\ 1 & -1 \end{pmatrix} \begin{pmatrix} 1 & 1 \\ 0 & 1 \end{pmatrix} = \begin{pmatrix} 0 & 1 \\ 1 & 0 \end{pmatrix}. \qquad \square$$

Corollary 7.10 *The only nontrivial involution in $\mathrm{SL}(2,\mathbb{Z})$ is -1.* $\qquad \square$

The next proposition can be proven by an explicit calculation.

Proposition 7.11 *If g is an involution in $\mathrm{GL}(2,\mathbb{Z})$ other than $\pm\mathbb{1}$, then the centraliser of g has just four elements:*

$$C_g(GL(2,\mathbb{Z})) = \{\pm\mathbb{1}, \pm g\} \simeq C_2 \times C_2. \qquad \square$$

It follows from the results of this subsection that $\mathrm{GL}(2,\mathbb{Z})$ has a rich supply of strongly-reversible elements, but $\mathrm{SL}(2,\mathbb{Z})$ has only $\pm\mathbb{1}$.

7.5.2 Conjugacy classification

Let us now broaden our attention to the conjugacy classes of other finite-order elements in $\mathrm{GL}(2,\mathbb{Z})$.

Proposition 7.12 *The conjugacy classification of the finite-order elements of $\mathrm{GL}(2,\mathbb{Z})$ is summarised in the following table.*

τ	Δ	order	class representative
2	1	1	$\mathbb{1}$
-2	1	1	$-\mathbb{1}$
0	-1	2	$\begin{pmatrix} 0 & 1 \\ 1 & 0 \end{pmatrix}$
0	-1	2	$\begin{pmatrix} 1 & 0 \\ 0 & -1 \end{pmatrix}$
-1	1	3	$\begin{pmatrix} 0 & -1 \\ 1 & -1 \end{pmatrix}$
0	1	4	$\begin{pmatrix} 0 & -1 \\ 1 & 0 \end{pmatrix}$
1	1	6	$\begin{pmatrix} 0 & -1 \\ 1 & 1 \end{pmatrix}$

Table 7.12: The elliptic conjugacy classes in $\mathrm{GL}(2,\mathbb{Z})$

There is just one conjugacy class of each order 3, 4, and 6.

Proof We have dealt with the involutions.

For g of order 3, 4, or 6, the characteristic polynomial is uniquely determined by the order, it is irreducible, and the representatives given in the table correspond to the rational canonical forms. Thus they represent up to conjugation in $\mathrm{GL}(2,\mathbb{Q})$. We have to see that we can always achieve the conjugation with an integer matrix. We apply Proposition 7.1.

For $g = \begin{pmatrix} a & b \\ c & d \end{pmatrix}$ of order 4, we have $d = -a$ and $-a^2 - bc = 1$, so it comes down to choosing $v_1 = (x,y) \in \mathbb{Z}^2$ such that $cx^2 - 2axy - by^2 = \pm 1$. The form here has discriminant -4. There is only one reduced positive form of discriminant -4, namely $[1, 0, 1]$, so all forms of this discriminant are properly equivalent either to this form or its negative, and represent ± 1.

For g of order 3, we have $d = -1 - a$ and $-a - a^2 - bc = 1$, so it comes down to choosing $x, y \in \mathbb{Z}^2$ such that $cx^2 - (1 + 2a)xy - by^2 = \pm 1$. This form has discriminant -3. Again, there are just two proper equivalence classes of this discriminant, represented by $\pm[1, 1, 1]$. So, again, ± 1 is represented.

For g of order 6, we have $d = 1 - a$ and $a - a^2 - bc = 1$, so it comes down to choosing $\xi, \eta \in \mathbb{Z}^2$ such that $cx^2 + (1 - 2a)xy - by^2 = \pm 1$. This form has also has discriminant -3, and we conclude as in the previous case. $\square$

Proposition 7.13

(i) *The elliptic elements of* SL$(2, \mathbb{Z})$ *have orders* 1, 2, 3, 4, *and* 6. *There is just one conjugacy class of order* 2, *and two each of orders* 3, 4 *and* 6.

(ii) *The elliptic elements of* PGL$(2, \mathbb{Z})$ *have orders* 1, 2, *and* 3. *There is just one conjugacy class of order* 3, *but two conjugacy classes of order* 2.

(iii) *The elliptic elements of* PSL$(2, \mathbb{Z})$ *have orders* 1, 2 *and* 3. *There is just one conjugacy class order* 2, *and two of order* 3.

Proof   (i) Starting from the conjugacy table for GL$(2, \mathbb{Z})$, we eliminate the elements having determinant -1. The remaining classes split as disjoint unions of classes in SL$(2, \mathbb{Z})$. There is just one proper involution, $-\mathbb{1}$. Applying Corollary 7.2 as before, we see that there are two classes of order 3, represented by $\omega = \begin{pmatrix} 0 & -1 \\ 1 & -1 \end{pmatrix}$ and $\omega^2 = \begin{pmatrix} -1 & 1 \\ -1 & 0 \end{pmatrix}$ (— the latter is not conjugate to ω because $-x^2 + xy - y^2$ does not represent $+1$). Similarly, there are two classes of order 4, represented by $\pm \begin{pmatrix} 0 & -1 \\ 1 & 0 \end{pmatrix}$, and two classes of order 6, represented by $-\omega$ and $-\omega^2$.

(ii) If g has finite order m in PGL$(2, \mathbb{Z})$, then g has order m or $2m$ in GL$(2, \mathbb{Z})$, so we just have to check the orders in PGL$(2, \mathbb{Z})$ of the finite-order elements of GL$(2, \mathbb{Z})$. It turns out that the elements of order 4 and 6 have orders 2 and 3 in the quotient. The two conjugacy classes of proper noncentral involutions are equivalent modulo $\pm \mathbb{1}$, but not equivalent to the elements of order 4. The classes of order 3 and 6 are equivalent in PGL$(2, \mathbb{Z})$.

(iii) Follows from (ii). The only nontrivial involutions in PSL$(2, \mathbb{Z})$ correspond to elements of order 4 in SL$(2, \mathbb{Z})$. The two classes (of ω and ω^2) of order 3 in SL$(2, \mathbb{Z})$ remain nonconjugate in PSL$(2, \mathbb{Z})$, but the the two classes of order 6 in SL$(2, \mathbb{Z})$ are equivalent to them. $\square$

7.6 Centralisers

Recall that we denote the centraliser of the element g in the group G by $C_g(G)$.

Proposition 7.14   *Suppose that $g \in \mathrm{GL}(2,\mathbb{Z})$ has an irreducible characteristic polynomial, and let λ be an eigenvalue. Then*

$$C_g(\mathrm{GL}(2,\mathbb{Z})) \simeq S \times C_2,$$

where S is a subgroup of the group of units in the maximal order of $\mathbb{Q}(\lambda)$.

Proof   Letting $\Delta = \det(g)$, as usual, the matrix g is conjugate over $\mathbb{C}$ to its Jordan form

$$g_1 = \begin{pmatrix} \lambda & 0 \\ 0 & \frac{\Delta}{\lambda} \end{pmatrix}.$$

The eigenvector equations have coefficients in $K = \mathbb{Q}(\lambda)$, so may be solved in that field, and hence there is a matrix U over K such that

$$U^{-1}gU = g_1.$$

Now the only matrices over K that commute with a given diagonal matrix having distinct diagonal entries are the scalar matrices. Thus if $h \in \mathrm{GL}(2,\mathbb{Z})$ commutes with g, then $U^{-1}hU$ commutes with g_1, and hence is a diagonal matrix over K. Thus $C_g(\mathrm{GL}(2,\mathbb{Z}))$ consists precisely of those $h \in \mathrm{GL}(2,\mathbb{Z})$ such that $U^{-1}hU$ is diagonal. But if

$$U^{-1}hU = \begin{pmatrix} \alpha & 0 \\ 0 & \beta \end{pmatrix},$$

then α and β are the roots of the characteristic polynomial of C, and hence are algebraic integers. Moreover, $\alpha\beta = \Delta = \pm 1$, so that α is a unit of the maximal order O of K. It follows that the set of eligible α is a subgroup S of the group of units of K.

For each $\alpha \in S$, there are two elements of the centraliser of g, and the map

$$C_g(\mathrm{GL}(2,\mathbb{Z})) \times \{\pm 1\} \to C_C(\mathrm{GL}(2,K))$$

$$(\alpha, s) \mapsto U \begin{pmatrix} \alpha & 0 \\ 0 & \pm\frac{s}{\alpha} \end{pmatrix} U^{-1}$$

is a group isomorphism.   □

Dirichlet elucidated the structure of the group of units of irrational quadratic fields. His units theorem [54, page 99] tells us that if $K = \mathbb{Q}(\sqrt{D})$, where D is a nonsquare integer, then the group of units U_K in the maximal order O_K of K

is generated by -1 and one other element u. Such a u may be found by solving Pell's equation, and this may be done by using continued fractions. The group of units of an imaginary quadratic field is finite.

Corollary 7.15 $C_g(\mathrm{GL}(2,\mathbb{Z})$ is abelian, for each hyperbolic g.

Proposition 7.14 applies, in particular, to g of order 3,4, and 6, and easily leads to the following result.

Proposition 7.16   If g has order 3, 4, or 6, then the centraliser of g just consists of the elements $\pm g^j$, for integral j. It is thus a cyclic group, of order 6, 4, or 6, respectively.

Note that for g of order 3, the centraliser is generated by $-g$, and for orders 4 and 6, it is generated by g.

The parabolic elements with $\tau = 2$ and $\Delta = 1$ are conjugate to one of

$$t_m = \begin{pmatrix} 1 & m \\ 0 & 1 \end{pmatrix} \tag{7.2}$$

with m an element of $\mathbb{N}$, the positive integers, whereas those with $\tau = -2$ and $\Delta = 1$ are conjugate to one of

$$-t_m = \begin{pmatrix} -1 & -m \\ 0 & -1 \end{pmatrix} \tag{7.3}$$

with $m \in \mathbb{N}$. The distinct $m \in \mathbb{N}$ give nonconjugate elements, so this is the full conjugacy classification.

The centraliser of t_m is generated by t_1 and $-\mathbb{1}$, and is the abelian group freely generated by these two. Thus the centraliser of any parabolic element is isomorphic to $C_2 \times \mathbb{Z}$.

7.7   Reversible elements

7.7.1   The group GL

Proposition 7.17   All elliptic elements of $\mathrm{GL}(2,\mathbb{Z})$ are strongly reversible.

Proof   Involutions are reversible in any group. There is just one conjugacy class for each order 3,4, and 6, so elements of these orders are conjugate to their inverses.

To see they are strongly reversible, observe that

$$\begin{pmatrix} 0 & -1 \\ 1 & 1 \end{pmatrix} = \begin{pmatrix} -1 & 0 \\ 1 & 1 \end{pmatrix} \begin{pmatrix} 0 & 1 \\ 1 & 0 \end{pmatrix},$$

so that each element of order 6 is reversed by an involution. It follows that each element of order 3 is also strongly reversible (since each such element has a square root). Also,

$$\begin{pmatrix} 0 & -1 \\ 1 & 0 \end{pmatrix} = \begin{pmatrix} 0 & 1 \\ 1 & 0 \end{pmatrix} \begin{pmatrix} 1 & 0 \\ 0 & -1 \end{pmatrix},$$

so that each element of order 4 is the product of two involutions.   □

Proposition 7.18 *All the parabolic elements of* $\mathrm{GL}(2,\mathbb{Z})$ *are strongly reversible.*

Proof The representatives $\pm t_m$ of formulas 7.2 and 7.3 are reversed by

$$\begin{pmatrix} 1 & 0 \\ 0 & -1 \end{pmatrix}.$$

□

From here on, we concentrate on the hyperbolic elements.

Lemma 7.19 *If F is a field and* $g \in \mathrm{GL}(2, F)$, *then*

$$\mathrm{trace}(g^{-1}) = \frac{\mathrm{trace}(g)}{\det(g)}.$$

Proof Direct calculation.   □

Proposition 7.20 *Let F be a field of characteristic other than* 2. *Let* $g \in \mathrm{GL}(2, F)$. *If* $\tau \neq 0$ *and* $\Delta = -1$, *then g is not reversible in* $\mathrm{GL}(2, F)$.

Proof By the lemma, $\mathrm{trace}(g^{-1}) = -\mathrm{trace}(g)$, so g cannot be reversible, since the trace is a conjugacy invariant.   □

Corollary 7.21 *Let* $g \in \mathrm{GL}(2, \mathbb{Z})$. *If* $\tau \neq 0$ *and* $\Delta = -1$, *then g is not reversible.*   □

Thus the only reversibles in $\mathrm{GL}(2, \mathbb{Z})$ with $\Delta = -1$ are the (two conjugacy classes of) involutions. All the elements of $\mathrm{GL}(2, \mathbb{Z})$ having determinant -1 and nonzero trace are hyperbolic.

It remains to consider hyperbolic elements with positive determinant, i.e. belonging to $\mathrm{SL}(2, \mathbb{Z})$. For each such g, both g and g^{-1} are rationally-conjugate to $\begin{pmatrix} 0 & -1 \\ 1 & \tau \end{pmatrix}$, and hence g is reversible in $\mathrm{GL}(2, \mathbb{Q})$.

Proposition 7.22 *If* $g \in \mathrm{SL}(2, \mathbb{Z})$ *has infinite order and* $h \in R_g(\mathrm{GL}(2, \mathbb{Z}))$, *then h has order 2 or 4.*

Proof Let $g = \begin{pmatrix} a & b \\ c & d \end{pmatrix}$ have infinite order, and be reversed by $h = \begin{pmatrix} \alpha & \beta \\ \gamma & \delta \end{pmatrix}$ in $GL(2, \mathbb{Z})$. Then comparing the $(1,2)$ entries in gh and hg^{-1}, we get

$$b(\alpha + \delta) = 0 = c(\alpha + \delta).$$

Since not both b and c are zero, $\text{trace}(h) = 0$. This implies that $h^4 = \mathbb{1}$. $\square$

Corollary 7.23 *Each element of $R(GL(2, \mathbb{Z}))$ is the product of at most four involutions.*

Proof Combining Propositions 7.22 and Corollary 7.21 with Corollary 2.23 we see that each reversible g of infinite order is the product of two elements of order dividing 4. Each element of order 4 is the product of two involutions, so that gives the result in this case.

But we have also seen (Proposition 7.17) that each element of finite order is the product of two involutions, so we are done. $\square$

Proposition 7.24 *Let g be a hyperbolic element of $SL(2, \mathbb{Z})$. Then the following are equivalent:*

(i) *g is reversible in $GL(2, \mathbb{Z})$*

(ii) *g is conjugate in $SL(2, \mathbb{Z})$ to a matrix $\begin{pmatrix} a & b \\ c & d \end{pmatrix}$ with $b = \pm c$ or $a = d$.*

Proof Let g be reversible. Then it is reversed by a matrix of order 2 or 4.

First, Suppose g is reversed by an element of order precisely 4. Each element of order 4 is conjugate to the matrix $J = \begin{pmatrix} 0 & 1 \\ -1 & 0 \end{pmatrix}$. Use the same conjugation to convert g to $\begin{pmatrix} a & b \\ c & d \end{pmatrix}$. Now look at the reversibility equation $JgJ^{-1} = g^{-1}$ and equate coefficients. We get that $b = c$. Thus g is conjugate to a symmetric matrix.

Next, consider strongly reversible g. Each noncentral element of order 2 is conjugate by a matrix with determinant $+1$ to one of

$$\sigma = \begin{pmatrix} 0 & 1 \\ 1 & 0 \end{pmatrix} \quad \text{or } \tau = \begin{pmatrix} 1 & 0 \\ 0 & -1 \end{pmatrix}.$$

Thus each element of $I^2(GL(2, \mathbb{Z}))$ is conjugate to a matrix reversed by one of these. Now look at the equation in each case. σ gives $b = -c$, and τ gives $a = d$.

Conversely, each symmetric element with $\Delta = +1$ is reversed by the above J. Thus each matrix with $\Delta = +1$, similar to a symmetric matrix, is reversible by an element of order 4. Similarly for the cases $b = -c$ and $d = a$. $\square$

Corollary 7.25　*Let $g \in \mathrm{GL}(2,\mathbb{Z})$. Then g is reversible in $\mathrm{GL}(2,\mathbb{Z})$ if and only if*

(i) *it has $\Delta = -1$ and is an involution, or*

(ii) *it has positive determinant and is similar to a matrix $\begin{pmatrix} a & b \\ c & d \end{pmatrix}$ with $b = \pm c$ or $a = d$.*

□

Corollary 7.26　*A hyperbolic element of $\mathrm{GL}(2,\mathbb{Z})$ is strongly reversible if and only if it is conjugate to a matrix $\begin{pmatrix} a & b \\ c & d \end{pmatrix}$ with $b = -c$ or $a = d$.*　□

But how do you determine whether g is conjugate to a matrix of one of the three types in condition (ii) of Corollary 7.25? This will lead us back to the first associated quadratic form. We will also see that a direct attack on the reversibility equation produces a connection with the representation of ± 1 by a second associated form.

Before taking this up, we record the corresponding results about reversibility in the related groups.

7.7.2　The group SL

Proposition 7.27　*Let $g \in \mathrm{SL}(2,\mathbb{Z})$. Then g is reversible in $\mathrm{SL}(2,\mathbb{Z})$ if and only if*

(i) *g is involutive, or*

(ii) *g is hyperbolic and is conjugate in $\mathrm{SL}(2,\mathbb{Z})$ to a symmetric matrix.*

Proof　Apart from the involution $-\mathbb{1}$, we have seen that the elliptic elements are all in different conjugacy classes of $\mathrm{SL}(2,\mathbb{Z})$ from their inverses. (Proposition 7.13).

No parabolic element is reversible in $\mathrm{SL}(2,\mathbb{Z})$. If a parabolic g were reversible, then, since it has infinite order, it would be conjugate to a symmetric matrix. But if $\begin{pmatrix} a & b \\ b & d \end{pmatrix}$ were parabolic, we would have $ad = b^2 + 1 > 0$ and $a + d = \pm 2$, forcing $a = d = \pm 1$ and $b = 0$.

In $\mathrm{SL}(2,\mathbb{Z})$ the only proper involution is $-\mathbb{1}$. Thus hyperbolic reversibles can only be reversed by elements of order 4. As a result, each hyperbolic reversible is conjugate to a symmetric matrix.　□

7.7.3 The group PGL

All elliptic and parabolic elements are reversible in $\mathrm{PGL}(2,\mathbb{Z})$.

For hyperbolic elements, we have to consider the possibility that $g^h = -g^{-1}$. This cannot occur for g with positive determinant, because of the trace formula, Lemma 7.19.

Proposition 7.28 *Let a hyperbolic $g \in \mathrm{GL}(2,\mathbb{Z})$ have $\Delta = -1$. Then g is not reversible in $\mathrm{PGL}(2,\mathbb{Z})$.*

Proof Reversibility in PGL amounts to $g^h = \pm g^{-1}$. We know already that $g^h = g^{-1}$ does not occur. Writing $g = \begin{pmatrix} a & b \\ c & d \end{pmatrix}$ and $h = \begin{pmatrix} \alpha & \beta \\ \gamma & \delta \end{pmatrix}$, the equation $g^h = -g^{-1}$ amounts to the system

$$\begin{cases} (a+d)\alpha + b\gamma - c\beta &= 0 \\ 2a\beta + b(\delta - \alpha) &= 0 \\ 2d\gamma + c(\alpha - \delta) &= 0 \\ (d+a)\delta + c\beta - b\gamma &= 0. \end{cases} \tag{7.4}$$

Combining the first and last, we get $(a+d)(\alpha+\delta) = 0$. But g has nonzero trace, so h has trace zero, hence its order divides 4.

Now considering in turn the conjugacy classes of elements of order 2 or 4, and using equations 7.4 after conjugating h to the representatives in Table 7.12 (as in the proof of Proposition 7.24), we find in every case that if g is reversed by an element h of of order dividing 4, then $g = \pm h$, contradicting the assumption that it is hyperbolic. $\qquad\qquad\square$

Corollary 7.29 *Let $g \in \mathrm{GL}(2,\mathbb{Z})$. Then g is reversible in $\mathrm{PGL}(2,\mathbb{Z})$ if and only if it is reversible in $\mathrm{GL}(2,\mathbb{Z})$.*

The elements of order 4 in $\mathrm{GL}(2,\mathbb{Z})$ become involutions in the quotient $\mathrm{PGL}(2,\mathbb{Z})$. Thus in $\mathrm{PGL}(2,\mathbb{Z})$, there are no elements of order 4, and we get that all reversible elements are strongly reversible.

7.7.4 PSL

The involutions of $\mathrm{PSL}(2,\mathbb{Z})$ are, as always, reversible, but the elements of order 3 are not.

No parabolic is reversible in $\mathrm{PSL}(2,\mathbb{Z})$. Since parabolics $g = \pm t_m$ $(m \neq 0)$ are not reversible in $\mathrm{SL}(2,\mathbb{Z})$, they could only be reversible in $\mathrm{PSL}(2,\mathbb{Z})$ if they could be conjugated to $-g^{-1}$ But this cannot happen, because of the trace.

Combining the results from the last two section, we deduce:

Proposition 7.30 *Let $g \in \mathrm{PSL}(2,\mathbb{Z})$. Then the following four conditions are equivalent:*

  (i) *g is reversible in $\mathrm{PSL}(2,\mathbb{Z})$*
 (ii) *g is reversible in $\mathrm{SL}(2,\mathbb{Z})$*
(iii) *g is involutive or is conjugate to a symmetric matrix*
 (iv) *g is strongly reversible.*

Proof Regarding the last condition, note that hyperbolic g can only be reversed by elements conjugate to $\begin{pmatrix} 0 & -1 \\ 1 & 0 \end{pmatrix}$, which is an involution in $\mathrm{PSL}(2,\mathbb{Z})$.

$\square$

7.8 Second associated form

Let's start again with the reversibility equation $g^h = g^{-1}$.

Throughout this section, we let $g = \begin{pmatrix} a & b \\ c & d \end{pmatrix} \in \mathrm{SL}(2,\mathbb{Z})$ and $h = \begin{pmatrix} \alpha & \beta \\ \gamma & \delta \end{pmatrix} \in \mathrm{GL}(2,\mathbb{Z})$, and we suppose g is hyperbolic. Then the equation $gh = hg^{-1}$ becomes

$$
\begin{aligned}
(a-d)\alpha &= -c\beta - b\gamma \\
b((\alpha+\delta) &= 0 \\
c((\alpha+\delta) &= 0 \\
(d-a)\delta &= c\beta + b\gamma.
\end{aligned}
\tag{7.5}
$$

We deduce:

Lemma 7.31 *Let $g \in \mathrm{SL}(2,\mathbb{Z})$, and $h \in \mathrm{GL}(2,\mathbb{Z})$, with g hyperbolic and $\mathrm{trace}(h) = 0$. Then $h \in R_g(\mathrm{GL}(2,\mathbb{Z}))$ if and only if*

$$
(a-d)\alpha + c\beta + b\gamma = 0.
\tag{7.6}
$$

$\square$

Corollary 7.32 *A hyperbolic $g \in \mathrm{SL}(2,\mathbb{Z})$ is reversible in $\mathrm{GL}(2,\mathbb{Z})$ if and only if there is a solution $(\alpha,\beta,\gamma,\delta) \in \mathbb{Z}^4$ to the Diophantine system*

$$
\left.
\begin{aligned}
(a)\qquad && \alpha + \delta &= 0 \\
(b)\qquad && \alpha^2 + \beta\gamma &= \pm 1 \\
(c)\qquad (a-d)\alpha + c\beta + b\gamma &= 0.
\end{aligned}
\right\}
\tag{7.7}
$$

Further, g is reversible in $\mathrm{SL}(2,\mathbb{Z})$ if and only if there is a solution that has the $-$ sign on the right-hand-side of equation (b), and g is strongly reversible if and only if there is a solution that has the $+$ sign.

We know from the proof of Proposition 7.24 that hyperbolic $g \in \mathrm{SL}(2,\mathbb{Z})$ with equal diagonal elements are strongly reversible in $\mathrm{GL}(2,\mathbb{Z})$. They may or may not be reversible in $\mathrm{SL}(2,\mathbb{Z})$:

Proposition 7.33   *If $g \in \mathrm{SL}(2,\mathbb{Z})$ is hyperbolic and has equal diagonal entries, then it is reversible in $\mathrm{GL}(2,\mathbb{Z})$ if and only if the equation*

$$x^2 - \frac{bc}{s^2}y^2 = -1$$

has a solution in integers x, y, where $s = \gcd(b,c)$.

Proof   Equation (c) of 7.7 reduces to $c\beta + b\gamma = 0$, so is solved by

$$\beta = bt/s, \qquad \gamma = -ct/s$$

for any integral t. Substituting this in Equation (b) (with the $-$ sign!), we get

$$\alpha^2 - \frac{bc}{s^2}t^2 = -1.$$

If we can solve this equation for α, t in integers, then g is reversible, and conversely. $\qquad\qquad\qquad\qquad\qquad\qquad\qquad\qquad\qquad\qquad\qquad\square$

Noting that $bc = a^2 + 1$ is a positive nonsquare, we have a standard 'negative Pell equation'. It is well-known when this may be solved [51]. In essence, for nonsquare κ, one looks at the continued fraction expansion of $\sqrt{\kappa}$, which is necessarily periodic. If the length of the period is odd, then the equation $x^2 - \kappa y^2 = -1$ can be solved in integers, and conversely.

So now we focus on hyperbolic $g = \begin{pmatrix} a & b \\ c & d \end{pmatrix}$ that have $a \neq d$. Let us say that g is an *ordinary hyperbolic* element if $\det(g) = +1$, and $a \neq d$.

Neither b nor c can be zero, for otherwise $ad = 1$, so $a = d = \pm 1$, and g is parabolic.

Let $g = \begin{pmatrix} a & b \\ c & d \end{pmatrix} \in \mathrm{SL}(2,\mathbb{Z})$ be an ordinary hyperbolic matrix. Take $r = \gcd(a-d,b,c)$, and $s = \gcd(a-d,c)$. Then $r|s$. Let $\mu = s/r$. Let (α_0, β_0) be any one solution of the linear Diophantine equation

$$(a-d)\alpha + c\beta = s.$$

Then the general solution of 7.7(c) is

$$\alpha = \frac{b}{r}t\alpha_0 - \frac{c}{s}u, \quad \beta = \frac{b}{r}t\beta_0 + \left(\frac{a-d}{s}\right)u, \quad \gamma = -\mu t, \qquad (u,t \in \mathbb{Z}).$$

Thus 7.7(b) becomes

$$At^2 + But + Cu^2 = \pm 1,$$

where

$$\begin{cases} A & = & \dfrac{\alpha_0^2 b^2}{r^2} - \dfrac{\mu\beta_0 b}{r}, \\ B & = & -\left(\dfrac{a-d}{r} + \dfrac{2\alpha_0 bc}{rs}\right), \\ C & = & \dfrac{c^2}{s^2}. \end{cases}$$

Thus g is reversible in $\mathrm{GL}(2,\mathbb{Z})$ if and only if $+1$ or -1 may be represented by the quadratic form $Ax^2 + Bxy + Cy^2$. It is reversible in $\mathrm{SL}(2,\mathbb{Z})$ if and only if -1 may be represented in this way. It is strongly reversible if and only if $+1$ may be so represented.

The construction associates a second quadratic form to each ordinary hyperbolic g. If we make a different choice of the solution α_0, β_0, then we get a different form, but it may be seen that it is properly equivalent. In fact the other solution must take the form

$$\alpha_0' = \alpha_0 - \left(\frac{c}{s}\right)v, \quad \beta_0' = \beta_0 + \left(\frac{a-d}{s}\right)v,$$

for some integer v, and this gives a new form $[A', B', C']$ with

$$\begin{pmatrix} A' & \frac{1}{2}B' \\ \frac{1}{2}B' & C' \end{pmatrix} = \begin{pmatrix} 1 & w \\ 0 & 1 \end{pmatrix}\begin{pmatrix} A & \frac{1}{2}B \\ \frac{1}{2}B & C \end{pmatrix}\begin{pmatrix} 1 & 0 \\ w & 1 \end{pmatrix},$$

where $w = bv/r$, as is readily checked.

Thus we have a well-defined map from the set of such g to the family of equivalence classes of forms. Denoting the class of any form induced by g by $[g]$, we can state:

Proposition 7.34 *An ordinary hyperbolic element g of $\mathrm{SL}(2,\mathbb{Z})$ is reversible in $\mathrm{GL}(2,\mathbb{Z})$ if and only if the forms of the class $[g]$ represent ± 1. It is reversible in $\mathrm{SL}(2,\mathbb{Z})$ if and only if -1 is represented, and is strongly reversible if and only if $+1$ is represented.*

Replacing g by g^{-1} changes the signs of $a-d$, b, and c, and hence we may use the negatives of α_0 and β_0 to solve the linear equation, so if g induces the form $[A,B,C]$, then g^{-1} induces the improperly equivalent form $[A,-B,C]$.

Example 7.35 The form induced by the reversible element $g = \begin{pmatrix} 5 & 8 \\ 8 & 13 \end{pmatrix}$ is $[-1,1,1]$. This shows that we may not confine attention to forms $[A,B,C]$ with *even B*.

Proposition 7.36 *The second form $[A,B,C]$ induced by an ordinary hyperbolic element $g \in \mathrm{SL}(2,\mathbb{Z})$ is primitive.*

Proof   Suppose the prime p divides A, B, and C. Then we deduce in turn that $p|\frac{c}{s}$, $p|(a-d)/r$, and so $p|b/r$ or $p|(\alpha_0(b/r)-\mu\beta_0)$. But if $p|(b/r)$, then we deduce that $p|\gcd\left(\frac{a-d}{r},\frac{b}{r},\frac{c}{r}\right)$, contradicting the fact that $r=gcd(a-d,b,c)$, so this does not occur. On the other hand, if p divides $(\alpha_0(b/r)-\mu\beta_0)$, then since α_0 and β_0 have no common factor, p must divide the gcd of b/r and μ. But μ is the gcd of $\frac{a-d}{r}$ and c/r, so again we find that p divides all three of $\frac{a-d}{r}$, b/r and c/r, which is impossible. $\square$

So to decide the reversibility question for such a g, we again require an algorithm for deciding which quadratic forms represent ± 1.

Proposition 7.37   *If g is an ordinary hyperbolic matrix, then its second associated class of forms $[g]$ has discriminant*

$$B^2 - 4AC = \frac{\tau^2 - 4\Delta}{r^2},$$

where $r=\gcd(a-d,b,c)$. In particular, the discriminant is positive.

Proof   Routine calculation. $\square$

The discriminant is never square, because $\tau^2 - 4\Delta$ is the discriminant of the characteristic polynomial, and this is irreducible.

We turn now to the problem of deciding when an indefinite form represents ± 1.

7.9   Gauss' method for indefinite forms

A form $[A,B,C]$ with positive nonsquare discriminant D is called *Gauss reduced* if $0 < B < \sqrt{D}$ and $\sqrt{D}-B < 2|A| < \sqrt{D}+B$. Gauss gave an algorithm for constructing from any given form having positive nonsquare discriminant, a reduced form that is properly equivalent to it. The algorithm is, as before, simple, and based on Lemma 7.3, part (iv). If $[A,B,A']$ is not yet reduced, take B' congruent to $-B$ mod $2A'$ and lying between $\sqrt{D}-2|A'|$ and $\sqrt{D}$. Then let $A'' = (B'^2 - D)/4A'$ (which is an integer!) and replace $[A,B,A']$ by $[A',B',A'']$. Continue until the form is reduced.

In Conway's terms, a Gauss reduced form is associated to a point on the river, because necessarily $AC < 0$.

In general, there is not a unique reduced form in a given proper equivalence class. What happens is that (1) there is a finite number, and crucially (2) they may be arranged in a cycle, where repeated application of the same step above takes you round and round the whole cycle. Of these two facts, (1) is obvious, but (2) seems harder — in Gauss' Disquisitiones its justification occupies 14 pages [95, Articles 188–193]. Moreover, Gauss works only with even B, and we need the result for all integers B. Fortunately, the Conway perspective makes (2) obvious as well:

Let us call a face that borders the river a *meadow*, and an edge that lies on the river a *stretch*. There are $+$ meadows and $-$ meadows, distinguished by the sign of the form. Each stretch separates two meadows, one of each sign, and each two adjacent stretches lie on the boundary of exactly one common meadow. Note that an edge of the complex is a stretch if and only if its adjacent faces have values of opposite signs. We may distinguish two kinds of stretch. The best way to think about the distinction is to imagine that we want to construct a bike-path along the river, running along meadows, using bridges

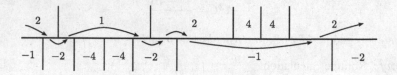

Figure 7.9 Bikepath along river

where necessary, but never cutting through the fence between two meadows of the same sign. If we do this as economically as possible, only crossing the river where essential, then we do not put a bridge across a stretch if it and both adjacent stretches lie along a single meadow (Figure 7.9). Let us call the stretches where we must put a bridge *bridge stretches*. Evidently, the step from one bridge stretch to the next corresponds precisely to a step of Gauss' reduction algorithm. So once you reach the river, the algorithm consists of skipping along from bridge to bridge. Now suppose that $f' = [A', B', C']$ is a reduced form properly equivalent to our original form $f = [A, B, C]$. Then there exists a basis $\{e_1, e_2\}$ for $\mathbb{Z}^2$ such that the matrix g with columns e_1, e_2 belongs to $SL(2, \mathbb{Z})$ and $f' = f \circ g$. Thus $f(e_1) = A'$ and $f(e_2) = B'$. Since A' and C' have opposite signs, the edge $\{e_1, e_2\}$ lies on the river, i.e. is a stretch. Moreover, it is readily seen that it is a bridge stretch, using the reduced form inequalities. Hence the reduction algorithm for f will eventually lead us to this bridge. Hence the form f' lies in the reduced form cycle of f.

Thus, to determine whether or not two forms are properly equivalent, one reduces them both, and checks whether or not they land in the same cycle. To determine equivalence, one should also compare one with some form improperly equivalent to the other.

Incidentally, the cycle of reduced forms always has an even number of terms, as may be seen from the fact that the sign of the first coefficient alternates (— you must cross the river an even number of times per cycle).

We cannot directly see from this Gauss cycle which numbers are properly represented by the form. However, we can see whether or not ± 1 is represented. We have:

Proposition 7.38 *If $[A,B,C]$ is a form of square-free discriminant $D > 0$, and if N is an integer with $|N| < \sqrt{D}/2$, then the form represents N primitively if and only if it is equivalent to a Gauss reduced form $[N,A',B']$.*

Proof The 'if' part is clear.

To see the other direction, suppose that $Ap^2 + Bpr + Cr^2 = N$, where $p, r \in \mathbb{Z}$ are relatively-prime. Choose $q, s \in \mathbb{Z}$ with $ps - qr = 1$. Then

$$\begin{pmatrix} p & q \\ r & s \end{pmatrix} [A,B,C] = [N,B',C'],$$

where

$$\begin{aligned} B' &= 2Apq + B(ps+qr) + Crs, \\ C' &= Aq^2 + Bqs + Cs^2. \end{aligned}$$

Now replace B by $B'' \equiv B \mod N$ so that $\sqrt{D} - 2|N| < B'' < \sqrt{D}$, and choose C'' to make $\mathrm{discr}[N,B'',C''] = D$. Then $[N,B'',C'']$ is Gauss reduced. $\square$

Corollary 7.39 *An indefinite form $[A,B,C]$ represents 1 or -1 if and only Gauss's reduction algorithm leads to a form $[1,B',C']$ or a form $[-1,B',C']$.*

Proof Discriminants are always congruent to 0 or 1 mod 4, so if $[A,B,C]$ has square-free discriminant, then its discriminant is at least 5, so the proposition always applies to $N = \pm 1$. $\square$

The beauty of this is that, compared with the slow walk along the Conway graph, Gauss' method takes you to the river on an express train (worst case time $O(\log(|A|+|B|+|C|))$ instead of $O(\sqrt{(|A|+|B|+|C|)})$), and , once at the river, the Gauss cycle skips along equally fast: there will usually be relatively few edges on the river that correspond to reduced forms.

One might wonder why Gauss used such a curious definition in the indefinite case, instead of something more straightforward, as in the definite case. However, the obvious version does not much constrain C, and it would no longer holds true (in the indefinite case) that the reduced forms would lie in a convenient cycle, and we could not so readily determine when ± 1 is represented. Gauss knew what he was doing.

Example 7.40 The form $[86898, 11202, 361]$ has discriminant 4092. Gauss' algorithm converts it in turn to:

$$[361, -372, 93], [93, 0, -11], [-11, 22, 82], [82, -22, -11], [-11, 22, 82].$$

Thus the cycle of reduced forms has length 2, and neither 1 nor -1 is represented.

Example 7.41 The form $[1806, 2466, 841]$ has discriminant 5772. Gauss' algorithm leads in turn to:

$$[841, -784, 181], [181, 60, -3], [-3, 6, 478], [478, -6, -3], [-3, 6, 478].$$

For this example, it takes 4 steps on Conway's graph to get to the river, and the river has a period of 27 edges, and shows that -3 and -26 are the two largest negative values primitively represented, while 49 is the smallest positive value represented. Gauss' cycle picks out just 3 points on the river.

Example 7.42 Let $S = \begin{pmatrix} 5 & 12 \\ 12 & 29 \end{pmatrix}$, $T = \begin{pmatrix} 5 & 8 \\ 8 & 13 \end{pmatrix}$, and take

$$g = T^{-1}ST = \begin{pmatrix} -763 & -1236 \\ 492 & 797 \end{pmatrix}.$$

Then $g \in \mathrm{SL}(2, \mathbb{Z})$ and is reversed by some element of order 4, by Corollary 7.25. Let's try the above methods on g. One finds that the second associated form is $[379967, -50546, 1681]$, with discriminant 8. Applying Gauss' algorithm to this yields a chain of properly equivalent forms, beginning with

$$[1681, -3246, 1567], [1567, -3022, 1457],$$

and after 25 further steps arriving at

$$[7, -6, 1], [1, 2, -1], [-1, 2, 1], [1, 2, -1].$$

The last two are Gauss reduced, and form the cycle of all reduced forms in the proper equivalence class. Of course, at least one of ± 1 is represented, as expected. In fact, both are represented.

Example 7.43 Taking $T = \begin{pmatrix} 5 & 8 \\ 8 & 13 \end{pmatrix}$, $\sigma = \begin{pmatrix} 0 & 1 \\ 1 & 0 \end{pmatrix}$, and

$$g = T^{-1}\sigma T\sigma = \begin{pmatrix} 105 & 64 \\ -64 & -39 \end{pmatrix},$$

we obtain a strongly-reversible element of $SL(2, \mathbb{Z})$. The second associated form is $[16, -23, 8]$, with discriminant 17. Applying Gauss's algorithm, we get the chain of properly equivalent forms

$$[8, -9, 2], [2, 1, -2], [-2, 3, 1], [1, 3, -2],$$
$$[-2, 1, 2], [2, 3, -1], [-1, 3, 2], [2, 1, -2],$$

so this time the cycle of reduced forms has length 6. Of course, to check reversibility, there is no need to continue beyond the form $[-2, 3, 1]$, since we know at that stage that 1 is represented. In general, whichever method we use, we stop as soon as we find ± 1 as a value. For this example, Conway's method wins by a lucky fluke at the very first step, because $[16, -23, 8](1, 1) = 1$! The river for this form is that shown in Figure 7.9.

Proposition 7.44 $R(GL(2, \mathbb{Z})) \neq I^2(GL(2, \mathbb{Z}))$

Proof   The hyperbolic element $g = \begin{pmatrix} 5 & 7 \\ 7 & 10 \end{pmatrix} \in SL(2, \mathbb{Z})$ is symmetric, and hence is reversible in $SL(2, \mathbb{Z})$. It has second associated form $[455, 299, 49]$. Gauss reduction gives the chain

$$[49, -5, -1], [-1, 13, 13], [13, -13, -1], [-1, 13, 13].$$

Thus the cycle of reduced forms has two elements, -1 is represented, but the smallest positive value represented is 13. Hence $+1$ is not represented, and this g is not strongly reversible in $GL(2, \mathbb{Z})$.   □

Remark 7.45   This shows that $R \neq I^2$ also in the group $Diffeo(\mathbb{T}^2)$ of diffeomorphisms of the torus. In fact, we may regard g as an (orientation-preserving) element of $Diffeo(\mathbb{T}^2)$, and if there existed an involution $h \in Diffeo(\mathbb{T}^2)$ that reversed g, then the image of h in the extended mapping class group $MCG(2, \mathbb{Z})$ would give an element of $GL(2, \mathbb{Z})$ that reversed g there.

7.10 Another way

We return to considering the first associated form. We denote it by $F(g)$. Note that if the first form associated to a hyperbolic $g \in SL(2, \mathbb{Z})$ represents ± 1, then, by Proposition 7.1, g and g^{-1} are surely conjugate, since both have the same characteristic polynomial, and their forms are equivalent. This does not characterise reversibility, but the first form may be used to give a characterisation.

First, we note the effect of conjugation on $F(g)$.

Proposition 7.46   *If $g \in \mathrm{SL}(2,\mathbb{Z})$ and $h \in \mathrm{GL}(2,\mathbb{Z})$, then*

$$F(g^h) = \det(h) \cdot (hF(g)).$$

Proof   Direct calculation.   □

Proposition 7.47   *Let g be a hyperbolic element of $\mathrm{SL}(2,\mathbb{Z})$. Then g is conjugate in $\mathrm{SL}(2,\mathbb{Z})$ to a symmetric matrix if and only if $F(g)$ is properly equivalent to $-F(g)$.*

Proof   Suppose first that g is conjugate in $\mathrm{SL}(2,\mathbb{Z})$ to a symmetric matrix. Then we can see that $F(g)$ is properly equivalent to $-F(g)$ using the previous proposition and basic properties of proper equivalence, because

$$[c, d-a, -c] \approx [-c, a-d, c] = -[c, d-a, -c].$$

For the converse, suppose that $F(g)$ is properly equivalent to $-F(g)$. Choose an $h \in \mathrm{SL}(2,\mathbb{Z})$ such that $hF(g) = -F(g)$. Then $F(g^h) = -F(g) = F(g^{-1})$, hence $g^h = g^{-1}$ or $g^h = -g$, by Proposition 7.5. But the second is ruled out, because g has nonzero trace. Thus g is reversible in $\mathrm{SL}(2,\mathbb{Z})$, and hence conjugate to a symmetric matrix, by Proposition 7.27.   □

This gives another algorithm for checking reversibility in $\mathrm{SL}(2,\mathbb{Z})$.

We could also give conditions for reversibility in $\mathrm{GL}(2,\mathbb{Z})$ by considering the condition (ii) of Proposition 7.24. However, since this condition involves three alternatives ($a = d$ or $b = \pm c$), this leads to a characterisation that takes the form: a hyperbolic g is reversible if and only if one of three conditions on the first associated form holds (— the condition that g be similar to a matrix having $b = -c$ translates into the form being equivalent one of the form $[A, B, A]$; that g is similar to a matrix having equal diagonal elements translates into the form being equivalent to one of the form $[A, 0, C]$). Thus it takes rather more work to verify nonreversibility using this test than that already given using the second associated form.

Returning to $\mathrm{SL}(2,\mathbb{Z})$, we can add some further useful equivalent criteria:

Proposition 7.48   *Let $g \in \mathrm{SL}(2,\mathbb{Z})$ be hyperbolic. Then the following are equivalent:*

(i) *g is reversible in $\mathrm{SL}(2,\mathbb{Z})$*
(ii) *there exists a lax basis $\{e_1, e_2\}$ with $F(g)e_1 = -F(g)e_2$.*

Proof   We know that reversibility is equivalent to conjugacy to a symmetric matrix. Writing this out in terms of matrix entries, we find that $g = \begin{pmatrix} a & b \\ c & d \end{pmatrix}$ is reversible if and only if there exist α, β, δ, $\gamma \in \mathbb{Z}$ with $\alpha\delta - \beta\gamma = 1$ and

$$[c, d - a, -b](\alpha, \gamma) = -[c, d - a, -b](\beta, \delta).$$

Letting $e_1 = (\alpha, \gamma)$ and $e_2 = (\beta, \delta)$, this just says that $F(g)(e_1) = -F(g)(e_2)$. Evidently, e_1 and e_2 are primitive vectors, and so we have a pair of faces on the river associated to the form $F(g)$. This shows that (i) implies (ii).

The converse is obvious, because we can define $h \in \mathrm{GL}(2, \mathbb{Z})$ taking $F(g)$ to $-F(g)$ by requiring that its matrix with respect to the basis $\{e_1, e_2\}$ be $\begin{pmatrix} 0 & -1 \\ 1 & 0 \end{pmatrix}$. A more picturesque way to understand this is to note that rotating the Conway complex through 180 deg about the middle of the edge between e_1 and e_2 obviously converts $F(g)$ to its negative.   □

Condition (ii) gives us the following algorithm: Start at any vertex of the Conway complex, and use Gauss reduction on the form $F(g)$ to find the river. Then travel the river until you find an edge such that the form has equal absolute values on the two adjacent faces (in which case g is reversible in $\mathrm{SL}(2, \mathbb{Z})$), or until you complete a period without finding such an edge (in which case g is not reversible in $\mathrm{SL}(2, \mathbb{Z})$), or until you reach a lake. In the latter case, the form is isotropic, so check the river in the other direction to the other lake. This algorithm reaches the river in logarithmic time, but then may take time $\sqrt{D}$ to traverse the river. We can speed it up to logarithmic time.

Proposition 7.49   *If the indefinite form f is properly equivalent to a form $[A, B, -A]$, then one of the properly equivalent forms $[A, B, -A]$ or $[-A, -B, A]$ is Gauss reduced.*

Proof   Without loss in generality, we may take $B > 0$, and we then must show that $[A, B, C]$ is reduced. We have $B^2 + 4A^2 = D$, so $B < \sqrt{D}$ and $4A^2 = (B - \sqrt{D})(B + \sqrt{D}$. Thus $2|A|$ lies between $B - \sqrt{D}$ and $B + \sqrt{D}$, as required.   □

Thus if we travel from bridge to bridge using Gauss reduction, we will hit the edge we seek, if it exists, within a single cycle of reduced forms.

7.11   Cyclically-reduced words

There is another way to distinguish the reversible elements in $\mathrm{PSL}(2, \mathbb{Z})$, based on the fact that the group is the free product $C_2 * C_3$. Denoting the generators by s (order 2) and t (order 3), as before, each conjugacy class has a unique representation as a cyclically reduced word in s and t (a reduced word in s and t is said to cyclically reduced if it does not take the form sws, twt^2, or t^2wt).

Such a word may be encoded, for hyperbolic elements, by a finite even-order cycle of ± 1's, and reversibility corresponds precisely to suitable symmetry of this sequence: it should be invariant under cyclic reversal followed by a sign change. It is possible to compute the cyclically reduced word representing a given element in an effective manner. For details, see Baake and Roberts [15].

Notes

Sources

Sources for this chapter include the survey by Baake and Roberts [15], besides the various classics cited already.

A useful identity

We have the identity

$$[A,B,C](m,n) \cdot [A,-B,C](v,\mu)$$

$$= \{\mu(mb+nc) - v(ma+nb)\}^2 - D(m\mu+nv)^2.$$

This tells us that if $[A,B,C]$ primitively represents $N \neq \pm 1$, then D is a quadratic residue of N: just pick relatively-prime m,n with $[A,B,C](m,n) = N$, and then pick μ, v with $m\mu + nv = 1$.

When $N = \pm 1$ it just tells us that D is a square $\pm$ a primitive value of $[A,-B,C]$, which is not all that helpful.

Reversible hyperbolic elements

All hyperbolic g with $3 \leqslant \tau \leqslant 19$ are reversible.

Pell's equation

If the negative Pell equation $x^2 - Dy^2 = -1$ has a solution, then *every* form f of discriminant D is improperly equivalent to $-f$.

Toral automorphisms

Let $\mathbb{T}^2$ denote the quotient of the plane by the group G generated by $(x,y) \mapsto (x+1,y)$ and $(x,y) \mapsto (x,y+1)$. Topologically, $\mathbb{T}^2$ is a torus. The group $\mathrm{GL}(2,\mathbb{Z})$

acts on $\mathbb{T}^2$ as follows:

$$\begin{pmatrix} a & b \\ c & d \end{pmatrix} \begin{bmatrix} x \\ y \end{bmatrix} = \begin{bmatrix} ax+by \\ cx+dy \end{bmatrix},$$

where $\begin{bmatrix} u \\ v \end{bmatrix}$ is the orbit of the point $\begin{pmatrix} u \\ v \end{pmatrix}$ under G. With this action, elements of $GL(2,\mathbb{Z})$ are described as *toral automorphisms* or *cat maps*. One of the most famous cat maps is *Arnold's cat map*, given by

$$\begin{pmatrix} 2 & 1 \\ 1 & 1 \end{pmatrix},$$

whose action on the torus is chaotic.

Reversibility problems for toral automorphisms have been studied in great detail by Baake, Neumärker, Roberts, and Weiss [15, 16, 20, 21]. They obtain most of the results about reversibility in this chapter, and much more. For instance, in [20, 21] they consider the finite groups $GL(2,\mathbb{Z}/n\mathbb{Z})$. They prove, among other theorems, that an element of $SL(2,\mathbb{Z})$, when reduced modulo n, is strongly reversible in the group $GL(2,\mathbb{Z}/n\mathbb{Z})$ (see [21, Theorem 1]). Also in [20, 21], they characterise group theoretic and dynamical features (such as centralisers, backwards orbits and connections to dynamical zeta functions) of toral endomorphisms that are not necessarily automorphisms. Most of their techniques generalise to automorphisms of higher dimensional tori, but do not give complete answers in this greater generality.

Poincaré's three-manifolds

The following construction was discovered by Poincaré [197, Tome VI, pages 236–239] (and rediscovered by Ghys and Sergiescu [97]).

Poincaré's purpose was to show that its list of Betti numbers (the list of dimensions of its rational homology groups) do not suffice to characterise the homeomorphism class of a compact manifold. His example T_g was of the form $\mathbb{T}^2 \times [0,1]$ with an identification of the end tori $\mathbb{T}^2 \times \{0\}$ and $\mathbb{T}^2 \times \{1\}$ induced by the toral automorphism associated to an element $g \in GL(2,\mathbb{Z})$.

In a little more detail, Consider the maps $T_i : \mathbb{R}^3 \to \mathbb{R}^3$ given by

$$\begin{aligned} T_1(x,y,t) &= (x+1,y,t) \\ T_2(x,y,t) &= (x,y+1,t) \\ T_3(x,y,t) &= (g(x,y),t+1). \end{aligned} \qquad (7.8)$$

(In the last equation, we identify $\mathbb{R}^3$ with $\mathbb{R}^2 \times \mathbb{R}$.) These three maps generate a group A_g that acts freely and discretely on $\mathbb{R}^3$, and T_g is the quotient $\mathbb{R}^3/A_g$. Naturally, its fundamental group is A_g.

Theorem 7.50 (Poincaré)   *Let $g, g' \in \mathrm{SL}(2, \mathbb{Z})$ be hyperbolic. Then the following are equivalent:*

(i) $\mathbb{T}_g^3$ *is homeomorphic to* $\mathbb{T}_{g'}^3$

(ii) $\pi_1(\mathbb{T}_g^3)$ *is isomorphic to* $\pi_1(\mathbb{T}_{g'}^3)$

(iii) g *is conjugate to* g' *in* $\mathrm{GL}(2, \mathbb{Z})$.

$\square$

Ghys–Sergiescu foliations

Ghys and Sergiescu were interested in the topological classification of foliations without Reeb components on compact 3-manifolds having solvable fundamental group. The key to this were two foliations of T_g^3.

Let λ_1 and $\lambda_2 = 1/\lambda_1$ be the two distinct real eigenvalues of a hyperbolic $g \in \mathrm{SL}(2, \mathbb{Z})$, and let e_1 and e_2 be corresponding eigenvectors.

For $i = 1, 2$, we define two foliations on $\mathbb{R}^2 \times \mathbb{R}$ by taking the leaves to be the family of planes parallel to $(e_i, 0)$ and $(0, 1)$. The maps T_j ($j = 1, 2, 3$) of Equation (7.8) map each leaf of each foliation bijectively onto another, so we get two foliations $\mathscr{F}_1$ and $\mathscr{F}_2$ of T_g^3 by passing to the quotient. The leaves of these foliations are, topologically, either planes or cylinders. All leaves are dense. The cyclindrical leaves of each foliation form a countable family. Each one meets the (quotiented) end torus $\mathbb{T}^2 \times \{0\}$ in a circle (cutting through a fundamental domain in $\mathbb{R}^2 \times \{0\}$ in a finite set of parallel line segments) containing a finite cycle of rational points, $(q, 0), (g(q), 0), \dots, (g^n(q), 0)$, with $q \in \mathbb{Q}^2 \mod \mathbb{Z}^2$ and $g^{n+1}(q) = q$. Conversely, all g-periodic rational points $q \in \mathbb{T}^2$ produce cylindrical leaves in this way; that is, there is a bijection between cycles of rational points on the torus under g and cyclindrical leaves.

We call these two foliations the G-S foliations corresponding to g. Ghys and Sergiescu showed [97, Theorem 1, page 188] that if g is hyperbolic, then each transversally-orientable codimension 1 C^r foliation of $\mathbb{T}_g^3$ is C^{r-2}-conjugate to one of the G-S foliations. (Here $r \in \mathbb{N}$, $r = \infty$ or $r = \omega$.)

Proposition 7.51   *Let g be a hyperbolic element of $\mathrm{SL}(2, \mathbb{Z})$. Then the following two conditions are equivalent.*

(i) *The two G-S foliations on $\mathbb{T}_g^3$ are topologically conjugate.*

(ii) *g is reversible in $\mathrm{GL}(2, \mathbb{Z})$.*

Hence the $\mathrm{GL}(2, \mathbb{Z})$-reversible elements of $\mathrm{SL}(2, \mathbb{Z})$ give torus bundles that have only one class of transversally-orientable foliations.

The authors are grateful to Étienne Ghys for aquainting us with material used in this note.

Unorientable geodesics

Let g be a hyperbolic element of $G = \mathrm{PSL}(2,\mathbb{Z})$. Thought of as a linear-fractional transformation $\frac{az+b}{cz+d}$ acting on the extended plane $\hat{\mathbb{C}}$, g has two real fixed points $\alpha < \beta$. The only circles in $\mathbb{C}$ that are mapped into themselves by g are those that pass through α and β. Just one of these circles intersects the upper half-plane $\mathbb{H}$ in a hyperbolic geodesic, namely the one, say γ, having $[\alpha, \beta]$ as diameter. Since all hyperbolic geodesics are Euclidean semicircles or vertical half-lines, we see that γ is *the unique geodesic* invariant under g.

There are two objects that are known as the *modular surface*. Most people use this term for the quotient space $M = \mathbb{H}/G$, obtained by identifying the orbits of G on $\mathbb{H}$ to points. However this is not strictly a surface (i.e. a 2-dimensional manifold). One obtains a true surface by forming $S = \mathbb{H}/[G,G]$. This surface is just a torus with one puncture.

The space M is a surface apart from two singular points p_2 and p_3, the images of i and $\omega = \frac{1}{2}(1+i\sqrt{3})$, respectively, which are, up to the group action, the fixed points of the elements of order 2 and 3, respectively. The map $\mathbb{H} \to M$ factors through S, and the factor $S \to M$ is an abelian branched covering. We shall call M the singular modular surface and S the nonsingular modular surface. The point p_2 has three preimages on S, and p_3 has two.

The hyperbolic metric on $\mathbb{H}$ induces metrics on S and M, and geodesics on $\mathbb{H}$ project to geodesics on both surfaces. The length on either surface of the geodesic invariant under a hyperbolic element g is of the form

$$\frac{1}{m} |\log |[\alpha, \beta, p, g(p)]||$$

where m is the largest integer such that g has a compositional m-th root in $\mathrm{PSL}(2,\mathbb{Z})$,

$$p = \frac{\alpha+\beta}{2} + i\left(\frac{\beta-\alpha}{2}\right)$$

and $[\alpha, \beta, \gamma, \delta]$ denotes the cross-ratio. We say that g is a *primitive element* of $\mathrm{PSL}(2,\mathbb{Z})$ if it has no roots in $\mathrm{PSL}(2,\mathbb{Z})$, i.e if $m = 1$. For primitive elements g, this length works out as

$$\log\left(\frac{\tau + \sqrt{\tau^2 - 4}}{2}\right)^2,$$

where, as usual, $\tau = \mathrm{trace}(g)$.

If we think of g as a matrix, $\begin{pmatrix} a & b \\ c & d \end{pmatrix}$, then its eigenvalues λ and fixed points z are related by $\lambda = cz + d$.

Let us denote the geodesic on M determined by a geodesic $\gamma \subset \mathbb{H}$ by $\tilde{\gamma}$. If we consider orientations on geodesics, then there is not a perfect correspondence between *oriented* geodesics on $\mathbb{H}$ and on M. A given orientation on $\gamma \subset \mathbb{H}$ determines an orientation on $\tilde{\gamma}$ only if there is no $h \in G$ that maps γ to itself and reverses orientation.

Let us call $\tilde{\gamma}$ *unorientable* if there exists an $h \in G$ that maps γ to itself and reverses orientation. The corresponding γ are called *reciprocal* geodesics by Sarnak[207].

Since the branched covering map is a covering map of Riemann surfaces away from the singularities, we see that an orientation on γ must determine an orientation on $\tilde{\gamma}$ unless γ passes through one of the singular points. We can be more precise:

Theorem 7.52 *Let $g \in \mathrm{PSL}(2, \mathbb{Z})$ be hyperbolic. Then g is reversible if and only if its invariant geodesic on the singular modular surface passes through p_2, or, equivalently, its invariant geodesic on the upper half-plane passes through one of the images of i under the action of $\mathrm{PSL}(2, \mathbb{Z})$.*

This provides another, geometric way to see that reversible elements of the group $\mathrm{PSL}(2, \mathbb{Z})$ are always strongly-reversible (Proposition 7.30). Alternatively, this geometric conclusion may also be obtained as a corollary of that algebraic proposition.

On the nonsingular modular surface, the unorientable geodesics lift to closed geodesics that pass through at least one of the three preimages of p_2, and are 'folded over' at one such preimage by the map $S \to M$.

Any lift of the geodesic associated to a hyperbolic element may be used to calculate the cyclically-reduced word representing its conjugacy class (see Chapter 7, Section 7.11).

Forms, matrices, and ideals

There is a three-way correspondence between the sets of (1) hyperbolic elements of $\mathrm{SL}(2, \mathbb{Z})$, (2) prime ideals in maximal orders of real quadratic fields and (3) binary integral quadratic forms. More precisely, one relates *primitive* elements of each set. To get a bijective correspondence, one has to impose a suitable equivalence relation on each set of primitive elements. For more information, see [29, 34, 51, 53, 54, 207, 222].

Open problems

Congruence subgroups

A *congruence subgroup* of $PSL(2,\mathbb{Z})$ is a subgroup defined by congruences involving the matrix coefficients. For example, the *level 2 congruence subgroup of* $PSL(2,\mathbb{Z})$ consists of those equivalence classes of matrices

$$\begin{pmatrix} a & b \\ c & d \end{pmatrix}$$

with a and d odd and b and c even. This group is isomorphic to the free group on two generators, so only the identity is reversible, by Proposition 3.24. The reversibility problem in other congruence subgroups is not always so simple, and remains open.

Hecke groups

The *Hecke group* Γ_q is the group generated by

$$z \mapsto -\frac{1}{z}, \qquad z \mapsto z + 2\cos\left(\frac{\pi}{q}\right),$$

where q is an integer greater than or equal to 3. The group Γ_3 is the modular group. In general, Γ_q is isomorphic to the free product $C_2 * C_q$. In this chapter we have seen a number of ways to tackle the reversibility problem in Γ_3, and it would be natural to adapt these methods for Γ_q.

The Picard group

The Picard group is the group of Möbius transformations

$$z \mapsto \frac{az+b}{cz+d},$$

where a, b, c, and d are Gaussian integers and $ad - bc = 1$. This group is discrete and it contains the modular group. Once again, the reversibility problem in this group has yet to be settled.

Higher dimensions

The conjugacy problem in $GL(n,\mathbb{Z})$, for $n > 2$ is difficult [219], and probably intractable, so it would be interesting to see whether the difficulties can be bypassed for the reversibility problem. Baake and Roberts [16] have made headway; they calculated centralisers and extended centralisers of elements of

$GL(n, \mathbb{Z})$ that are diagonalisable over $\mathbb{C}$ with distinct eigenvalues. Ishibashi [139] proved that each element of $GL(n, \mathbb{Z})$ can be expressed as a product of $3n + 9$ involutions for every $n > 2$, and Laffey [158] proved that, provided $n \geqslant 84$, this number can be reduced to 41.

Ternary forms

It is not a problem about reversibility, but it is worth mentioning that there is nothing like so satisfactory a theory about the representation of integers by ternary quadratic forms.

8

Real homeomorphisms

In this chapter, we consider the group $\text{Homeo}(\mathbb{R})$ of homeomorphisms of the real line $\mathbb{R}$ onto itself, under composition. We also consider the subgroup $\text{Homeo}^+(\mathbb{R})$ comprised of orientation-preserving homeomorphisms (that is, order-preserving homeomorphisms).

These are our first examples of 'big' groups. Hitherto, we have considered groups that were finite, countable, or were, at worst, finite-dimensional manifolds. With its natural topology[1], $\text{Homeo}(\mathbb{R})$ is homeomorphic to two disjoint copies of the separable infinite-dimensional Hilbert space [30, Proposition VI.8.1]. We shall not be concerned with $\text{Homeo}(\mathbb{R})$ as a topological group.

The full conjugacy classification of $\text{Homeo}(\mathbb{R})$ is understood. We shall begin by describing this classification, and then apply it to address questions of reversibility in $\text{Homeo}(\mathbb{R})$ and $\text{Homeo}^+(\mathbb{R})$. We shall see (in Subsection 8.2.4) that typical orientation-preserving homeomorphisms have uncountably many square roots. Thus, by Corollary 2.23, there is a rich supply of reversible elements. These results on conjugacy and reversibility will be important in later chapters about special groups of homeomorphisms of the line and circle.

Let us establish some notation used throughout the chapter. As usual, the identity element (the map $x \mapsto x$) is denoted by $\mathbb{1}$. The group $\text{Homeo}^+(\mathbb{R})$ is a subgroup of index two of $\text{Homeo}(\mathbb{R})$. The second coset is denoted by $\text{Homeo}^-(\mathbb{R})$: it consists of orientation-reversing homeomorphisms (that is,

---

[1] The group $\text{Homeo}(\mathbb{R})$ becomes a topological group when given the topology of uniform convergence on compact sets. A complete metric ρ inducing this topology is defined by

$$\rho = \sum_{n=1}^{\infty} \frac{2^{-n} d_n}{1 + d_n},$$

where

$$d_n(f,g) = \sup\{|f(x) - g(x)| + |f^{-1}(x) - g^{-1}(x)| : -n \leqslant x \leqslant n\},$$

for $n = 1, 2, \ldots$. For more information about this metric, see, for example, [68].

order-reversing homeomorphisms). For example, the homeomorphism $x \mapsto -x$ is orientation-reversing. The *degree* of a homeomorphism f is the number $\deg(f)$ given by

$$\deg(f) = \begin{cases} 1 & \text{if } f \in \text{Homeo}^+(\mathbb{R}), \\ -1 & \text{if } f \in \text{Homeo}^-(\mathbb{R}). \end{cases}$$

The map deg is a homomorphism from $\text{Homeo}(\mathbb{R})$ onto the cyclic group C_2.

Let $-\infty \leqslant a < b \leqslant +\infty$. We denote the group of homeomorphisms of the open interval (a, b) onto itself by $\text{Homeo}(a, b)$, and the subgroup of orientation-preserving elements by $\text{Homeo}^+(a, b)$. We use similar notation for homeomorphism groups of other types of intervals (such as closed intervals). Each orientation-preserving homeomorphism of one interval onto another induces an isomorphism of the corresponding groups. Using this observation, one sees that $\text{Homeo}(a, b)$ is (noncanonically) isomorphic to $\text{Homeo}(\mathbb{R})$, and the group $\text{Homeo}^+(a, b)$ is isomorphic to $\text{Homeo}^+(\mathbb{R})$.

We denote the set of fixed points of a homeomorphism f by $\text{fix}(f)$. Homeomorphisms that reverse orientation have exactly one fixed point, whereas orientation-preserving homeomorphisms may have any number of fixed points. The fixed-point set of a homeomorphism f is a closed set, and conversely, every closed subset of $\mathbb{R}$ arises as the fixed-point set of a homeomorphism.

It follows from Proposition 2.19 and these observations about fixed points that both groups $\text{Homeo}(\mathbb{R})$ and $\text{Homeo}^+(\mathbb{R})$ have trivial centres. The only automorphisms of $\text{Homeo}(\mathbb{R})$ are inner automorphisms (see [89]).

8.1 Involutions

Apart from the identity, all the elements of finite order in $\text{Homeo}(\mathbb{R})$ have order two, and belong to $\text{Homeo}^-(\mathbb{R})$.

Proposition 8.1 *The only homeomorphisms of $\mathbb{R}$ that have finite order are the orientation-reversing involutions and the identity map.*

Proof Given a nontrivial element f of $\text{Homeo}^+(\mathbb{R})$, there is a point x in $\mathbb{R}$ for which $f(x) \neq x$. Either $x < f(x)$, in which case $x, f(x), f^2(x), \dots$ is a strictly increasing sequence, or $x > f(x)$, in which case $x, f(x), f^2(x), \dots$ is a strictly decreasing sequence. In both cases, f is of infinite order. If f is an orientation-reversing homeomorphism of finite order, then f^2 is a an element of $\text{Homeo}^+(\mathbb{R})$ of finite order, so $f^2 = 1$. $\qquad\square$

There is just one conjugacy class of proper (that is, nonidentity) involutions in $\text{Homeo}(\mathbb{R})$.

Proposition 8.2 *Each proper involution is conjugate to the map $x \mapsto -x$.*

Proof By Proposition 8.1, each proper involution τ must reverse orientation. Let $h(x) = x - \tau(x)$. Then h is an orientation-preserving homeomorphism and $h\tau(x) = -h(x)$ for each real number x. $\qquad\qquad\square$

There are many of these orientation-reversing, proper involutions. In fact, each homeomorphism σ of $[0, +\infty)$ onto itself determines a unique proper involution τ by

$$\tau(x) = \begin{cases} \sigma^{-1}(-x) & \text{if } x < 0, \\ -\sigma(x) & \text{if } x \geqslant 0. \end{cases}$$

All proper involutions that fix 0 arise in this way. The other proper involutions are obtained from those that fix 0 by conjugating by translations.

One may describe the involutions in geometric terms as those homeomorphisms whose graphs are symmetric under reflection in the diagonal $y = x$ (see the left-hand graph of Figure 1.8).

8.2 Conjugacy

In this section we describe the conjugacy classes of $\text{Homeo}(\mathbb{R})$. The classification is described in terms of a concept known as the *topological signature*.

8.2.1 Topological signatures

Let sign denote the function from $\mathbb{R}$ to $\{-1, 0, 1\}$ given by

$$\text{sign}(x) = \begin{cases} -1 & \text{if } x < 0, \\ 0 & \text{if } x = 0, \\ 1 & \text{if } x > 0. \end{cases}$$

To each homeomorphism f we associate the continuous function

$$s(f) : \mathbb{R} \to \{-1, 0, 1\}, \quad x \mapsto \text{sign}(f(x) - x).$$

The function $s(f)$ records where the graph of f is below, on, or above, the line $y = x$. In Figure 8.1, the graph of a function f is shown and the corresponding values of $s(f)$ are labelled. The fixed-point set $\text{fix}(f)$ consists of those real numbers x with $s(f)(x) = 0$.

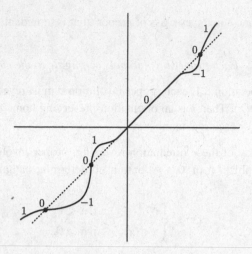

Figure 8.1 A graph of a homeomorphism f with values of $s(f)$ labelled

We endow $\{-1,0,1\}$ with the largest topology that makes the function sign continuous. Thus $\{1\}$ and $\{-1\}$ are open, but $\{0\}$ is not. Let Σ denote the family of all continuous maps from $\mathbb{R}$ to $\{-1,0,1\}$. One may think of an element k of Σ as the specification of a closed set $k^{-1}(0)$, and a choice of sign (± 1) on each connected component of the complement of $k^{-1}(0)$.

Proposition 8.3 *We have $s(\mathrm{Homeo}^+(\mathbb{R})) = \Sigma$.*

Proof Let $k \in \Sigma$. We choose an orientation-preserving homeomorphism g whose fixed-point set is equal to the closed set $k^{-1}(0)$. For each point x in the open set $k^{-1}(1)$, we choose g to satisfy $g(x) > x$, and for each point x in the open set $k^{-1}(-1)$, we choose g to satisfy $g(x) < x$. Then $s(g) = k$. $\square$

The next lemma gives two elementary properties of the function s.

Lemma 8.4 *Suppose that $g, h \in \mathrm{Homeo}(\mathbb{R})$. Then*

(i) $s(h^{-1}gh) = \deg(h)(s(g) \circ h)$
(ii) $s(g^{-1}) = -\deg(g)s(g)$. $\square$

Motivated by formula (i), we say that two maps s and t in Σ are *topologically equivalent* if there exists a map h in $\mathrm{Homeo}(\mathbb{R})$ such that

$$t = \deg(h)(s \circ h).$$

This defines an equivalence relation on Σ. We call the equivalence classes *topo-*

logical signatures. For a homeomorphism g, we refer to the equivalence class of $s(g)$ as the *topological signature*, or just *signature*, of g.

Informally, one may think of a topological signature as a homeomorphism class of pairs $(\mathbb{R}, F)$, with F closed, and a *pattern of signs* (± 1) on the complement of F. One is allowed to 'reverse' the set F, but then the pattern of signs must also be reversed.

To give a simple example, in the case where F is a singleton, with two complementary components, there are three distinct (inequivalent) patterns of signs, namely $(+1, +1)$, $(-1, +1)$, and $(+1, -1)$. The pattern of signs $(-1, -1)$ is equivalent to $(+1, +1)$. Examples of homeomorphisms that fix only 0 and have these patterns of signs are $x \mapsto x + |x/2|$, $x \mapsto 2x$, and $x \mapsto x/2$, respectively.

As an even simpler example, letting 1 stand for the constant map from $\mathbb{R}$ to 1, and similarly for -1 and 0, we find that 1 and -1 have the same topological signature. This is the signature of any fixed-point-free homeomorphism. Only the identity map has signature 0.

8.2.2 Oriented signatures

We shall see that the topological signature classifies the elements of the group $\mathrm{Homeo}^+(\mathbb{R})$ up to conjugacy in $\mathrm{Homeo}(\mathbb{R})$. To characterise the conjugacy classes of elements of $\mathrm{Homeo}^+(\mathbb{R})$ in $\mathrm{Homeo}^+(\mathbb{R})$ itself, we introduce the idea of *oriented topological equivalence*. We say that two elements s and t of Σ are oriented topologically equivalent if there is an orientation-*preserving* homeomorphism h such that

$$t = s \circ h.$$

We call the corresponding equivalence classes *oriented topological signatures*. Obviously, each topological signature is the union of one or two oriented topological signatures (two, unless it has suitable 'symmetry'; see Section 8.3). We shall see that the conjugacy classes in $\mathrm{Homeo}^+(\mathbb{R})$ are in bijective correspondence with the oriented topological signatures.

8.2.3 Conjugacy of fixed-point-free homeomorphisms

Proposition 8.5 *Two orientation-preserving homeomorphisms f and g with either $f(x) > x$ and $g(x) > x$, or $f(x) < x$ and $g(x) < x$, for each real number x are conjugate in* $\mathrm{Homeo}^+(\mathbb{R})$.

Proof By replacing f and g with f^{-1} and g^{-1} if necessary, we can assume that $f(x) > x$ and $g(x) > x$ for each real number x. Choose any increasing homeomorphism $k : [0, f(0)] \to [0, g(0)]$. Both sequences

$$\ldots, f^{-2}(0), f^{-1}(0), 0, f(0), f(0), f^2(0), \ldots$$

and

$$\ldots, g^{-2}(0), g^{-1}(0), 0, g(0), g(0), g^2(0), \ldots$$

are strictly increasing and unbounded above and below. Therefore we can define a function $h : \mathbb{R} \to \mathbb{R}$ by the equations

$$h(x) = g^n k f^{-n}(x), \quad x \in [f^n(0), f^{n+1}(0)],$$

for $n \in \mathbb{Z}$. This map h belongs to $\mathrm{Homeo}^+(\mathbb{R})$ and it satisfies $hfh^{-1} = g$. $\qquad\square$

Because an open interval (a, b) is homeomorphic to $\mathbb{R}$, we have the following corollary of Proposition 8.5.

Corollary 8.6 *Two maps f and g in $\mathrm{Homeo}^+(a, b)$ with either $f(x) > x$ and $g(x) > x$, or $f(x) < x$ and $g(x) < x$, for each element x of (a, b) are conjugate in $\mathrm{Homeo}^+(a, b)$.* $\qquad\square$

The two conjugacy classes of fixed-point-free maps in $\mathrm{Homeo}^+(\mathbb{R})$ combine to make a single conjugacy class in $\mathrm{Homeo}(\mathbb{R})$.

Corollary 8.7 *Each fixed-point-free element g of $\mathrm{Homeo}^+(\mathbb{R})$ is conjugate in $\mathrm{Homeo}(\mathbb{R})$ to $x \mapsto x + 1$.*

Proof It suffices to note that the homeomorphism $x \mapsto -x$ conjugates $x \mapsto x - 1$ to $x \mapsto x + 1$. $\qquad\square$

8.2.4 Conjugacy in the order-preserving homeomorphism group

We define a *bump domain* of an orientation-preserving homeomorphism f to be a connected component of the complement of $\mathrm{fix}(f)$.

Proposition 8.8 *Suppose that f and g are orientation-preserving homeomorphisms with $s(f) = s(g)$. Then there is a function h in $\mathrm{Homeo}^+(\mathbb{R})$ such that $g = hfh^{-1}$, and h fixes each of the fixed points of f and g.*

Proof Define h as follows. For each bump domain (a, b) of f, we use Corollary 8.6 and choose a homeomorphism k in $\mathrm{Homeo}^+(a, b)$ such that $kfk^{-1} = g$ on (a, b). Define $h(x) = k(x)$ on (a, b). If x is a fixed point of f and g then let $h(x) = x$. The resulting orientation-preserving homeomorphism h satisfies $hfh^{-1} = g$. $\qquad\square$

Theorem 8.9   *Two orientation-preserving homeomorphisms f and g are conjugate in* $\mathrm{Homeo}^+(\mathbb{R})$ *if and only if they have the same oriented topological signature.*

Proof   If there exists an orientation-preserving homeomorphism h such that $s(g) = s(f) \circ h$, then, by Lemma 8.4(i), $s(h^{-1}fh) = s(g)$. Proposition 8.8 then shows that $h^{-1}fh$ and g are conjugate, so f and g are conjugate. The converse follows directly from Lemma 8.4(i).   □

A *flow* in $\mathrm{Homeo}^+(\mathbb{R})$ is a continuous homomorphism

$$(\mathbb{R}, +) \to \mathrm{Homeo}^+(\mathbb{R}), \qquad t \mapsto g_t.$$

The simplest example of a flow is given by $g_t(x) = x + t$ for each real number t. Using Proposition 8.8, we can embed any orientation-preserving homeomorphism g in a flow; that is, we can find a flow g_t with $g_1 = g$, in which case we say that g is *flowable*. One can do this by working with each bump domain separately, and conjugating to homeomorphisms such as $x \mapsto x + 1$ that we know are flowable. In fact, because of the freedom in choosing conjugating maps (this freedom is demonstrated in the choice of k in the proof of Proposition 8.5), g can be embedded in many flows. It can be shown [89] that the intersection of two such flows is the image of a discrete action, and is generically just $\{g^n : n \in \mathbb{Z}\}$. Thus $C_g(\mathrm{Homeo}^+(\mathbb{R}))$ is enormous, and nonabelian. For example, if g is an orientation-preserving homeomorphism without fixed points, then it is not hard to see that $C_g(\mathrm{Homeo}^+(\mathbb{R}))$ is isomorphic to the group of orientation-preserving homeomorphisms of the circle, which we study in the next chapter. The extended centraliser $E_g(\mathrm{Homeo}^+(\mathbb{R}))$ is isomorphic to the full group of homeomorphisms of the circle. For a more general orientation-preserving homeomorphism g, one can describe the centraliser (and extended centraliser) by first working with the bump domains and fixed-point set of g separately, and then considering elements of the centraliser that permute the bump domains and fixed-point set.

A consequence of this rich collection of flows is that each order-preserving homeomorphism g has numerous square roots (and, more generally, nth roots). In particular, if we choose two distinct square roots h and k of g, then $k^{-1}h$ is reversible, by Corollary 2.23, so there are plenty of reversible elements. We discuss reversible elements in more detail later.

Since all positive powers and roots of an orientation-preserving homeomorphism g have the same signature as g, we have the following corollary of Theorem 8.9.

Corollary 8.10   *If $g \in \mathrm{Homeo}^+(\mathbb{R})$, then all positive powers and roots of g are conjugate to one another in* $\mathrm{Homeo}^+(\mathbb{R})$.   □

The negative roots and powers are also conjugate to one another in the group $\mathrm{Homeo}^+(\mathbb{R})$, and they are conjugate to the positive powers and roots in the full homeomorphism group $\mathrm{Homeo}(\mathbb{R})$.

Corollary 8.10 has an interesting structural consequence (which was observed in greater generality in the Notes of Chapter 2).

Corollary 8.11   *Each element of* $\mathrm{Homeo}^+(\mathbb{R})$ *is a commutator.*

Proof   Given an element g of $\mathrm{Homeo}^+(\mathbb{R})$, we choose another element h of $\mathrm{Homeo}^+(\mathbb{R})$ with $hgh^{-1} = g^2$. Then $g = hgh^{-1}g^{-1}$, a commutator.   □

8.2.5 Conjugacy in the full homeomorphism group

Elements of $\mathrm{Homeo}^+(\mathbb{R})$ can never be conjugate to elements of $\mathrm{Homeo}^-(\mathbb{R})$, but some orientation-preserving homeomorphisms that are not conjugate in $\mathrm{Homeo}^+(\mathbb{R})$ can yet be conjugate in $\mathrm{Homeo}(\mathbb{R})$.

Theorem 8.12   *Two orientation-preserving homeomorphisms f and g are conjugate in* $\mathrm{Homeo}(\mathbb{R})$ *if and only if they have the same topological signature.*

Proof   Suppose there is a homeomorphism h such that $s(g) = \deg(h)(s(f) \circ h)$. By Lemma 8.4(i), we have that $s(h^{-1}fh) = s(g)$. By Proposition 8.8, there is a homeomorphism k such that $k^{-1}h^{-1}fhk = g$, so f is conjugate to g in $\mathrm{Homeo}(\mathbb{R})$. The converse follows directly from Lemma 8.4(i).   □

It remains to consider conjugacy of orientation-*reversing* homeomorphisms.

Theorem 8.13   *Two orientation-reversing homeomorphisms f and g are conjugate in* $\mathrm{Homeo}(\mathbb{R})$ *if and only if f^2 and g^2 are conjugate in* $\mathrm{Homeo}(\mathbb{R})$ *by a map that takes the fixed point of f to the fixed point of g.*

Proof   If $hfh^{-1} = g$ for some homeomorphism h, then h must map the fixed point of f to the fixed point of g, and $hf^2h^{-1} = g^2$. Conversely, let f have fixed point p and let g have fixed point q, and suppose there is a homeomorphism h that satisfies $h(p) = q$ and $hf^2h^{-1} = g^2$. By replacing h with hf if necessary, we can assume that $h \in \mathrm{Homeo}^+(\mathbb{R})$. Define a map k by

$$k(x) = \begin{cases} g^{-1}hf(x) & \text{if } x \leqslant p, \\ h(x) & \text{if } x > p. \end{cases}$$

The map k is a homeomorphism and satisfies $kfk^{-1} = g$.   □

Theorem 8.13 fails if we remove the condition that the conjugating map from f^2 to g^2 takes the fixed point of f to the fixed point of g. To see this, consider the orientation-preserving homeomorphism k whose graph is shown in Figure 8.2.

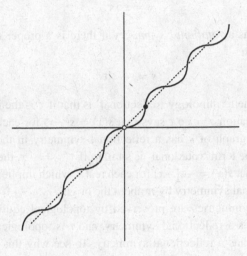

Figure 8.2 The graph of a homeomorphism with nonconjugate orientation-reversing square roots

The fixed-point set of k is the integers. The pattern of signs alternates between ± 1 from one interval in the complement of fix(k) to the next. The graph k has a rotational symmetry by π about each of the points $(0,0)$ (marked by a white spot) and $(1,1)$ (marked by a black spot). We will see in the next section that there are orientation-reversing homeomorphisms f and g with fixed points 0 and 1, respectively, such that $f^2 = g^2 = k$. (In fact, we can see this now: choose an orientation-*preserving* square root h of k. We can assume, after conjugation if need be, that h also rotational symmetries about $(0,0)$ and $(1,1)$. Now define $f = -h$. Since h commutes with the map $x \mapsto -x$, the homeomorphism f is an orientation-reversing square root of k that fixes 0. We can construct g in a similar way.) The two maps f^2 and g^2 are conjugate in Homeo($\mathbb{R}$) as they are equal, and yet it is impossible to find a conjugating map that takes 0 to 1 because such a map would reverse the pattern of signs. Therefore, by Theorem 8.13, f and g are not conjugate in Homeo($\mathbb{R}$).

8.3 Reflectional and rotational symmetries

Recall that Σ denotes the collection of continuous maps from $\mathbb{R}$ to $\{-1,0,1\}$ (where the nontrivial open sets of $\{-1,0,1\}$ are $\{-1\}$, $\{1\}$, and $\{-1,1\}$). We say that an element s of Σ has a *reflectional symmetry* if there is a proper involution τ in $\mathrm{Homeo}^-(\mathbb{R})$ with

$$s = s \circ \tau.$$

We say that s has a *rotational symmetry* if there is a proper involution τ in $\mathrm{Homeo}^-(\mathbb{R})$ with

$$s = -s \circ \tau.$$

The reason for the terminology 'reflectional' is that if τ is the involution $x \mapsto -x$, then the equation $s = s \circ \tau$ says that $s(x) = s(-x)$ for each real x, which implies that the graph of s has a reflectional symmetry in the vertical axis. The reason for the term 'rotational' is similar: if $\tau(x) = -x$, then the equation $s = -s \circ \tau$ says that $s(x) = -s(-x)$ for each real x, which implies that the graph of s has a rotational symmetry by π about the origin.

Reflectional symmetries are preserved by topological equivalence, in the sense that if s has a reflectional symmetry, and t is topologically equivalent to s, then t also has a reflectional symmetry. To see why this is so, suppose that $s = s \circ \tau$ for a proper involution τ in $\mathrm{Homeo}^-(\mathbb{R})$, and suppose that there is a homeomorphism h with $t = \deg(h)(s \circ h)$. Then $h^{-1}\tau h$ is also a proper involution in $\mathrm{Homeo}^-(\mathbb{R})$, and

$$t \circ (h^{-1}\tau h) = \deg(h)(s \circ h) \circ (h^{-1}\tau h) = \deg(h)(s \circ h) = t.$$

In a similar way you can show that rotational symmetries are preserved by topological equivalence.

The simplest functions with reflectional symmetry are constant functions. The simplest functions with rotational symmetry are those functions s in Σ for which $s^{-1}(0)$ is a singleton and the pattern of signs is either $(-1,+1)$ or $(+1,-1)$. Figure 8.3 shows graphs of homeomorphisms f and g for which $s(f)$ has a reflectional symmetry and $s(g)$ has a rotational symmetry.

The concept of reflectional symmetry can be used to classify the strongly-reversible elements of $\mathrm{Homeo}(\mathbb{R})$. We saw earlier that the only nontrivial involutions in $\mathrm{Homeo}(\mathbb{R})$ are orientation *reversing*. Therefore the only strongly-reversible element of the group $\mathrm{Homeo}^+(\mathbb{R})$ is the identity element. Since the product of two orientation-reversing maps is an orientation-preserving map, we see that the strongly-reversible elements of $\mathrm{Homeo}(\mathbb{R})$ are either involutions or else orientation-preserving maps that are the product of two proper

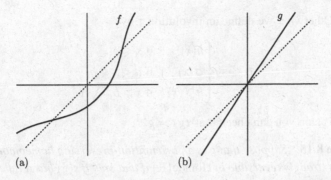

Figure 8.3 (a) The graph of a homeomorphism f for which $s(f)$ has a reflectional symmetry (b) The graph of a homeomorphism g for which $s(g)$ has a rotational symmetry

involutions. It is this second class of maps that can be characterised using reflectional symmetries, as we demonstrate shortly. Before that, we make the following important observation.

Theorem 8.14 *If an orientation-preserving homeomorphism is reversed by an orientation-reversing homeomorphism, then it is also reversed by a proper involution.*

Proof Suppose that $g \in \mathrm{Homeo}^+(\mathbb{R})$, $h \in \mathrm{Homeo}^-(\mathbb{R})$, and $hgh^{-1} = g^{-1}$. Let p be the fixed point of h. Suppose that p is a fixed point of g. Then we define an involution τ by

$$\tau(x) = \begin{cases} h(x) & \text{if } x \leqslant p, \\ h^{-1}(x) & \text{if } x > p. \end{cases}$$

It is straightforward to check that $\tau g \tau = g^{-1}$. Suppose now that p is not a fixed point of g. Then p lies in a bump domain (a, b) of g. Lemma 2.17 tells us that h permutes the fixed points of g, so h also permutes the bump domains of g. However, since $p \in (a, b)$, h must fix (a, b) as a set. It follows that either (a, b) is the whole real line, or else it is an interval of finite width and h interchanges a and b. Since g is free of fixed points in (a, b), we can choose an orientation-reversing involution σ of (a, b) such that $\sigma g \sigma = g^{-1}$. (For example, if (a, b) is $\mathbb{R}$ and $g(x) = x + 1$, then we could choose $\sigma(x) = -x$, and by Corollary 8.6 we can conjugate to this special situation.) If $(a, b) = \mathbb{R}$, then the result is now proven. Otherwise, we define an involution τ by

$$
\tau(x) = \begin{cases} h(x) & \text{if } x \leqslant a, \\ \sigma(x) & \text{if } a < x \leqslant b, \\ h^{-1}(x) & \text{if } x > b, \end{cases}
$$

and once again you can check that $\tau g \tau = g^{-1}$.   □

Theorem 8.15 *Suppose that g is an orientation-preserving homeomorphism. Then g is strongly reversible in $\mathrm{Homeo}(\mathbb{R})$ if and only if $s(g)$ has a reflectional symmetry.*

Proof Suppose first that g is strongly reversible. Then there is a proper involution τ such that $\tau g \tau = g^{-1}$. Using Lemma 8.4(i) and (ii) we see that

$$
s(g) = -s(g^{-1}) = -s(\tau g \tau) = s(g) \circ \tau,
$$

so $s(g)$ has a reflectional symmetry.

Conversely, suppose that $s(g)$ has a reflectional symmetry; that is, suppose there is a proper involution τ with $s(g) = s(g) \circ \tau$. Then, applying Lemma 8.4(i) and (ii) again, we see that $s(\tau g \tau) = s(g^{-1})$. It follows from Proposition 8.8 that there is an orientation-preserving homeomorphism h such that $h \tau g \tau h^{-1} = g^{-1}$. Since $h\tau$ reverses orientation, we see from Theorem 8.14 that g is strongly reversible.   □

Proposition 8.2 tells us that all nontrivial involutions are conjugate to the map $x \mapsto -x$. It follows that an orientation-preserving homeomorphism that is strongly reversible is conjugate to a map that is reversible by $x \mapsto -x$. As we proved in Chapter 1, such maps have graphs that are symmetric about the line $y = -x$, like the graph shown in Figure 8.3(a).

We have now finished with reflectional symmetries, so we move on to consider rotational symmetries instead. Theorem 8.15 tells us that, given a homeomorphism g that preserves orientation, there is a proper involution τ such that

$$
\tau g \tau = g^{-1}
$$

if and only if $s(g)$ has a reflectional symmetry. There is a similar result about rotational symmetries, which states that there is a proper involution τ such that

$$
\tau g \tau = g
$$

if and only if $s(g)$ has a rotational symmetry. We will prove this in Theorem 8.17 below; in fact, we will prove more than this, as we will also show that g has a rotational symmetry if and only if it has an orientation-reversing square root. Before stating this theorem we need a result that is similar in type, but simpler than, Theorem 8.14.

Theorem 8.16   *If an orientation-preserving homeomorphism commutes with an orientation-reversing homeomorphism, then it also commutes with a proper involution.*

Proof   Suppose that $g \in \text{Homeo}^+(\mathbb{R})$, $h \in \text{Homeo}^-(\mathbb{R})$, and $hgh^{-1} = g$. Let p be the fixed point of h. Then Corollary 2.18 tells us that $g(p) = p$ (this is trivial to verify). We define an involution τ by

$$\tau(x) = \begin{cases} h(x) & \text{if } x \leqslant p, \\ h^{-1}(x) & \text{if } x > p. \end{cases}$$

It is straightforward to check that $\tau g \tau = g$.   □

Theorem 8.17   *Let g be an orientation-preserving homeomorphism. The following are equivalent:*

(i) $s(g)$ *has a rotational symmetry*
(ii) g *commutes with a proper involution*
(iii) g *has an orientation-reversing square root.*

Proof   First we show that (i) implies (ii). Suppose there is a proper involution τ such that $s(g) = -s(g) \circ \tau$. Then Lemma 8.4(i) tells us that $s(g) = s(\tau g \tau)$. It follows from Proposition 8.8 that there is an orientation-preserving homeomorphism h such that $h \tau g \tau h^{-1} = g$. Since $h\tau$ reverses orientation, we see from Theorem 8.16 that g commutes with a proper involution.

Next we show that (ii) implies (iii). Suppose that g commutes with a proper involution. Because g preserves orientation, it has an orientation-preserving square root f, which is conjugate to g. Therefore f also commutes with a proper involution, which we denote by τ. The homeomorphism $f\tau$ is a square root of g that reverses orientation.

Finally we show that (iii) implies (i). Suppose that k is an orientation-reversing homeomorphism with $k^2 = g$. Let τ be the proper involution given by

$$\tau(x) = \begin{cases} k(x) & \text{if } x \leqslant p, \\ k^{-1}(x) & \text{if } x > p, \end{cases}$$

where p is the unique fixed point of k. Then $\tau g \tau = g$, so, by Lemma 8.4(i),

$$s(g) = -s(g) \circ \tau.$$ □

We remark that the orientation-reversing maps τ and k in this proof can be chosen to have the same fixed point.

An orientation-preserving homeomorphism that commutes with a proper involution is conjugate to a homeomorphism that commutes with the map $x \mapsto -x$. The graph of a homeomorphism that commutes with $x \mapsto -x$ has a rotational symmetry of π about 0, like the graphs shown in Figures 8.2 and 8.3(b).

A consequence of Theorems 8.14 and 8.16 (and Proposition 8.3) is that the definitions of reflectional and rotational symmetries are unchanged if we replace the proper involution τ by any orientation-reversing homeomorphism.

8.4 Reversible elements

In this section we first study the reversible homeomorphisms that preserve orientation, and then study the reversible homeomorphisms that reverse orientation.

8.4.1 Orientation-preserving reversible elements

Suppose that g is an orientation-preserving homeomorphism. Theorem 8.14 tells us that if g is reversed by an orientation-*reversing* homeomorphism, then in fact g is strongly reversible in Homeo($\mathbb{R}$) (that is, it is reversed by a proper involution). However, it could be that g is reversed by an orientation-*preserving* homeomorphism, but is not strongly reversible in Homeo($\mathbb{R}$). Here we will characterise the reversible elements of Homeo$^+(\mathbb{R})$, and then supply examples of the various types of reversible homeomorphisms.

We begin by characterising those elements of Homeo$^+(\mathbb{R})$ that are reversed by fixed-point-free maps.

Proposition 8.18 *An orientation-preserving homeomorphism g is reversed by a fixed-point-free homeomorphism if and only if g is conjugate in* Homeo$^+(\mathbb{R})$ *to a map f in* Homeo$^+(\mathbb{R})$ *that fixes each integer and satisfies*

$$f(x+1) = f^{-1}(x) + 1, \qquad x \in \mathbb{R}.$$

Proof Suppose that g is reversed by a fixed-point-free map h. By Proposition 8.5, one of h or h^{-1} is conjugate in Homeo$^+(\mathbb{R})$ to $x \mapsto x+1$. Since h and h^{-1} both reverse g, we can conjugate g to give a map f that is reversed by $x \mapsto x+1$. Therefore f satisfies $f(x+1) = f^{-1}(x) + 1$ for each real number x. The map f must have a fixed point, and by conjugating by a translation we may assume that the fixed point is 0. Then the equation $f(x+1) = f^{-1}(x)+1$ ensures that all integers are fixed by f. The converse is immediate. $\quad\square$

For more general reversible elements of $\text{Homeo}^+(\mathbb{R})$, we need the following elementary lemma.

Lemma 8.19 *Suppose that g and h are order-preserving homeomorphisms with $hgh^{-1} = g^{-1}$. Then each fixed point of h is also a fixed point of g.*

Proof Suppose that h fixes the point p. From the equation $hgh^{-1} = g^{-1}$ we deduce that $hg(p) = g^{-1}(p)$ and $hg^{-1}(p) = g(p)$. Order-preserving homeomorphisms such as h have no periodic points other than fixed points; thus $g(p) = g^{-1}(p)$. Therefore $g^2(p) = p$, so $g(p) = p$. $\quad\square$

Suppose that $hgh^{-1} = g^{-1}$, where g and h are elements of $\text{Homeo}^+(\mathbb{R})$. We see from Lemma 8.19 that g fixes each fixed point of h, and therefore maps each connected component J of the complement of $\text{fix}(h)$ onto itself. Restricted to this open interval J, the equation $hgh^{-1} = g^{-1}$ still holds, and h is free of fixed points on J; these are the circumstances of Proposition 8.18 (but with an interval J instead of $\mathbb{R}$). This gives us the following theorem.

Theorem 8.20 *Suppose that $g, h \in \text{Homeo}^+(\mathbb{R})$ and $hgh^{-1} = g^{-1}$. Then for each connected component J of $\mathbb{R} \setminus \text{fix}(h)$, there is a homeomorphism $\phi : J \to \mathbb{R}$ such that g fixes $\phi^{-1}(\mathbb{Z})$ pointwise and*

$$(\phi g \phi^{-1})(x+1) = (\phi g \phi^{-1})^{-1}(x)+1, \qquad x \in \mathbb{R}. \qquad \square$$

We finish this subsection with examples of orientation-preserving homeomorphisms that are

(a) reversible by an order-preserving map but not by an order-reversing map
(b) reversible by an order-reversing map but not by an order-preserving map
(c) reversible by both order-preserving and order-reversing maps.

For (a), consider a homeomorphism g that has fixed-point set equal to

$$(-\infty, 0] \cup \left\{ \tfrac{1}{n} : n \in \mathbb{N} \right\} \cup \mathbb{N}.$$

On the complement of the fixed-point set, $s(g)$ alternates between ± 1. The graph of such a homeomorphism is shown in Figure 8.4. Since g^{-1} has the same oriented topological signature as g, we see from Theorem 8.9 that g and g^{-1} are conjugate in $\text{Homeo}^+(\mathbb{R})$. However, g and g^{-1} not conjugate by an orientation-reversing homeomorphism because $s(g)$ does not have a reflectional symmetry.

For (b), we can choose a homeomorphism such as the one whose graph is

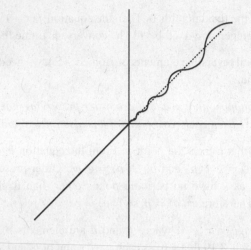

Figure 8.4 The graph of a homeomorphism that is reversible by an orientation-preserving map but not by an orientation-reversing map

shown in Figure 8.3(a), which has a reflectional symmetry (so is reversed by an orientation-reversing map). It is not reversed by an orientation-preserving map because it has only finitely many fixed points. The simplest example for (c) is the identity map. For a nontrivial example, consider the homeomorphism k whose graph is shown in Figure 8.2. Since $s(k)$ has a reflectional symmetry, k is strongly reversible, and it is also reversed by a translation.

8.4.2 Orientation-reversing reversible elements

If an orientation-reversing homeomorphism g is reversible, then we can find reversing maps of degrees both 1 and -1. This is because if g is reversed by a map h, then g is also reversed by hg, and the degrees of g and hg have opposite signs. The following theorem characterises the orientation-reversing reversible elements of Homeo($\mathbb{R}$).

Theorem 8.21   Let $g \in$ Homeo$^-(\mathbb{R})$. *The following are equivalent:*

(i) *g is reversible*
(ii) *g^2 is reversed by a homeomorphism that fixes the fixed point of g*
(iii) *g^2 is reversed by an involution that fixes the fixed point of g.*

Proof   That (i) and (ii) are equivalent follows immediately from Theorem 8.13. Clearly, statement (iii) implies (ii). To finish the proof, we show that (ii) implies (iii). Suppose then that $h^{-1}g^2h = g^{-2}$, where h is a homeomorphism that fixes the fixed point p of g. As we observed at the start of this subsection, we can assume that h is orientation reversing. Let τ be the proper involution defined by

$$\tau(x) = \begin{cases} h(x) & \text{if } x \leqslant p, \\ h^{-1}(x) & \text{if } x > p. \end{cases}$$

Then τ reverses g^2 and fixes p.   □

Theorem 8.21 has an interesting corollary. Let us define a *point of symmetry* of a function s in Σ that has a reflectional symmetry to be the fixed point of a proper involution τ that satisfies $s = s \circ \tau$. We define a point of symmetry of a function with a rotational symmetry in a similar way.

Corollary 8.22   *An orientation-preserving homeomorphism f has a reversible orientation-reversing square root if and only if $s(f)$ has a reflectional symmetry and a rotational symmetry that share the same point of symmetry.*

Proof   This result is certainly true if f is the identity map, so let us assume that this is not so. Suppose first that $f = g^2$, where is g reversible and orientation reversing, with fixed point p. Theorem 8.21(iii) says that f is reversed by a proper involution τ with fixed point p. Therefore $s(f)$ has a reflectional symmetry with point of symmetry p. Also, f commutes with the proper involution σ given by the usual formula

$$\sigma(x) = \begin{cases} g(x) & \text{if } x \leqslant p, \\ g^{-1}(x) & \text{if } x > p, \end{cases}$$

so $s(f)$ has a rotational symmetry with point of symmetry p.

Conversely, suppose that $s(f)$ has a reflectional symmetry and a rotational symmetry both with point of symmetry p. Theorem 8.17 tells us that f has an orientation-reversing square root g. In fact, in proving Theorem 8.17 we found an orientation-reversing square root g that *has fixed point p*. Notice that f fixes the point p too. Next, Theorem 8.15 tells us that f is strongly reversible. In fact, in proving Theorem 8.15 we found a proper involution τ that reverses f and *has fixed point p*.[2] The implication (iii) implies (i) from Theorem 8.21 then shows that g is reversible.   □

We finish here with an example of a function s in Σ that has a reflectional symmetry and a rotational symmetry that share the same point of symmetry.

[2]   Verifying these claims about the proofs of Theorems 8.15 and 8.17 requires you to reexamine several earlier results carefully.

Let s have fixed-point set equal to

$$\mathbb{Z} \cup \left\{ \tfrac{1}{n} : n \in \mathbb{Z} \setminus \{0\} \right\},$$

and suppose that the pattern of signs of s alternates between ± 1 on the complement of $\mathrm{fix}(s)$. Then 0 is a point of symmetry of both a reflectional symmetry and a rotational symmetry. We shall see in Chapter 11 that this kind of thing cannot occur for the signatures of nontrivial orientation-preserving diffeomorphisms.

8.5 Products of involutions and reversible elements

Let us denote by I^n the elements of $\mathrm{Homeo}(\mathbb{R})$ that can be expressed as a product of n involutions.

Theorem 8.23

 (i) *Each element of* $\mathrm{Homeo}^-(\mathbb{R})$ *belongs to* I^3.
 (ii) *Each element of* $\mathrm{Homeo}^+(\mathbb{R})$ *belongs to* I^4.
 (iii) *Thus* $I^4 = \mathrm{Homeo}(\mathbb{R})$.

Proof To prove (i), first choose an element g of $\mathrm{Homeo}^-(\mathbb{R})$. Let σ be an orientation-reversing involution such that $\sigma(x) < g(x)$ for each real number x. Then $g^{-1}\sigma(x) > x$ for each real x. Therefore, by Proposition 8.5, $g^{-1}\sigma$ is conjugate to the fixed-point-free map $x \mapsto x + 1$, which is reversed by the involution $x \mapsto -x$. Hence $g^{-1}\sigma \in I^2$, which implies that $\sigma \in I^3$.

To prove (ii), choose an element g of $\mathrm{Homeo}^+(\mathbb{R})$. Then $-g \in \mathrm{Homeo}^-(\mathbb{R})$ (where $-g$ denotes the map $x \mapsto -g(x)$). Hence $-g \in I^3$, which implies that $g \in I^4$.

Statement (iii) is an immediate consequence of (i) and (ii).   □

Clearly, 4 is the smallest integer n such that $I^n = \mathrm{Homeo}(\mathbb{R})$, because I^3 consists only of orientation-reversing homeomorphisms (products of one or three proper involutions) or strongly-reversible homeomorphisms (products of one or two proper involutions).

A corollary of Theorem 8.23 is that each element of $\mathrm{Homeo}(\mathbb{R})$ can be expressed as a product of two reversible elements of $\mathrm{Homeo}(\mathbb{R})$ (because I^2 consists of strongly-reversible elements). There is a similar result for $\mathrm{Homeo}^+(\mathbb{R})$.

Theorem 8.24 *Each element of* $\mathrm{Homeo}^+(\mathbb{R})$ *can be expressed as a product of two reversible elements of* $\mathrm{Homeo}^+(\mathbb{R})$.

Proof   Let $f(x) = x + \frac{1}{4}$ and $g(x) = x + \frac{1}{2}\sin x$. The map f has no fixed points, and g and fg are both reversible in $\text{Homeo}^+(\mathbb{R})$. Since $f = (fg)g^{-1}$, we can express f, and hence any fixed-point-free homeomorphism, as a product of two reversible elements of $\text{Homeo}^+(\mathbb{R})$.

Now consider an arbitrary element f of $\text{Homeo}^+(\mathbb{R})$. We will define two reversible homeomorphisms g and h such that $f = gh$. For each element x of $\text{fix}(f)$, we let $g(x) = h(x) = x$. On a component I in the complement of $\text{fix}(f)$, we observe that f has no fixed points, so, by the observation at the start of this proof, we can choose two reversible orientation-preserving homeomorphisms g_I and h_I of I such that $f = g_I h_I$ on I. We define $g = g_I$ and $h = h_I$ on I. The maps g and h have now been constructed such that $f = gh$, and one can check that they are both reversible in $\text{Homeo}^+(\mathbb{R})$. $\square$

Notes

Sources

The study of $\text{Homeo}(\mathbb{R})$ as a group began with Schreier and Ulam [208], who proved Proposition 8.5. The classification of homeomorphisms up to topological conjugacy goes back to Fine and Schweigert [89]. Fine and Schweigert also made the observation (given in the introduction) that $\text{Homeo}(\mathbb{R})$ has a trivial centre, and proved Corollary 8.11. See also [132, Section 4.2] and [220].

The results about products of involutions in Sections 8.3 and 8.5 are due to Fine and Schweigert [89] and Young [251]. See also [141, 188]. Part of Theorem 8.17 is proven in [179]. For Subsection 8.4.1, see [45]. Theorem 8.24 can be found in [99].

The methods used in this chapter may be used to determine the normal subgroup structure of $\text{Homeo}(\mathbb{R})$.

Piecewise-affine homeomorphisms

Let $\text{PLF}(\mathbb{R})$ denote the group of homeomorphisms of $\mathbb{R}$ that are locally affine at all but a *finite* number of points, and let $\text{PL}(\mathbb{R})$ denote the group of homeomorphisms of $\mathbb{R}$ that are locally affine at all but a *discrete* set of points. (This is standard notation; 'P' stands for 'piecewise', 'L' stands for 'linear', and 'F' stands for 'finite'.) Brin and Squier [39, 40] dealt with the conjugacy problem in these groups. Later, Gill and the second author [99] addressed the reversibility problem.

Higher dimensions

There are scattered results related to reversibility in homeomorphism groups of higher-dimensional manifolds. Anderson [6] proved that for a variety of Hausdorff topological spaces X (including $\mathbb{R}^n$ and the Cantor set), if we are given a nontrivial element h of $\mathrm{Homeo}(X)$, then every other element of $\mathrm{Homeo}(X)$ is the product of six conjugates of h and h^{-1}. This implies, in particular, that $\mathrm{Homeo}(X)$ is simple. If $\mathrm{Homeo}(X)$ contains an involution, then it also implies that $\mathrm{Homeo}(X) = I^6$.

Open problems

Products of piecewise-affine reversibles

Let $\mathrm{PL}^+(\mathbb{R})$ denote the subgroup of those members of $\mathrm{PL}(\mathbb{R})$ that preserve orientation. The problem of determining the least integer m for which

$$R^m(\mathrm{PL}^+(\mathbb{R})) = \mathrm{PL}^+(\mathbb{R})$$

remains unsolved. In [99] it is proven that $m \leqslant 4$.

Thompson's groups

Thompson's group F can be realised as a subgroup of $\mathrm{PL}(\mathbb{R})$. The reversibility problem in this group is trivial, but it is not for the other Thompson groups T and V [38, 47]. The conjugacy problem in these groups has been solved (see, for example, [124, 129]), and it is unclear whether the reversibility problem in these groups is of special interest.

Higher dimensions

The structure of the group $\mathrm{Homeo}(\mathbb{R})$ is well understood; however, our knowledge of $\mathrm{Homeo}(\mathbb{R}^n)$, for $n > 1$, is far from complete. In particular, most reversibility questions in these groups remain open.

Products of conjugators

In [6], Anderson asks what is the least positive integer m such that, for any nonidentity element h of $\mathrm{Homeo}(\mathbb{R}^n)$, every other element of $\mathrm{Homeo}(\mathbb{R}^n)$ is a product of m conjugates of h and h^{-1}. Anderson proved that $m \leqslant 6$. He asks the same question for homeomorphism groups of other topological spaces.

Subgroups

Problems abound about various subgroups of Homeo($\mathbb{R}$), and groups of homeomorphisms in higher dimensions, only a few of which are dealt with in later chapters. For related problems, see the notes on Chapters 9 and 11.

9

Circle homeomorphisms

The line and the circle are the only connected real manifolds of dimension one. In the previous chapter we considered reversibility in the group $\text{Homeo}(\mathbb{R})$ of homeomorphisms of the line (the noncompact case), and in this chapter we consider reversibility in the group $\text{Homeo}(\mathbb{S})$ of homeomorphisms of the unit circle $\mathbb{S}$ onto itself, under composition (the compact case).

Let us introduce some concepts and notation that are used throughout the chapter. Many sources (such as [41, 96, 63, 145, 239]) deal with the basic theory of circle homeomorphisms that we outline here. We can think of $\mathbb{S}$ either as the unit circle in $\mathbb{R}^2$, or else as the quotient of the real line $\mathbb{R}$ by the group of translations by integers. For the most part, we use the first representation of $\mathbb{S}$, however, for this paragraph only we use the second representation. Using the second representation, we see that elements of $\text{Homeo}(\mathbb{S})$ can be classified as orientation-preserving or reversing depending on whether their lifts to $\text{Homeo}(\mathbb{R})$ are orientation-preserving or reversing, respectively. We denote by $\text{Homeo}^+(\mathbb{S})$ the subgroup of orientation-preserving homeomorphisms, and we denote by $\text{Homeo}^-(\mathbb{S})$ the coset of orientation-reversing homeomorphisms. We can characterise orientation-preserving and reversing homeomorphisms using the representation of $\mathbb{S}$ as the unit circle as follows: given any three points a, b, and c that occur in anticlockwise order around $\mathbb{S}$, and $f \in \text{Homeo}(\mathbb{S})$, then $f(a)$, $f(b)$, and $f(c)$ also occur in anticlockwise order around $\mathbb{S}$ if and only if $f \in \text{Homeo}^+(\mathbb{S})$.

The identity element of $\text{Homeo}(\mathbb{S})$ will, as usual, be denoted by $\mathbb{1}$. The fixed-point set of a circle homeomorphism f will be denoted by $\text{fix}(f)$. As for real-line homeomorphisms, we can define the *degree* $\deg(f)$ of f by

$$\deg(f) = \begin{cases} 1 & \text{if } f \in \text{Homeo}^+(\mathbb{S}), \\ -1 & \text{if } f \in \text{Homeo}^-(\mathbb{S}). \end{cases}$$

166

More generally, we can define the degree of any continuous map g from $\mathbb{S}$ to itself, as follows. The map g induces an endomorphism of the fundamental group $\pi_1(\mathbb{S})$, and since $\pi_1(\mathbb{S}) \cong \mathbb{Z}$, this endomorphism corresponds to multiplication by some integer d, which is called the *degree* of g.

For distinct points a and b in $\mathbb{S}$, we denote by (a,b) the open anticlockwise interval from a to b in $\mathbb{S}$. Let $[a,b]$ denote the closure of (a,b). For a proper open interval J in $\mathbb{S}$, we say that $u < v$ in J if $(u,v) \subset J$.

We can use this notation to show that each orientation-reversing circle homeomorphism f has exactly two fixed points, as follows. Choose a point x in $\mathbb{S}$ that is not fixed by f. If t is sufficiently close to (and anticlockwise from) x, then the image of the interval (x,t) under f is disjoint from (x,t). Let p be the point such that $I = (x,p)$ is the largest interval with this property. Since I and $f(I)$ lie in the same component of $\mathbb{S} \setminus \{x, f(x)\}$ we see, by maximality, that $f(p) = p$. Similarly, there is exactly one other fixed point of f, in the other component of $\mathbb{S} \setminus \{x, f(x)\}$.

The group $\mathrm{Homeo}(\mathbb{S})$ contains $\mathrm{Homeo}(\mathbb{R})$ as a subgroup: $\mathrm{Homeo}(\mathbb{R})$ is the stabiliser of a point. Thus $\mathrm{Homeo}(\mathbb{S})$ inherits the orientation-reversing involutions from $\mathrm{Homeo}(\mathbb{R})$; in particular, $\mathrm{Homeo}(\mathbb{S})$ contains reflections in lines through the origin. There are also orientation-preserving involutions in $\mathrm{Homeo}(\mathbb{S})$; for example, the rotation by π. Since the group $\mathrm{Homeo}^+(\mathbb{S})$ of orientation-preserving circle homeomorphisms is simple (see, for example, [96, Theorem 4.3]), and it contains nontrivial involutions such as the rotation by π, we see that it is generated by its involutions. In Theorem 9.17 we show that, in fact, each member of $\mathrm{Homeo}^+(\mathbb{S})$ can be expressed as a product of three orientation-preserving involutions.

Like $\mathrm{Homeo}(\mathbb{R})$, the group of homeomorphisms of the circle has trivial centre (by Proposition 2.19) and the only automorphisms are inner automorphisms [244].

9.1 Involutions

Consider an orientation-reversing involution τ with fixed points p and q. Let σ denote reflection in the line through p and the origin. The stabiliser of p in $\mathrm{Homeo}(\mathbb{S})$ is a subgroup isomorphic to $\mathrm{Homeo}(\mathbb{R})$. By Proposition 8.2, the images of τ and σ are conjugate in $\mathrm{Homeo}(\mathbb{R})$. Since all reflections in lines through the origin are conjugate by suitably chosen rotations, we have proved the next proposition.

Proposition 9.1 *Each orientation-reversing involution in* Homeo($\mathbb{S}$) *is conjugate in* Homeo($\mathbb{S}$) *to the reflection in the horizontal axis (the x-axis).*   □

We move on to consider orientation-preserving homeomorphisms.

Proposition 9.2 *Each orientation-preserving involution in* Homeo($\mathbb{S}$) *other than the identity is conjugate in* Homeo$^+$($\mathbb{S}$) *to the rotation by π.*

Proof   Let τ be a nontrivial orientation-preserving involution. Were τ to have a fixed point then, after identifying the stabiliser of this fixed point with the real-line homeomorphism group Homeo($\mathbb{R}$), we would have a nontrivial order-preserving homeomorphism of $\mathbb{R}$ with finite order, which, by Proposition 8.1, is impossible. Let σ denote rotation by π. Given a point p in $\mathbb{S}$, choose any orientation-preserving homeomorphism k from $[p, \tau(p)]$ to $[p, \sigma(p)]$ (so that $k(p) = p$ and $k(\tau(p)) = \sigma(p)$), and define an element h of Homeo$^+$($\mathbb{S}$) by

$$h(x) = \begin{cases} k(x) & \text{if } x \in [p, \tau(p)], \\ \sigma k \tau(x) & \text{if } x \in [\tau(p), p]. \end{cases}$$

One can easily check that h is an element of Homeo$^+$($\mathbb{S}$), and $h\tau h^{-1} = \sigma$, as required.   □

9.2 Conjugacy

9.2.1 Topological signatures

For orientation-preserving homeomorphisms with fixed points we extend the concept of signature from the previous chapter (Subsection 8.2.1), as follows. If f is an element of Homeo$^+$($\mathbb{S}$) with fixed points then each point x in $\mathbb{S}$ is either a fixed point of f or else it lies in an open interval component J in the complement of the fixed-point set of f. The signature $s(f)$ of f is the function from $\mathbb{S}$ to $\{-1, 0, 1\}$ given by the equation

$$s(f)(x) = \begin{cases} 1, & \text{if } x < f(x) \text{ in } J, \\ 0, & \text{if } f(x) = x, \\ -1, & \text{if } f(x) < x \text{ in } J. \end{cases}$$

Basic properties of the signature are summarised in the next elementary lemma, which is almost equivalent to Lemma 8.4.

Lemma 9.3 *If f is a member of* Homeo$^+$($\mathbb{S}$) *with a fixed point, and h is a member of* Homeo($\mathbb{S}$), *then*

(i) $s(h^{-1}fh) = \deg(h)(s(f) \circ h)$

(ii) $s(f^{-1}) = -s(f)$. $\square$

9.2.2 Rotation number

The rotation number is a quantity defined for every member of $\mathrm{Homeo}^+(\mathbb{S})$ (not just those with fixed points), which remains invariant under conjugation. We give a brief description of rotation numbers; for more details see [96].

Given an orientation-preserving homeomorphism f, choose any point x in $\mathbb{S}$, and let θ_n be the angle in $[0, 2\pi)$ measured anticlockwise between $f^{n-1}(x)$ and $f^n(x)$. The *rotation number* of f, denoted $\rho(f)$, is the unique number in $[0,1)$ such that the expression

$$(\theta_1 + \cdots + \theta_n) - 2\pi n \rho(f)$$

is bounded for all integers n. The quantity $\rho(f)$ is independent of x. The rotation number is invariant under conjugation in $\mathrm{Homeo}^+(\mathbb{S})$, and, if $h \in \mathrm{Homeo}^-(\mathbb{S})$, then $\rho(hfh^{-1}) \equiv -\rho(f) \pmod 1$. Observe also that $\rho(f^n) \equiv n\rho(f) \pmod 1$, for each integer n.

A straightforward consequence of the definition of $\rho(f)$ is that $\rho(f) = 0$ if and only if f has a fixed point. In this case, $\mathbb{S}$ can be partitioned into a closed set $\mathrm{fix}(f)$, consisting of fixed points of f, and a countable collection of open intervals, on each of which f is free of fixed points.

Now suppose that $\rho(f) = p/q$, where p and q are coprime positive integers. Then $\rho(f^q) = 0$, so f has periodic points with period q. The integer q is known as the *minimal period* of f, because it is the smallest positive integer n for which f^n has fixed points. In future we denote the minimal period by n_f.

The remaining possibility is that $\rho(f)$ is irrational. In this case we define K_f to be the unique nonempty minimal set in the poset consisting of f-invariant compact subsets of $\mathbb{S}$ ordered by inclusion. We describe K_f as the *minimal set* of f. Either $K_f = \mathbb{S}$ or else K_f is a perfect subset of $\mathbb{S}$ with empty interior – a Cantor set. In the latter case there is a sequence of open intervals (a_i, b_i), for $i = 1, 2, \ldots$, such that $[a_i, b_i] \cap [a_j, b_j] = \emptyset$ when $i \neq j$, and K_f is the complement of $\bigcup_{i=1}^\infty (a_i, b_i)$. The set I_f of *inaccessible points* of K_f is the complement of $\bigcup_{i=1}^\infty [a_i, b_i]$. If $K_f = \mathbb{S}$ then we define $I_f = \mathbb{S}$. There is a continuous, surjective map w_f of $\mathbb{S}$ to itself of degree 1 such that $w_f f = R_\theta w_f$, where $\theta = 2\pi\rho(f)$. The map w_f is chosen such that it is a homeomorphism when restricted to I_f, and it maps each interval $[a_i, b_i]$ to a single point. This map w_f is unique up to post-composition by rotations.

9.2.3 Conjugacy

We state four theorems which together determine the conjugacy classes in each of the groups $\mathrm{Homeo}^+(\mathbb{S})$ and $\mathrm{Homeo}(\mathbb{S})$. The first result is on homeomorphisms with fixed points.

Theorem 9.4 *Two orientation-preserving circle homeomorphisms f and g, each of which has a fixed point, are conjugate by a homeomorphism of degree ε if and only if there is a homeomorphism h of degree ε such that $s(g) = \varepsilon \cdot (s(f) \circ h)$.*

Proof Suppose first that $h^{-1}fh = g$, where h has degree ε. By Lemma 9.3(i), $s(g) = \varepsilon \cdot (s(f) \circ h)$. Conversely, suppose that $s(g) = \varepsilon \cdot (s(f) \circ h)$, where h has degree ε. Let p be a fixed point of f and q a fixed point of g. Choose k in $\mathrm{Homeo}^+(\mathbb{S})$ with $k(p) = q$, and define $g_1 = k^{-1}gk$. Then g_1 fixes p and, by Lemma 9.3, $s(g_1) = s(g) \circ k$. Hence $s(g_1) = \varepsilon \cdot (s(f) \circ h \circ k)$. Since g_1 and f share a fixed point we can apply Theorem 8.12 to deduce that g_1 and f, and hence also g and f, are conjugate by a homeomorphism of degree ε. $\square$

Note in particular that a map in $\mathrm{Homeo}^+(\mathbb{S})$ with fixed points is conjugate in $\mathrm{Homeo}^+(\mathbb{S})$ to all of its positive powers.

Next we look at orientation-preserving circle homeomorphisms whose rotation numbers are rational numbers.

Theorem 9.5 *Two orientation-preserving circle homeomorphisms f and g, both with rational rotation numbers, are conjugate by a homeomorphism of degree ε if and only if $\rho(f) \equiv \varepsilon\rho(g) \pmod{1}$ and f^n is conjugate to g^n (where $n = n_f = n_g$) by a homeomorphism of degree ε that maps one of the n-point periodic orbits of f to an n-point periodic orbit of g.*

Before we prove this theorem, let us analyse what it says in more detail. The maps f and g are assumed to have rational rotation numbers; say $\rho(f) = r/n_f$ and $\rho(g) = s/n_g$, where n_f and n_g are the minimal periods of f and g, respectively, and $0 \leqslant r < n_f$ and $0 \leqslant s < n_g$. The statement $\rho(f) \equiv \varepsilon\rho(g) \pmod{1}$ implies that either $r/n_f = s/n_g$ (if $\varepsilon = 1$ or $\rho(g) = 0$) or $r/n_f = 1 - s/n_g$ (if $\varepsilon = -1$ and $\rho(g) \neq 0$). In both cases $n_f = n_g$, and that is why we can define an integer $n = n_f = n_g$. The final statement about the conjugacy between f^n and g^n means that there is a periodic point p of f and a homeomorphism h of degree ε such that $hfh^{-1} = g$ and such that h maps the orbit $\{p, f(p), \ldots, f^{n-1}(p)\}$ to the orbit $\{q, g(q), \ldots, g^{n-1}(q)\}$, where $h(p) = q$.

Proof Suppose that $hfh^{-1} = g$, where h has degree ε. Then $\rho(f) \equiv \varepsilon\rho(g)$ $\pmod{1}$ and $hf^nh^{-1} = g^n$. From the equation $hf^i(p) = g^ih(p)$ we see that h maps any periodic orbit of f to a periodic orbit of g.

Conversely, suppose that $\rho(f) \equiv \varepsilon\rho(g)$ (mod 1) and $kf^nk^{-1} = g^n$, where k is a circle homeomorphism of degree ε that maps a periodic orbit of f to a periodic orbit of g. After replacing f by kfk^{-1} we may assume that $\rho(f) = \rho(g) = q/n$, $f^n = g^n$, and f and g share a common periodic orbit; we must prove that f and g are conjugate in $\mathrm{Homeo}^+(\mathbb{S})$.

Let the common periodic orbit be $\{p, f(p), \ldots, f^{n-1}(p)\}$, for some point p. Since $\rho(f) = \rho(g)$, we deduce that the points $p, f(p), \ldots, f^{n-1}(p)$ occur in the same order around $\mathbb{S}$ as the points $p, g(p), \ldots, g^{n-1}(p)$. Therefore $f^i(p) = g^i(p)$ for all integers i. Let $f^m(p)$ be the first point in this orbit anticlockwise from p. Let $J = [p, f^m(p)]$. We define, for $i = 0, 1, \ldots, n-1$,

$$h(x) = g^i f^{-i}(x), \qquad x \in f^i(J).$$

The map h fixes each of the points $f^i(p)$ and each of the intervals $f^i(J)$, so it is an orientation-preserving homeomorphism. One can check that it satisfies $hf = gh$ on each of the intervals $f^i(J)$. The calculation is straightforward on all but the interval $f^{n-1}(J)$. On this interval we see that

$$hf = f \quad \text{and} \quad gh = gg^{n-1}f^{-(n-1)} = g^n f^{-(n-1)} = f^n f^{-(n-1)} = f,$$

so $hf = gh$.   □

Our third theorem is on orientation-preserving homeomorphisms with irrational rotation number. We sketch a proof; see [175, Theorem 2.3] for details. Recall that an orthogonal map of the circle of degree 1 is a rotation, and an orthogonal map of the circle of degree -1 is a reflection in a line through the origin.

Theorem 9.6   *Two orientation-preserving circle homeomorphisms f and g, both with irrational rotation numbers, are conjugate by a homeomorphism of degree ε if and only if $\rho(f) \equiv \varepsilon\rho(g)$ (mod 1) and there is an orthogonal map of degree ε that maps $w_f(I_f)$ to $w_g(I_g)$.*

Proof   It suffices to prove the theorem with $\varepsilon = 1$, because the $\varepsilon = -1$ case reduces to the $\varepsilon = 1$ case once we replace g by $\tau g\tau$, where $\tau(x) = -x$.

Suppose that $hfh^{-1} = g$, where h has degree 1. Then $h(I_f) = I_g$. Let $\theta = 2\pi\rho(f) = 2\pi\rho(g)$. Since $w_f f = R_\theta w_f$ we have $w_f h^{-1}g = R_\theta w_f h^{-1}$. By uniqueness of w_g up to post-composition by rotations we deduce that $R_\phi w_g = w_f h^{-1}$ for some real number ϕ. Thus

$$w_f(I_f) = R_\phi w_g h(I_f) = R_\phi w_g(I_g).$$

Conversely, suppose that $\rho(f) = \rho(g)$, and $w_f(I_f)$ and $w_g(I_g)$ are equivalent by a rotation. Redefining w_g we can assume that $w_f(I_f) = w_g(I_g)$. Let

$\theta = 2\pi\rho(f)$. The map w_f is a homeomorphism when restricted to I_f (and similarly for w_g), which implies that we can define a homeomorphism h on I_f by $h = w_g^{-1} w_f$. The identity $hfh^{-1} = g$ is satisfied on I_g. We can extend the definition of h from I_f to its closure, namely K_f (the minimal set of f), and the identity remains intact. Now, $K_f = \mathbb{S} \setminus \bigcup(a_i, b_i)$ for some collection of open intervals (a_i, b_i), where $[a_i, b_i] \cap [a_j, b_j] = \emptyset$ for $i \neq j$. These open intervals can be presented as a collection $\{f^n(u_i, v_i) : n \in \mathbb{Z}, i \in \beta\}$, where $\beta \in \mathbb{N} \cup \{\infty\}$, in such a way that no two sets of the form $f^n(u_i, v_i)$ are equal. Either $h(u_i)$ and $h(v_i)$ are equal, or else they are the end points of an open interval in the complement of K_g. In the latter case, define h_i to be an orientation-preserving homeomorphism of $[u_i, v_i]$ to $[h(u_i), h(v_i)]$. Finally, we define

$$k(x) = \begin{cases} h(x) & \text{if } x \in K_f, \\ hf^n(u_i) & \text{if } x \in f^n(u_i, v_i) \text{ and } h(u_i) = h(v_i), \\ g^n h_i f^{-n}(x) & \text{if } x \in f^n(u_i, v_i) \text{ and } h(u_i) \neq h(v_i). \end{cases}$$

The map k belongs to $\text{Homeo}^+(\mathbb{S})$ and it satisfies $kfk^{-1} = g$. $\qquad \square$

It remains to consider conjugacy between orientation-reversing maps. The next theorem is similar to Theorem 8.13.

Theorem 9.7 *Two orientation-reversing circle homeomorphisms f and g are conjugate in $\text{Homeo}(\mathbb{S})$ if and only if f^2 and g^2 are conjugate in $\text{Homeo}(\mathbb{S})$ by a homeomorphism that maps the pair of fixed points of f to the pair of fixed points of g.*

Proof Suppose that $hf^2h^{-1} = g^2$, and h maps the fixed points p and q of f to the fixed points of g. Define

$$k(x) = \begin{cases} h(x) & \text{if } x \in [p, q], \\ g^{-1} hf(x) & \text{if } x \in [q, p]. \end{cases}$$

Then one can check that $k \in \text{Homeo}(\mathbb{S})$ and $kfk^{-1} = g$. The converse is trivial. $\qquad \square$

9.3 Reversible elements

In the previous section we solved the conjugacy problem in $\text{Homeo}(\mathbb{S})$ (and $\text{Homeo}^+(\mathbb{S})$). The reversibility problem is just a special case. Here we examine the relationship between reversible elements and rotation numbers.

Suppose that f and h are members of $\text{Homeo}^+(\mathbb{S})$ such that $hfh^{-1} = f^{-1}$. Then

$$\rho(f^{-1}) \equiv \rho(hfh^{-1}) \equiv \rho(f) \pmod 1.$$

But $\rho(f^{-1}) \equiv -\rho(f) \pmod 1$, hence $\rho(f)$ is equal to either 0 or $\frac{1}{2}$. On the other hand, if h reverses orientation and still $hfh^{-1} = f^{-1}$, then

$$\rho(hfh^{-1}) \equiv -\rho(f) \pmod 1,$$

so, in this case, the rotation number tells us nothing about reversibility.

We shall see later (at the end of Subsection 9.4.4) that a map f that is orientation-preserving map can be reversed by an orientation-*reversing* map if and only if f is strongly reversible by an orientation-*reversing* involution. There are, however, some orientation-preserving homeomorphisms that are not strongly reversible in $\text{Homeo}(\mathbb{S})$, but are nevertheless reversible by orientation *preserving* maps; one example is given at the end of Subsection 9.4.4.

9.4 Strongly-reversible elements

9.4.1 Strongly-reversible elements in $\text{Homeo}^+(\mathbb{S})$

In this section we prove the following theorem, which gives a full classification of the strongly-reversible maps in $\text{Homeo}^+(\mathbb{S})$.

Theorem 9.8 *An element f of* $\text{Homeo}^+(\mathbb{S})$ *is strongly reversible if and only if either it is an involution or else it has a fixed point and there is an orientation-preserving homeomorphism h of rotation number $\frac{1}{2}$ such that $s(f) = -s(f) \circ h$.*

We saw in Section 9.3 that the reversible elements of $\text{Homeo}^+(\mathbb{S})$ must have rotation number 0 or $\frac{1}{2}$. To prove Theorem 9.8, let us first investigate homeomorphisms with rotation number 0.

Theorem 9.9 *An element f of* $\text{Homeo}^+(\mathbb{S})$ *with a fixed point is strongly reversible in* $\text{Homeo}^+(\mathbb{S})$ *if and only if there is a homeomorphism h belonging to* $\text{Homeo}^+(\mathbb{S})$ *with rotation number $\frac{1}{2}$ such that $s(f) = -s(f) \circ h$.*

Recall that by identifying the fixed point of a map f in $\text{Homeo}^+(\mathbb{S})$ with ∞ we can consider f as a member of $\text{Homeo}^+(\mathbb{R})$. Thus in the proof of Theorem 9.9 we are able to apply Proposition 8.8 to such a map f.

Proof If $\sigma f \sigma = f^{-1}$ for an orientation-preserving involution σ, then $s(f) = -s(f) \circ \sigma$, by Lemma 9.3. Conversely, suppose that there is a homeomorphism h in $\text{Homeo}^+(\mathbb{S})$ with rotation number $\frac{1}{2}$ such that $s(f) = -s(f) \circ h$.

By Lemma 9.3, $s(h^{-1}fh) = s(f^{-1})$. Using Proposition 8.8 we can construct a map k belonging to $\text{Homeo}^+(\mathbb{S})$ that fixes each fixed point of f, and satisfies $k^{-1}h^{-1}fhk = f^{-1}$.

Now choose a fixed point q of f. The equation $s(f) = -s(f) \circ h$ tells us that all iterates of q under h also lie in the closed set $\text{fix}(f)$. Since h^2 has fixed points, the sequence $h^2(q), h^4(q), h^6(q), \ldots$ converges; its limit is another point p in $\text{fix}(f)$, and $h^2(p) = p$. The points p and $h(p)$ are distinct elements of $\text{fix}(f)$ because h has no fixed points. Define

$$\mu(x) = \begin{cases} hk(x) & \text{if } x \in [p, h(p)], \\ k^{-1}h^{-1}(x) & \text{if } x \in [h(p), p]. \end{cases}$$

One can check that μ is an involution in $\text{Homeo}^+(\mathbb{S})$ and $\mu f \mu = f^{-1}$. $\qquad\square$

Next we consider orientation-preserving homeomorphisms that have rotation number $\frac{1}{2}$.

Theorem 9.10 *An element of* $\text{Homeo}^+(\mathbb{S})$ *with rotation number* $\frac{1}{2}$ *is strongly reversible in* $\text{Homeo}^+(\mathbb{S})$ *if and only if it is an involution.*

Proof All involutions are strongly reversible by the identity map. Conversely, let f be a homeomorphism with rotation number $\frac{1}{2}$, and let σ be an involution in $\text{Homeo}^+(\mathbb{S})$ such that $\sigma f \sigma = f^{-1}$. Choose an element x of $\text{fix}(f^2)$. Then $f(x)$ is also an element of $\text{fix}(f^2)$, and by interchanging x and $f(x)$ if necessary, we may assume that $\sigma(x) \in (x, f(x)]$. Suppose that $\sigma(x) \neq f(x)$. Since σ maps $(\sigma(x), x)$ onto $(x, \sigma(x))$, we have that $\sigma f(x) \in (x, \sigma(x))$. Likewise, f maps $(x, f(x))$ onto $(f(x), x)$, therefore $f\sigma(x) \in (f(x), x)$. However, $f\sigma(x) = \sigma f^{-1}(x) = \sigma f(x)$, and yet $(x, \sigma(x)) \cap (f(x), x) = \emptyset$. This is a contradiction, therefore $\sigma(x) = f(x)$. This implies that σf, which is an orientation-preserving involution, fixes x; hence it is the identity map. Therefore $f = \sigma$, as required. $\qquad\square$

We will see later that there are reversible homeomorphisms with rotation number $\frac{1}{2}$ that are not strongly reversible. An example is given in Subsection 9.4.4.

Together Theorems 9.9 and 9.10 can be used to prove Theorem 9.8.

Proof of Theorem 9.8 If f is a strongly-reversible element of $\text{Homeo}^+(\mathbb{S})$, then it is reversible, so it must have rotation number equal to either 0 or $\frac{1}{2}$. If it has rotation number 0, then there is an orientation-preserving homeomorphism h with rotation number $\frac{1}{2}$ such that $s(f) = -s(f) \circ h$, by Theorem 9.9. If f has rotation number $\frac{1}{2}$, then it is an involution, by Theorem 9.10. The converse implication follows immediately from Theorem 9.9. $\qquad\square$

9.4.2 **Strongly reversible and orientation preserving in** Homeo($\mathbb{S}$)

Each strongly-reversible element of Homeo$^+$($\mathbb{S}$) is also a strongly-reversible member of Homeo($\mathbb{S}$); however, there may be other element of Homeo$^+$($\mathbb{S}$) that are strongly reversible in Homeo($\mathbb{S}$), but not in Homeo$^+$($\mathbb{S}$). The following theorem classifies all the strongly-reversible orientation-preserving elements of Homeo($\mathbb{S}$).

Theorem 9.11 *Let f be an orientation-preserving circle homeomorphism. Either*

(i) $\rho(f) = 0$, *in which case f is strongly reversible if and only if there is a homeomorphism h such that $s(f) = -\deg(h)(s(f) \circ h)$ and either h preserves orientation and has rotation number $\frac{1}{2}$ or else reverses orientation*

(ii) $\rho(f)$ *is nonzero and rational, in which case f is strongly reversible if and only if f^{n_f} is strongly reversible by an orientation-reversing involution*

(iii) $\rho(f)$ *is irrational, in which case f is strongly reversible if and only if $w_f(I_f)$ has a reflectional symmetry.*

We divide our analysis of strongly-reversible maps in Homeo($\mathbb{S}$) (and the proof of Theorem 9.11) between three cases corresponding to when the rotation number is 0, rational, or irrational. The first case is Theorem 9.11(i). We need a preliminary lemma.

Lemma 9.12 *If f is a fixed-point-free homeomorphism of an open proper arc A in $\mathbb{S}$, then f is conjugate to f^{-1} on A by an orientation-reversing involution of A.*

Proof The situation is topologically equivalent to the situation when f is a fixed-point-free homeomorphism of the real line, and so the result follows from Theorem 8.12.   □

Proof of Theorem 9.11(i) If $\tau f \tau = f^{-1}$ for an involution τ, then

$$s(f) = -\deg(\tau)(s(f) \circ \tau),$$

by Lemma 9.3. For the converse, we are given a homeomorphism h that satisfies $s(f) = -\deg(h)(s(f) \circ h)$. Either h preserves orientation and satisfies $\rho(h) = \frac{1}{2}$, in which case the result follows from Theorem 9.9, or else h reverses orientation.

In the latter case, by Proposition 8.8 and Lemma 9.3 there is an orientation-preserving homeomorphism k that fixes the fixed points of f, and satisfies $k^{-1}h^{-1}fhk = f^{-1}$. Define $s = k^{-1}h^{-1}$. As s is an orientation-reversing homeomorphism, it has exactly two fixed points p and q. Let a denote the point in fix(f) that is clockwise from p, and closest to p. Possibly $a = p$. Define b to be the point in fix(f) that is anticlockwise from p and closest to p. Let $J = (a,b)$. Possibly $J = \emptyset$. Similarly we define an interval J' about q. Since a point z in (p,q) is a fixed point of f if and only if the point $s(z)$ in (q,p) is a fixed point of f, we see that J and J' do not intersect. Now, f fixes J so we can, by Lemma 9.12, choose an orientation-reversing involution τ_J of J such that $\tau_J f \tau_J(x) = f^{-1}(x)$ for x in J. Similarly we define $\tau_{J'}$. From the equation $sfs^{-1} = f^{-1}$ and the definition of the intervals J and J' we deduce that s fixes J and J'. Hence we can define

$$\mu(x) = \begin{cases} \tau_J(x) & \text{if } x \in J, \\ \tau_{J'}(x) & \text{if } x \in J', \\ s(x) & \text{if } x \in [p,q] \setminus (J \cup J'), \\ s^{-1}(x) & \text{if } x \in [q,p] \setminus (J \cup J'). \end{cases} \tag{9.1}$$

One can check that μ is an orientation-reversing involution that satisfies $\mu f \mu = f^{-1}$.   □

To prove Theorem 9.11(ii) we use a lemma that enables us to deal with rational rotation numbers of the form $1/n$, rather than m/n. Recall that n_f denotes the minimal period of f.

Lemma 9.13 *Let f be an element of* Homeo$^+(\mathbb{S})$ *with a periodic point. If f^d is strongly reversible for an integer d in $\{1, 2, \ldots, n_f - 1\}$ that is coprime to n_f, then f is strongly reversible.*

Proof   There exist integers u and t such that $dt = 1 + un_f$. Let q be the positive integer between 0 and n_f, and coprime to n_f, such that $\rho(f) = q/n_f$. Observe that

$$\rho(f^{dt}) \equiv (1 + un_f)\rho(f) \equiv \rho(f) + uq \equiv \rho(f) \pmod 1.$$

Recall that a map g in Homeo$^+(\mathbb{S})$ with fixed points is conjugate in Homeo$^+(\mathbb{S})$ to all its powers. Let $g = f^{n_f}$. Then g is conjugate to g^{dt}. In other words, f^{n_f} is conjugate to $(f^{dt})^{n_f}$. Apply Theorem 9.5 to the maps f and f^{dt} to see that these two maps are conjugate. The second map f^{dt} is strongly reversible, because f^d is strongly reversible. Conjugacy preserves the property of being strongly reversible, therefore f is also strongly reversible.   □

We first prove a special case of Theorem 9.11(ii).

Lemma 9.14 *Let f be an orientation-preserving homeomorphism of $\mathbb{S}$ with rotation number $\frac{1}{n}$, for a positive integer n. Suppose that there is an open interval J such that the intervals $J, f(J), \ldots, f^{n-1}(J)$ are pairwise disjoint, and such* *that* $\text{fix}(f^n)$ *is the complement of* $J \cup f(J) \cup \cdots \cup f^{n-1}(J)$. *Then f is strongly reversible by an orientation-reversing involution that maps $f^k(J)$ to $f^{-k+1}(J)$ for each integer k.*

Proof Let R denote an anticlockwise rotation by $2\pi/n$. After conjugating f suitably, the function f and interval J may be adjusted so that $f^k(J) = R^k(J)$ for all integers k. Note that f^n has the same signature on each of the intervals $R^k(J)$ because, by Lemma 9.3 (1),

$$s(f^n)(f(x)) = s(f^{-1}f^n f)(x) = s(f^n)(x).$$

Let τ denote reflection in a line ℓ through the origin that bisects J. Thus τ fixes J as a set. Let σ denote reflection in a line through the origin that is π/n anticlockwise from ℓ. Then $R = \sigma\tau$. Orient J in an anticlockwise sense, and choose an increasing homeomorphism ϕ of J, without fixed points, such that $\tau\phi\tau = \phi^{-1}$ on J. This is possible because we can, by conjugation, consider J to be the real line and consider τ still to be a reflection, in which case ϕ can be chosen to be a translation. Now define an orientation-preserving circle homeomorphism g to satisfy $g(x) = f(x)$, for x in $\text{fix}(f^n)$, and $g(x) = R^{k+1}\phi R^{-k}(x)$, for x in $R^k(J)$. This implies that g^n has the same signature on all of the intervals $R^k(J)$ (our choice of ϕ determines whether this signature is -1 or 1). By Theorem 9.5, g is conjugate to f. Also, one can check that $\sigma g\sigma = g^{-1}$ and $\sigma(g^k(J)) = g^{-k+1}(J)$ for each integer k. Since the property of being strongly reversible is preserved under conjugation, the result follows. $\square$

Proof of Theorem 9.11(ii) Suppose first that f is reversed by an involution τ. Then f^{n_f} is also strongly reversible by τ. We must show that f^{n_f} is reversed by an orientation-reversing involution; thus we may assume that τ is orientation preserving. In this case, by Theorem 9.8, f is an involution. Therefore f^{n_f} is the identity, and as such it is reversible by any orientation-reversing involution.

Conversely, suppose that there is an orientation-reversing involution τ such that $\tau f^{n_f}\tau = f^{-n_f}$. Let p be a fixed point of f^{n_f}. There is an integer d that is coprime to n_f such that the distinct points $p, f^d(p), \ldots, f^{(n_f-1)d}(p)$ occur in that order anticlockwise around $\mathbb{S}$. The function $(f^d)^{n_f}$ is strongly reversible because it equals $(f^{n_f})^d$. If we can deduce that f^d is strongly reversible then it follows from Lemma 9.13 that f is strongly reversible. In other words, it is sufficient to prove the theorem when the points $p, f(p), \ldots, f^{n_f-1}(p)$ occur in that order around $\mathbb{S}$.

The map τf has a fixed point q which, by replacing τ with $f^k\tau f^{-k}$ for an appropriate integer k, we can assume lies in the interval $(f^{-1}(p), p]$. Either q is a fixed point of f^{n_f} (that is, a periodic point of f) or it is not. In the former case let $J = [q, f(q)]$ and define, for integers $k = 0, 1, \ldots n_f - 1$,

$$\mu(x) = f^k \tau f^k(x), \quad x \in f^{-k}(J). \tag{9.2}$$

This is a well-defined homeomorphism because $f^{n_f} \tau f^{n_f} = \tau$. One can check that μ is an involution and satisfies $\mu f \mu = f^{-1}$. In Figure 9.1 the action of μ on certain f-iterates of q is shown in the case $n_f = 2m$.

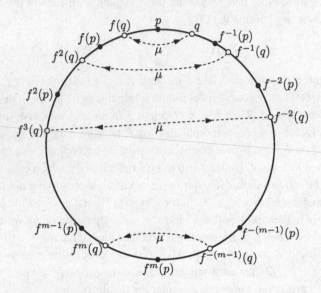

Figure 9.1 The action of f and μ.

If q is not a fixed point of f^{n_f}, then it lies in a unique component $J' = (s, t)$ in the complement of $\mathrm{fix}(f^{n_f})$. The interval J' is contained in $(f^{-1}(p), p)$, and both s and t lie in $\mathrm{fix}(f^{n_f})$. From the equation $\tau f^{n_f} \tau = f^{-n_f}$, we can deduce that τ maps J' to another open interval component in $\mathbb{S} \setminus \mathrm{fix}(f^{n_f})$. Also, $f(J')$ is a component of $\mathbb{S} \setminus \mathrm{fix}(f^{n_f})$. But $\tau(J')$ and $f(J')$ both contain the point $\tau(q) = f(q)$; therefore $\tau(J') = f(J')$. This implies that $\tau(s) = f(t)$ and $\tau(t) = f(s)$.

We define an involution μ in a similar fashion to (9.2), but this time we do not, yet, define μ on the intervals $J', f(J'), \ldots, f^{n_f-1}(J')$. Specifically, let $J = [t, f(s)]$ and define, for each integer k,

$$\mu(x) = f^k \tau f^k(x), \quad x \in f^{-k}(J).$$

We can extend μ to $J', f(J'), \ldots, f^{n_f-1}(J')$ using Lemma 9.14, as follows. Let f_0 be any homeomorphism with the same rotation number as f, which coincides with f on each interval $f^k(J')$, and which satisfies $f_0^{n_f}(x) = x$ for all other points x. By Lemma 9.14, f_0 is reversed by an orientation-reversing involution

μ_0 that maps $f^k(J')$ to $f^{-k+1}(J')$ for each integer k. We define $\mu(x) = \mu_0(x)$ for each point x in $f^k(J')$. The now fully defined map μ is an orientation-reversing involutive homeomorphism of $\mathbb{S}$ that satisfies $\mu f \mu = f^{-1}$. □

To prove Theorem 9.11(iii) we need a lemma. We prove the lemma explicitly, although it can be deduced quickly from Lemma 9.14.

Lemma 9.15 *Let J and J' be two disjoint nontrivial closed intervals in $\mathbb{S}$, and let g be an orientation-preserving homeomorphism from J to J'. Then there exists an orientation-reversing homeomorphism γ from J to J' such that $\gamma g^{-1} \gamma = g$.*

Proof Let a and b be points such that $J = [a,b]$. Choose a point q in (a,b). Choose an orientation-reversing homeomorphism α from $[a,q]$ to $[g(q),g(b)]$. The map γ defined by

$$\gamma(x) = \begin{cases} \alpha(x) & \text{if } x \in [a,q], \\ g\alpha^{-1}g(x) & \text{if } x \in [q,b], \end{cases}$$

has the required properties. □

Proof of Theorem 9.11(iii) If f is strongly reversible, then Theorem 9.8 tells us that any involution that reverses f must reverse orientation. Since $I_{f^{-1}} = I_f$, we see from Theorem 9.6 that $w_f(I_f)$ has a reflectional symmetry.

Conversely, suppose that there is a reflection τ in a line through the origin of $\mathbb{R}^2$ that fixes $w_f(I_f)$. We define an involution μ from I_f to I_f by the equation $\mu(x) = w_f^{-1}\tau w_f(x)$. In this equation, w_f^{-1} is the inverse of the function $w_f : I_f \to w_f(I_f)$. For x in I_f we have

$$\mu f \mu(x) = (w_f^{-1}\tau w_f)(w_f^{-1}R_\theta w_f)(w_f^{-1}\tau w_f)(x) = f^{-1}(x).$$

Recall that K_f is the complement in $\mathbb{S}$ of the union of a countable collection of disjoint open intervals (a_i, b_i), and I_f is the complement in $\mathbb{S}$ of the union of the intervals $[a_i, b_i]$. We can extend the definition of μ to K_f by defining $\mu(a_i)$ to be the limit of $\mu(x_n)$, where x_n is a sequence in I_f that converges to a_i. Similarly for b_i. The extended map μ is a homeomorphism from K_f to itself. Notice that μ has the property that for each integer i there is an integer j such that μ interchanges a_i and b_j, and also interchanges a_j and b_i.

We can extend the definition of μ to the whole of $\mathbb{S}$ by introducing, for each i, an orientation-reversing homeomorphism $\phi_i : [a_i, b_i] \to [a_j, b_j]$ (where $\mu(a_i) = b_j$ and $\mu(b_i) = a_j$), and defining $\mu(x) = \phi_i(x)$ for $x \in [a_i, b_i]$. The resulting map μ will be a homeomorphism. It remains only to show how to choose particular maps ϕ_i such that μ is an involution that satisfies $\mu f \mu = f^{-1}$.

Let $J = (a_i, b_i)$ and let $J' = (a_j, b_j)$. We have two collections

$$\{\ldots, f^{-1}(J), J, f(J), f^2(J), \ldots\}, \quad \{\ldots, f^{-1}(J'), J', f(J'), f^2(J'), \ldots\}, \quad (9.3)$$

each consisting of pairwise disjoint intervals. These two collections either share no common members, or else they coincide. In the first case, choose an arbitrary orientation-reversing homeomorphism γ from J to J'. In the second case, there is an integer m such that $f^m(J) = J'$. Apply Lemma 9.15 with $g = f^m$ to deduce the existence of an orientation-reversing homeomorphism γ from J to J' satisfying $\gamma f^{-m} \gamma = f^m$. In both cases we define, for each integer n,

$$\mu(x) = \begin{cases} f^n \gamma f^n(x) & \text{if } x \in f^{-n}(J), \\ f^n \gamma^{-1} f^n(x) & \text{if } x \in f^{-n}(J'). \end{cases}$$

One can check that μ is well defined for points x in one of the intervals of (9.3), and that $\mu^2(x) = x$ and $\mu f \mu(x) = f^{-1}(x)$. In this manner μ can be defined on each of the intervals (a_k, b_k). The resulting map is a homeomorphism of $\mathbb{S}$ that is an involution and satisfies $\mu f \mu = f^{-1}$. $\qquad\square$

It follows from Theorem 9.11 and the results about conjugacy in Section 9.2 that an orientation-preserving circle homeomorphism is reversed by a homeomorphism that is orientation-*reversing* if and only if it is reversed by an involution that is orientation -*reversing*. In Subsection 9.4.4 we sketch the details of an example to show that there are orientation-preserving circle homeomorphisms that are reversible by orientation-preserving homeomorphisms, but are *not* strongly reversible in Homeo($\mathbb{S}$). This implies that the properties of being reversible and strongly reversible are not equivalent in either Homeo($\mathbb{S}$) or Homeo$^+$($\mathbb{S}$).

9.4.3 Strongly reversible and orientation reversing in Homeo($\mathbb{S}$)

Recall that each orientation-reversing homeomorphism has exactly two fixed points. The following theorem classifies the strongly-reversible orientation-reversing elements of Homeo($\mathbb{S}$).

Theorem 9.16 *An orientation-reversing homeomorphism f is strongly reversible if and only if there is an orientation-reversing homeomorphism s that interchanges the pair of fixed points of f and satisfies $s f^2 s^{-1} = f^{-2}$.*

Proof If f is strongly reversible, then it is expressible as a composite of two involutions, one of which, τ, must reverse orientation. From the equation $\tau f \tau = f^{-1}$ we see that $\tau f^2 \tau = f^{-2}$, and that τ preserves the pair of fixed points of f as a set. Either τ interchanges the fixed points of f, which is the condition stated in Theorem 9.16, or it fixes each of them. In the latter case, τf is an orientation-preserving involution with fixed points. Hence it is the identity map. Therefore $f = \tau$, and we can choose the map s from Theorem 9.16 to be any orientation-reversing homeomorphism that swaps the fixed points of f.

For the converse, suppose that a and b are the two fixed points of f. By Theorem 9.11(i) we can construct from s an orientation-reversing involution μ such that $\mu f^2 \mu = f^{-2}$. We require that μ, like s, interchanges a and b; this is not given by the statement of Theorem 9.11, however, it is immediate from the definition of μ in (9.1). Now define

$$\tau(x) = \begin{cases} \mu(x) & \text{if } x \in [a,b], \\ f\mu f(x) & \text{if } x \in [b,a]. \end{cases}$$

Then τ is an involution in $\mathrm{Homeo}^-(\mathbb{S})$ and $\tau f \tau = f^{-1}$. $\square$

The hypothesis that s interchanges the pair of fixed points of f cannot be dropped from Theorem 9.16; one can construct an example of an orientation-reversing homeomorphism f and involutions τ such that $\tau f^2 \tau = f^{-2}$ even though f is not strongly reversible. There are also examples of orientation-reversing circle homeomorphisms that are reversible, but are not strongly reversible.

9.4.4   Reversible but not strongly reversible

We construct a map that is reversible in $\mathrm{Homeo}^+(\mathbb{S})$, but not strongly reversible in $\mathrm{Homeo}(\mathbb{S})$.

Let a be the point on $\mathbb{S}$ with coordinates $(0,1)$ and let b be the point with coordinates $(0,-1)$. Let $\cdots < a_{-2} < a_{-1} < a_0 < a_1 < a_2 < \cdots$ be an infinite sequence of points in (b,a) that accumulates only at a and b. We construct a doubly-infinite sequence s consisting of 1s and -1s as follows. Let u represent the string of six numbers $1,1,1,-1,-1,1$. Let $-u$ represent the string $-1,-1,-1,1,1,-1$. Then s is given by $\ldots, u, -u, u, -u, \ldots$. We say that two doubly-infinite sequences (x_n) and (y_n) are equal if and only if there is an integer k such that $x_{n+k} = y_n$ for all n. Our sequence s has been constructed such that $s = -s$, and the sequence formed by reversing s is not equal to s.

Let g be an orientation-preserving homeomorphism from $[b,a]$ to $[b,a]$ that fixes only the points a_i, a, and b. Suppose that g is defined in any fashion on the intervals

$$\ldots, (a_{-2}, a_{-1}), (a_{-1}, a_0), (a_0, a_1), (a_1, a_2), \ldots$$

such that the signature of g on these intervals is determined by s. For x in (a,b) we define $g(x) = R_\pi g R_\pi(x)$, where R_π is the rotation by π. A diagram of the homeomorphism g is shown in Figure 9.2.

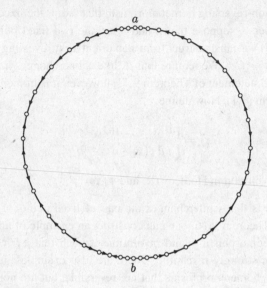

Figure 9.2 The flow of g.

Since we can embed g in a flow, we can certainly choose a square root $\sqrt{g}$ of g (which shares the same signature as g). Define another circle homeomorphism f by the formula

$$f(x) = \begin{cases} R_\pi \sqrt{g}(x) & \text{if } x \in [b,a], \\ \sqrt{g} R_\pi(x) & \text{if } x \in (a,b). \end{cases}$$

This map f has rotation number $\frac{1}{2}$ and it satisfies $f^2 = g$. The map f is not reversed by an orientation-preserving involution, by Theorem 9.10, because it is not an involution. Nor is f reversed by an orientation-reversing involution; for it were then g would also be reversed by the same orientation-reversing involution, and from Theorem 9.11 one can deduce that this would mean that the sequence formed by reversing s is equal to s. Finally, f is reversible since g is reversible; a conjugation from g to g^{-1} can be constructed that maps a_i to a_{i+6} for each i.

9.5 Products of involutions

We began this chapter with the observation that Homeo$^+(\mathbb{S})$ is generated by involutions, and now we can prove a stronger result.

Theorem 9.17 *Each element of* Homeo$^+(\mathbb{S})$ *can be expressed as a composite of three orientation-preserving involutions.*

There are elements of Homeo$^+(\mathbb{S})$ that cannot be expressed as a composite of two orientation-preserving involutions; for example, all those maps without rotation number 0 or $\frac{1}{2}$.

A simple corollary of Theorem 9.17 is that Homeo$^+(\mathbb{S})$ is uniformly perfect, meaning that there is a positive integer N such that each element of Homeo$^+(\mathbb{S})$ can be expressed as a composite of N or fewer commutators. Since it is easy to express the rotation by π as a commutator, and each involution in Homeo$^+(\mathbb{S})$ is conjugate to the rotation by π, it follows from Theorem 9.17 that Homeo$^+(\mathbb{S})$ is uniformly perfect with $N = 3$. In fact, Eisenbud, Hirsch, and Neumann [74] proved that Homeo$^+(\mathbb{S})$ is uniformly perfect with $N = 1$.

Proof of Theorem 9.17   Choose an element f of Homeo$^+(\mathbb{S})$ that is not an involution. There exists a point x in $\mathbb{S}$ such that x, $f(x)$, and $f^2(x)$ are three distinct points. By replacing f with f^{-1} and x with $f^2(x)$ if necessary we can assume that x, $f(x)$, and $f^2(x)$ occur in that order anticlockwise around $\mathbb{S}$. Notice that $f^{-1}(x)$ lies in $(f(x),x)$. Select a point y in $(x,f(x))$ that is sufficiently close to $f(x)$ that $f^{-1}(y) \in (f^2(x),x)$. Choose an orientation-preserving homeomorphism g from $[x,f(x)]$ to $[f(x),x]$ such that $g(y) = f(y)$, $g(t) < \min(f(t),f^{-1}(t))$ in (x,y), and $f(t) < g(t) < f^{-1}(t)$ in $(y,f(x))$. A graph of such a function is shown in Figure 9.3.

Define an involution σ in Homeo$^+(\mathbb{S})$ by the equation

$$\sigma(t) = \begin{cases} g(t) & \text{if } t \in [x,f(x)], \\ g^{-1}(t) & \text{if } t \in [f(x),x]. \end{cases}$$

Let us determine the fixed points of σf. For a point t in $[x,f(x)]$ we have that $\sigma f(t) = t$ if and only if $f(t) = g(t)$. This implies that either $t = x$ or $t = y$. For a point t in $(f(x),x)$ we have that $\sigma f(t) = t$ if and only if $\sigma f g(w) = g(w)$, where $w = g^{-1}(t)$ is a point in $(x,f(x))$. Therefore $g(w) = f^{-1}(w)$. This equation has no solutions in $(x,f(x))$, hence σf has no fixed points in $(f(x),x)$. The direction of flow of σf on the complement of $\{x,y\}$ can also easily be deduced

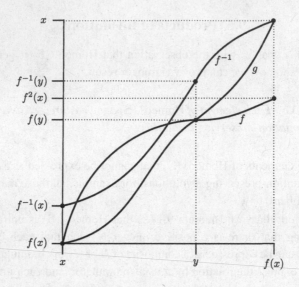

Figure 9.3 Graphs of f, f^{-1}, and g.

from the construction of g, and we find that

$$s(\sigma f)(t) = \begin{cases} 0 & \text{if } t = x, y, \\ 1 & \text{if } t \in (x, y), \\ -1 & \text{if } t \in (y, x). \end{cases}$$

By Theorem 9.9, σf can be expressed as a composite of two involutions in $\mathrm{Homeo}^+(\mathbb{S})$. Therefore f can be expressed as a composite of three involutions in $\mathrm{Homeo}^+(\mathbb{S})$. $\square$

Corollary 9.18 *Each element of* $\mathrm{Homeo}(\mathbb{S})$ *can be expressed as a composite of three involutions.*

There are elements of $\mathrm{Homeo}(\mathbb{S})$ that cannot be expressed as a composite of two involutions; these are the elements that are not strongly reversible.

Proof of Corollary 9.18 Because of Theorem 9.17 we need only prove that an arbitrarily chosen orientation-reversing homeomorphism f can be expressed as a product of three involutions. Let f have fixed points p and q. Choose an orientation-reversing involution τ with fixed points p and q such that on (q, p) both $\tau(x) > f(x)$ and $\tau(x) > f^{-1}(x)$. Therefore $\tau f(x) > x$ on (p, q) and $\tau f(x) > x$ on (q, p). Theorem 9.11(i) now shows that τf is reversed by an orientation-reversing involution. That is, there are involutions α and β such that $\tau f = \alpha \beta$. Hence $f = \tau \alpha \beta$, and the result is proven. $\square$

Since the composite of two involutions is a reversible map, and involutions are reversible maps, we see from Theorems 9.17 and Corollary 9.18 that

$$\mathrm{Homeo}^{+}(\mathbb{S}) = R^2(\mathrm{Homeo}^{+}(\mathbb{S})) \quad \text{and} \quad \mathrm{Homeo}(\mathbb{S}) = R^2(\mathrm{Homeo}(\mathbb{S})).$$

Notes
Sources

The elementary theory of circle homeomorphisms is developed in [41, 96, 63, 145, 239]. Ghys sketches sketches details of a conjugacy classification in [96]. Theorem 9.6 was first proven by Markley [175, Theorem 2.3]. All the theorems on reversibility can be found in [102]. Corollary 9.18 was proven in [89, Theorem 25].

Subgroups

There are many interesting subgroups of $\mathrm{Homeo}(\mathbb{R})$ and $\mathrm{Homeo}(\mathbb{S})$, and one may ask about their reversible elements. For instance, there are Möbius groups, discrete groups, and groups of piecewise-affine maps. We shall consider the full diffeomorphism group of the line in Chapter 11. For the rest, the reader who would like to study reversibility in homeomorphism groups should start by familiarising him or herself with the seminal work of masters such as Smale, Mather, Epstein, Markovic, and Ghys. A useful starting point for diffeomorphism groups is the survey by Navas [187]. For piecewise-affine groups, see [39, 40, 99].

Higher dimensions

Little is known about reversibility in $\mathrm{Homeo}(\mathbb{S}^n)$, for $n > 1$. Anderson [6] has shown that, for $n = 2, 3$, each element of $\mathrm{Homeo}(\mathbb{S}^n)$ can be expressed as a product of six involutions. (See similar remarks at the end of Chapter 8.) Anderson's methods extend to n-dimensions provided that the annulus property holds on $\mathbb{S}^n$ (and this is unknown).

A well known-theorem due to Brouwer and Kerékjártó says that a periodic homeomorphism of the closed unit disc or the plane is conjugate by a homeomorphism to an orthogonal map (a reflection or a rotation). Likewise a periodic homeomorphism of the two-dimensional sphere is conjugate to an orthogonal map [56].

We continue the discussion of reversibility in higher dimensions in the Notes of Chapter 11.

Manifolds

The homeomorphism groups of general manifolds pose formidable challenges. For any manifold M, the mapping class group $C = \mathrm{MCG}(M)$ (the group of free isotopy classes of homeomorphisms) is a quotient of $\mathrm{Homeo}(M)$, so one should start by studying the simpler problem of identifying the reversible elements of C. The two-dimensional torus is the simplest manifold with nontrivial mapping class group, and we have covered it in Chapter 7. For higher dimensional tori, one would have to solve the reversibility problem in $\mathrm{GL}(n, \mathbb{Z})$.

For compact orientable surfaces of finite genus greater than 2, six involutions always suffice to generate C [37]. Thus $I^\infty(C) = C$. However, Korkmaz has observed [156] that $I^n(C)$ is always proper in C, whenever the genus exceeds 2.

Open problems

Higher dimensions

Investigate reversibility in $\mathrm{Homeo}(\mathbb{S}^n)$, for $n > 1$. We suggest that you begin with the work of Anderson [6], and then move on to manuscripts that cite Anderson.

Piecewise-affine circle homeomorphisms

The reversibility problem is well understood in $\mathrm{Homeo}(\mathbb{S})$, as we have seen, and it is also well understood in the group $\mathrm{PL}(\mathbb{R})$ of piecewise-affine homeomorphisms of the real line. An obvious next step is to investigate reversibility in the group of piecewise-affine homeomorphisms of the circle (viewed as $\mathbb{R}/\mathbb{Z}$).

Closed, transitive subgroups of $\mathrm{Homeo}^+(\mathbb{S})$

Ghys raises many open problems about $\mathrm{Homeo}^+(\mathbb{S})$ in [96]. Among these is Problem 4.4, which asks whether each closed, transitive proper subgroup of $\mathrm{Homeo}^+(\mathbb{S})$ is conjugate to either $\mathrm{SO}(2, \mathbb{R})$, $\mathrm{PSL}(2, \mathbb{R})$, the group $\mathrm{Homeo}_k^+(\mathbb{S})$ of orientation-preserving homeomorphisms that commute with the rotation of order k, or a cyclic cover of $\mathrm{PSL}(2, \mathbb{R})$ or $\mathrm{Homeo}_k^+(\mathbb{S})$. This proposal was confirmed by Giblin and Markovic [98] when the closed, transitive subgroup contains a nonconstant continuous path. Without the assumption that there is such a path, the problem remains open.

10

Formal power series

In this chapter, we consider the group of formally-invertible power series in one indeterminate (under formal composition). One is led to consider the group of such series as a necessary preliminary to the study of diffeomorphisms in one real variable, or biholomorphic germs in one complex variable.

10.1 Power series structures

In this section we define the group of formally-invertible power series and some important subgroups.

10.1.1 The coefficient field

We work with power series that have coefficients in an arbitrary commutative field F of characteristic zero. Some aspects of reversibility depend on algebraic features of the field F, and in particular on the presence of primitive fourth-roots of unity

For each positive integer p, *which is not necessarily a prime number*, we let S_p denote the multiplicative group of pth roots of unity of F. This is a cyclic group whose order divides p. We let $F^\times$ denote the multiplicative group of nonzero elements of F, and we let F_p denote the subgroup of pth powers of elements of $F^\times$. There is a short exact sequence of groups associated to the field F:

$$\{1\} \to S_p \to F^\times \to F_p \to \{1\},$$

where the map $S_p \to F^\times$ is inclusion, and the map $F^\times \to F_p$ is $x \mapsto x^p$.

For an element a in $F^\times$, we let $[a]_p$ denote the coset aF_p, an element of the quotient $F^\times/F_p$. Take, for instance, $F = \mathbb{R}$. Then when p is even, $[a]_p$ is essentially the sign of a, whereas when p is odd, there is a single equivalence class $[a]_p$, which is independent of a.

We denote by $M(F)$ the set of positive integers p such that F has a solution to $\omega^p = -1$. Thus $p \in M(F)$ if and only if S_p is a proper subset of S_{2p}. Note that $p \in M(F)$ if and only if $q \in M(F)$, where q is the highest power of 2 that divides p. Clearly, $M(F)$ always contains all odd natural numbers, and when $F = \mathbb{R}$, $M(F)$ is precisely the set of odd natural numbers. In contrast, when F is algebraically closed, $M(F)$ is the set of all positive integers.

10.1.2 The algebra of formal power series

The set $F[[X]]$ of formal power series forms a commutative F-algebra with unit $1 = 1 + 0X + 0X^2 + \cdots$ under coefficientwise sums and convolution product:

$$\sum_{n=0}^{\infty} a_n X^n + \sum_{n=0}^{\infty} b_n X^n = \sum_{n=0}^{\infty} (a_n + b_n) X^n$$

$$\sum_{n=0}^{\infty} a_n X^n \cdot \sum_{n=0}^{\infty} b_n X^n = \sum_{n=0}^{\infty} \left(\sum_{k=0}^{n} a_k b_{n-k} \right) X^n.$$

It is an integral domain, a local ring, and a complete valuation ring with respect to the order valuation

$$f \mapsto \mathrm{ord}(f),$$

where the order $\mathrm{ord}(\sum_{n=0}^{\infty} a_n X^n)$ of a nonzero series is the least integer n with $a_n \neq 0$. The algebra $F[X]$ of polynomials is dense in $F[[X]]$ with respect to the metric associated to the valuation.

We remark that an infinite sequence (f_n) of power series converges with respect to this metric if and only if, for each nonnegative integer k, the coefficient of X^k in f_n is eventually constant. We interpret the convergence of infinite sums and products of power series accordingly, as the convergence in this sense of the sequence of partial sums or products, respectively.

10.1.3 The semigroup of formal power series

The maximal ideal $\mathcal{M}$ that consists of power series with $a_0 = 0$ has additional structure: its elements can be composed formally, and it forms a composition semigroup. The formal composition is defined by

$$\sum_{n=1}^{\infty} a_n X^n \circ \sum_{n=1}^{\infty} b_n X^n = \sum_{n=1}^{\infty} a_n \left(\sum_{m=1}^{\infty} b_m X^m \right)^n. \tag{10.1}$$

Here, by

$$\left(\sum_{m=1}^{\infty} b_m X^m \right)^n,$$

we understand the nth power with respect to the convolution product. Notice that when this expression is expanded, the coefficient of X^k is 0 for $k < n$. It follows that the coefficient of X^k in the sum (10.1) is the the same as in the *finite* sum

$$\sum_{n=1}^{k} a_n \left(\sum_{m=1}^{k} b_m X^m \right)^n.$$

Therefore the formal composition makes sense for all pairs of series

$$f = \sum_{n=1}^{\infty} a_n X^n \quad \text{and} \quad g = \sum_{n=1}^{\infty} b_n X^n$$

belonging to $\mathcal{M}$. The first three terms of the composition are

$$a_1 b_1 X + \left(a_1 b_2 + a_2 b_1^2 \right) X^2 + \left(a_1 b_3 + 2 a_2 b_1 b_2 + a_3 b_1^3 \right) X^3, \qquad (10.2)$$

but the expression for the nth term rapidly becomes complicated. It is known as (the power series version of) Faà di Bruno's formula, and can be expressed in terms of (or used to define) the so-called Bell polynomials. The formula takes a somewhat tidier form if we write the series in the form

$$f = \sum_{n=1}^{\infty} \frac{a_n}{n!} X^n \quad \text{and} \quad g = \sum_{n=1}^{\infty} \frac{b_n}{n!} X^n.$$

Then

$$f \circ g = \sum_{n=1}^{\infty} \frac{c_n}{n!} X^n, \quad \text{where} \quad c_n = \sum_{k=1}^{n} B_{n,k} a_k,$$

and $B_{n,k}$ depends on the coefficients b_n. In fact, $B_{n,k}$ is a polynomial in the coefficients $b_1, \ldots, b_{n-k+1}$, given explicitly by

$$B_{n,k} = \sum_{\{B_1, \ldots, B_k\}} b_{|B_1|} \cdots b_{|B_k|}, \qquad (10.3)$$

where the sum is taken over the set of all partitions of the set $\{1, \ldots, n\}$ into pairwise-disjoint blocks $B_1, \ldots, B_k$, and $|B_j|$ is the number of elements in B_j.

The composition is associative, and the series

$$\mathbb{1} = X = X + 0 X^2 + 0 X^3 + \cdots$$

is the identity for composition. We emphasise that $\mathbb{1} \neq 1$.

10.1.4 The group of formally-invertible formal power series

A series

$$\sum_{n=1}^{\infty} a_n X^n$$

in $\mathcal{M}$ is invertible under formal composition if and only if $a_1 \neq 0$. (For proof of this statement, and more detail on the basic algebra of formal series, see [49, Chapter 1].) The collection of invertible elements of $\mathcal{M}$ forms a group, the group of formally-invertible formal power series over the field F, which we denote by $\mathcal{G}$.

To avoid confusion with the convolution product, we shall consistently denote compositions by $f \circ g$, rather than the usual fg found in all other chapters. We denote compositional powers $g \circ \cdots \circ g$ by $g^{\circ n}$. However, we shall, as is usual in group theory, sometimes refer to the group operation as the 'product'. We will have occasion to consider multiples $\lambda \sum_{n=1}^{\infty} a_n X^n$, with $\lambda \in F$. These should be read as equal to $\sum_{n=1}^{\infty} \lambda a_n X^n$ (or, equivalently, as the convolution product of the series $\lambda + 0X + 0X^2 + \cdots$ with $\sum_{n=1}^{\infty} a_n X^n$).

We shall characterise the sets $I^n(\mathcal{G})$, $R^n(\mathcal{G})$, and $R_g(\mathcal{G})$ quite explicitly. The most interesting result is that membership in I^∞ is determined by congruence 'modulo X^4'. This allows us to deduce that just four involutions suffice to represent a general series with $a_1 = \pm 1$.

10.1.5 Multipliers

The coefficient a_1 of an invertible series

$$f = \sum_{n=1}^{\infty} a_n X^n$$

in $\mathcal{G}$ is a conjugacy invariant (that is, it depends only on the conjugacy class of the series f in $\mathcal{G}$). This coefficient is usually called the *multiplier* of the series. We denote the multiplier of f by $m(f)$.

We consider the normal subgroups

$$\mathcal{H} = \{f \in \mathcal{G} : f \text{ has multiplier } \pm 1\},$$
$$\mathcal{G}_1 = \{f \in \mathcal{G} : f \text{ has multiplier } 1\}$$

of $\mathcal{G}$. The subgroup $\mathcal{G}_1$ has index 2 in $\mathcal{H}$, and we denote the second coset by

$$-\mathcal{G}_1 = \{f \in \mathcal{H} : f \text{ has multiplier } - 1\}.$$

We remark that all the sets $I(\mathcal{G})$, $\mathcal{H}$, $\mathcal{G}_1$, $I^\infty(\mathcal{G})$, and $R^\infty(\mathcal{G})$ are closed in $F[[X]]$, and hence complete with respect to the induced metric.

10.1.6 Series modulo X^{k+1}

We also consider the corresponding sets and groups 'modulo X^{k+1}'. To be precise, given elements f and g of $\mathcal{G}$ and a positive integer k, we say that

$$f \equiv g \quad (\text{mod } X^{k+1})$$

if f and g have the same coefficients up to and including that of X^k.

For each positive integer k, we let

$$J_k = \{f \in \mathcal{G} : f \equiv X \quad (\text{mod } X^{k+1})\} = \left\{ X + \sum_{n=k+1}^{\infty} a_n X^n : a_n \in F \right\},$$

and observe that J_k is a normal subgroup of $\mathcal{G}$. We denote the sets of involutions, reversibles, and products of them in the quotient group $\mathcal{G}/J_k$ by

$$I_k = I(\mathcal{G}/J_k), \quad R_k = R(\mathcal{G}/J_k), \quad I_k^n = I^n(\mathcal{G}/J_k), \quad R_k^n = R^n(\mathcal{G}/J_k),$$

where $n \in \mathbb{N} \cup \{\infty\}$. Those that are groups are normal subgroups of $\mathcal{G}/J_k$.

Note that in this chapter we use the notation I^n and R^n (in which the relevant group is not indicated) only to refer to the sets associated to the full power series group $\mathcal{G}$.

10.2 Elements of finite order

There are two order concepts for an element

$$f = \sum_{n=1}^{\infty} a_n X^n$$

of $\mathcal{G}$. The first is the order valuation $\text{ord}(f)$ mentioned earlier, defined to be the least integer n with $a_n \neq 0$. This is always equal to 1 for elements of $\mathcal{G}$. The second is the group-theoretic concept of order, namely the least positive integer n with $f^{\circ n} = \mathbb{1}$. It is this second type of order that we discuss here.

Proposition 10.1 *Let f be a nontrivial element of $\mathcal{G}_1$, that is,*

$$f = X + aX^{p+1} + \cdots,$$

with $a \neq 0$. Then f has infinite order.

Proof For any positive integer n,

$$f^{\circ n} = X + naX^{p+1} + \cdots,$$

so f cannot have finite order.   □

Corollary 10.2 *The only involution in $\mathcal{G}$ with multiplier 1 is $\mathbb{1}$.*

The multiplier of an involution must be ± 1. Thus the *proper* involutions (elements of order exactly 2 in $\mathcal{G}$) all have multiplier -1.

Proposition 10.3 *Suppose that an element f of $\mathcal{G}$, given by*

$$f = \omega X + \cdots, \quad \omega \in F,$$

has finite order m. Then ω is a primitive mth root of unity in F, and f is conjugate in $\mathcal{G}$ to its linear part ωX.

Proof Since

$$X = f^{\circ m} = \omega^m X + \mathrm{O}(X^2),$$

we see that $\omega^m = 1$.

Let ϕ be the element of $\mathcal{G}$ given by

$$\phi = X + \omega^{-1} f + \omega^{-2} f^{\circ 2} + \cdots + \omega^{1-m} f^{\circ(m-1)}.$$

Then one calculates that $\phi \circ f = \omega \phi$, so $\phi \circ f \circ \phi^{-1} = \omega X$. Finally, if ω were not a *primitive* root of unity, then ωX, and hence f, would not have order m. □

Corollary 10.4 *Each involution τ in $\mathcal{G}$ is conjugate to the series $-X$.* □

The condition $\tau^{\circ 2} = \mathbb{1}$ translates into an infinite chain of polynomial identities in the coefficients. These may be put in the form of recursion relations specifying the odd-index coefficients in terms of the preceding coefficients. The even-index coefficients are unconstrained.

10.3 Conjugacy

We denote the compositional product $f_1 \circ \cdots \circ f_n$ by $\prod_{j=1}^{\circ n} f_j$, and use $\prod_{j=1}^{\circ \infty} f_j$ for the limit $\lim_{n \to \infty} \prod_{j=1}^{\circ n} f_j$ (if it exists with respect to the valuation metric).

Lemma 10.5 *Each element ϕ of $\mathcal{G}$ can be written as a convergent product*

$$\phi = (\lambda X) \circ \prod_{j=2}^{\circ \infty} (X + a_j X^j),$$

where λ and the coefficients a_j belong to F, and $\lambda \neq 0$.

Proof We give only a sketch of the proof of this straightforward computation. One converts the inverse $\phi^{\circ -1}$ to the form $X + \sum_{n=j}^{\infty} b_n X^n$, for progressively larger integers j, by using right composition with the factors of the infinite product, for suitable constants λ and a_j. □

Corollary 10.6 *A continuous function* ν *mapping* $\mathscr{G}$ *to some topological space is a conjugacy invariant (that is, it is constant on each conjugacy class) if and only if*

$$\nu((\lambda^{-1}X) \circ \phi \circ (\lambda X)) = \nu(\phi), \quad \nu\left((X + aX^j)^{\circ -1} \circ \phi \circ (X + aX^j)\right) = \nu(\phi),$$

whenever $\phi \in \mathscr{G}$, $\lambda \in F^\times$, $a \in F$, *and* j *is a positive integer greater than 1.*

We note that the coefficient projection maps from $F[[X]]$ to F are all continuous (with respect to *any* metric on F), as are all polynomials depending on a finite number of variables, and indeed all sequences of such polynomials (regarded as maps to the product $\prod_{n=1}^\infty F$).

The conjugacy classes of $\mathscr{G}$ are well understood. For our purposes, it suffices to consider the classes contained in $\mathscr{H}$.

Theorem 10.7

(i) *Each nontrivial element* ϕ *of* $\mathscr{G}_1$ *is conjugate in* $\mathscr{G}$ *to one of the polynomials*

$$X + aX^{p+1} + \alpha a^2 X^{2p+1},$$

    where $a \in F^\times$, p *is a positive integer, and* $\alpha \in F$. *The integer* p, *the coset* $[a]_p$, *and the coefficient* α *are conjugacy invariants.*

(ii) *Each element* ϕ *of* $-\mathscr{G}_1$ *is conjugate to* $-X$ *or one of the polynomials*

$$-X + aX^{p+1} - \alpha a^2 X^{2p+1},$$

    where $a \in F^\times$, p *is a positive even integer, and* $\alpha \in F$. *Again, the integer* p, *the coset* $[a]_p$, *and the coefficient* α *are conjugacy invariants.*

$\square$

In the sequel, we denote the conjugacy invariant p associated to an element ϕ of $\mathscr{H}$ by $p(\phi)$.

When $F = \mathbb{C}$ (or any algebraically closed field), then $[a]_p$ is independent of a, so p and α are the only conjugacy invariants.

Proof of Theorem 10.7   We sketch a proof of (i); case (ii) is similar. One conjugates ϕ by generators of the form λX and $(X + aX^j)$ to fix the coefficient of X^{p+1}, and then eliminate the coefficients of $X^{p+2}, \ldots, X^{2p}$ as well as the coefficients of X^{2p+2} and higher powers. The inverse of $(X + aX^j)$ is given by an explicit formula:

$$(X + aX^j)^{\circ -1} = X + \sum_{t=1}^\infty c_t X^{1+(j-1)t},$$

where the coefficients c_t are given by the recursion

$$c_t = (-a)^t - \sum_{r=1}^{t-1} \binom{(j-1)r}{t-r} a^{t-r} c_r.$$

After some manipulation, one obtains the explicit conjugation formula

$$(X+aX^j)^{\circ-1} \circ \phi \circ (X+aX^j) = X + \sum_{m=p+1}^{\infty} a_m X^m \left[\frac{(1+aX^{j-1})^m}{1+ajX^{j-1}} \right] + O(X^{2p+j}).$$

From this we see that, for $k < p+j$, the effect on the $(p+k)$th coefficient when ϕ is conjugated by $X+aX^j$ is to change it from a_{p+k} to a_{p+k} plus a sum of terms of the form $sa^u a_m$, where s is an integer, u is a positive integer, $\dot{a}_m \neq 0$, $m \geqslant p+1$, and $m+u(j-1) = p+k$. Indeed, s is the coefficient of y^u in $(1+ay)^m/(1+ajy)$.

The integer p, the coset $[a]_p$, and the coefficient α can be shown to be conjugacy invariants with the help of Corollary 10.6. $\qquad \square$

Apart from facilitating the proof of Theorem 10.7, the explicit conjugation formula for $(X+aX^j)$ given above allows us to deduce the following proposition, which makes it easier to evaluate the invariant α.

Proposition 10.8 *Suppose that an element ϕ of $\mathcal{G}_1$, given by*

$$\phi = X + \sum_{n=2}^{\infty} a_n X^n,$$

has invariants p, $[a_{p+1}]_p$, and α.

(i) *The coefficients $a_2, \ldots, a_p$ are zero, and $a_{p+1} \neq 0$.*

(ii) *The invariant α is of the form $a_{2p+1}/a_{p+1}^2 + A/a_{p+1}^b$, where A is a polynomial in $a_{p+1}, \ldots, a_{2p-2}$ with rational coefficients, and b is a positive integer.*

(iii) *If p is odd, and a_n is 0 whenever n is odd and $n < 2p+1$, then $A = 0$, so*

$$\alpha = \frac{a_{2p+1}}{a_p^2}.$$

Proof Once again, we only sketch a proof. The reason for the second part is that for $p+1 < p+j < 2p+1$, one removes a_{p+j} by conjugating with $X+aX^j$, where

$$a = \frac{a_{p+j}}{(j-1-p)a_{p+1}}.$$

For $j > p$, the same conjugation removes a_{p+j}, but has no further affect on a_{2p+1}.

The reason for the third part is that in the special case where p is odd and

$a_m = 0$ for *odd* m between $p+1$ and $2p$, the conjugations required to remove the remaining a_m (for $m \neq p+1$ or $2p+1$) have no effect on the coefficient of X^{2p+1}: only even-indexed coefficients are affected. Thus, in this case, a_{2p+1} already has its 'final' value.   □

The representative form $\pm X + aX^p + bX^{2p+1}$ for an element of $\mathcal{H}$ is referred to as a *normal form*. Theorem 10.7 says that each element of $\mathcal{H}$ is conjugate to one in this normal form. An alternative collection of normal forms, representing each conjugacy class at least once is

$$\frac{\pm X}{(1 + aX^p + bX^{2p})^{\frac{1}{p}}},$$

where p is a positive integer, $a, b \in F$, and the expression should be expanded as a power series in the most obvious way.

10.4 Centralisers

The centralisers of elements of $\mathcal{G}$ are understood. We are interested mainly in the centralisers of elements of $\mathcal{G}_1$.

10.4.1 Special elements

Recall that S_p denotes the multiplicative group of pth roots of unity of the field F. The next proposition is straightforward.

Proposition 10.9 *Let p be a positive integer and $\omega \in S_p$. Then the series ωX commutes with a series f in $\mathcal{G}$ if and only if f takes the form*

$$f = \sum_{k=0}^{\infty} a_k X^{kp+1}.$$ □

The normal forms of Section 10.3 are all of this type, so we draw a corollary, for future use.

Corollary 10.10 *If*

$$f = \pm X + aX^{p+1} + bX^{2p+1} \quad or \quad f = \frac{\pm X}{(1 + aX^p + bX^{2p})^{\frac{1}{p}}},$$

with p a positive integer and $a, b \in F$, then ωX belongs to C_f, whenever ω is a pth root of unity.   □

Let us define an even more special class of series of the type in Proposition 10.9 given by

$$g_{\alpha,\lambda} = \frac{\alpha X}{(1+\lambda X^p)^{1/p}} = \alpha X - \left(\frac{\alpha\lambda}{p}\right) X^{p+1} + \frac{(p+1)\alpha\lambda^2}{2p^2} X^{2p-1} + \mathrm{O}\left(X^{3p-1}\right),$$

where $\alpha \in F^\times$ and $\lambda \in F$. An elementary calculation gives

$$g_{\alpha,\lambda} \circ g_{\beta,\mu} = g_{\gamma,\nu},$$

where

$$\gamma = \alpha\beta, \qquad \nu = \lambda\beta^p + \mu.$$

Thus $g_{\alpha,\lambda}$ commutes with $g_{\beta,\mu}$ if and only if

$$\lambda(\beta^p - 1) = \mu(\alpha^p - 1).$$

In particular, we obtain the following proposition.

Proposition 10.11   *The series $g_{\alpha,\lambda}$ commutes with $g_{\beta,\mu}$ whenever $\alpha, \beta \in S_p$ and $\lambda, \mu \in F$.*   □

10.4.2 A class of subgroups of $\mathscr{G}$

Proposition 10.12   *Suppose that an element f of $\mathscr{G}_1$ is given by*

$$f = X + f_{p+1}X^{p+1} + \mathrm{O}(X^{p+2}),$$

with $f_{p+1} \in F^\times$. Then each element g in $C_f(\mathscr{G})$ satisfies $m(g)^p = 1$.

Proof   Let

$$g = \sum_{n=1}^\infty g_n X^n.$$

Comparing the coefficients of X^{p+1} in $f \circ g = g \circ f$, we find that

$$g_{p+1} + f_{p+1}g_1^{p+1} = g_1 f_{p+1} + g_{p+1},$$

hence $g_1^p = 1$.   □

Let $\mathscr{K}_p$ denote the set of those elements f of $\mathscr{G}$ that satisfy

$$f \equiv \lambda X + bX^{p+1} \quad (\mathrm{mod}\ X^{p+2})$$

for constants λ in S_p and b in F.

Proposition 10.13   *The group $\mathscr{K}_p$ is a subgroup of $\mathscr{G}$, for each positive integer p.*

Proof We calculate

$$\left(\lambda X + bX^{p+1} + \cdots\right) \circ \left(\lambda'X + b'X^{p+1} + \cdots\right) = \lambda\lambda'X + (b\lambda' + b'\lambda)X^{p+1} + \cdots,$$

and

$$\left(\lambda X + bX^{p+1} + \cdots\right)^{\circ -1} = \frac{X}{\lambda} - \frac{bX^{p+1}}{\lambda^2} + \cdots.$$

Thus $\mathscr{K}_p$ is closed under composition and inverses. $\square$

We denote by $S_p \times F$ the direct product of the multiplicative group $(S_p, \times)$ and the additive group $(F, +)$. We define $\Psi : \mathscr{K}_p \to S_p \times F$ by setting

$$\Psi(\lambda X + bX^{p+1} + \cdots) = \left(\lambda, \frac{b}{\lambda}\right). \tag{10.4}$$

One can readily check that Ψ is a group homomorphism.

Proposition 10.14 *Suppose that an element f of $\mathscr{G}_1$ takes the form*

$$f \equiv X + f_{p+1}X^{p+1} + f_{2p+1}X^{2p+1} \pmod{X^{2p+2}},$$

with $f_{p+1} \neq 0$. Then the centraliser $C_f(\mathscr{G})$ is a subgroup of $\mathscr{K}_p$.

Proof Let g be an element of the centraliser $C_f(\mathscr{G})$ given by

$$g = \sum_{n=1}^{\infty} g_n X^n.$$

By Proposition 10.12, $g_1 \in S_p$. So it remains to show that $g_n = 0$, for $2 \leqslant n \leqslant p$.

Suppose this is not so, and pick a least integer n with $2 \leqslant n \leqslant p$ and $g_n \neq 0$. Calculating $f \circ g$ and $g \circ f$, and comparing the coefficients of X^{p+n}, we obtain

$$g_{n+p} + (p+1)f_{p+1}g_n = ng_nf_{p+1} + g_{n+p}.$$

Therefore $(p+1-n)f_{p+1}g_n = 0$, which is impossible. $\square$

10.4.3 Formal iterates

We describe a way to find a large collection of elements of $C_f(\mathscr{G})$, when $f \in \mathscr{G}_1$. Let f be given by

$$f = X + \sum_{n=2}^{\infty} f_n X^n.$$

Then for each positive integer m, the iterate $f^{\circ m}$ takes the form

$$f^{\circ m} = X + \sum_{n=2}^{\infty} P_n(m; f_2, \ldots, f_n)X^n,$$

where P_n are polynomials in $f_2, \ldots, f_n$, with integral coefficients. We observe a crucial fact about the dependence of P_n on m, as m varies.

Lemma 10.15 *For a fixed series f in $\mathcal{G}_1$ and a fixed positive integer n, the function P_n is a polynomial in m (over F), of degree at most $n-1$*

Proof For

$$f = X + \sum_{k=2}^{\infty} f_k X^k \quad \text{and} \quad g = X + \sum_{k=2}^{\infty} g_k X^k,$$

let us assign weight $k-1$ to the terms f_k and g_k. Then to each monomial in the coefficients f_k and g_k, where $k \in \mathbb{N}$, (that is, each product of an element of F and various coefficients f_k and g_k), let us assign weight equal to the sum of the weights of the coefficients f_k and g_k that occur in it. Let us say that a polynomial in the coefficients f_k and g_k is *isobaric* of weight i if each constituent monomial has weight exactly i. Looking at the expression (10.3) for the coefficients in Faà di Bruno's formula, we see that the coefficient of X^k ($k > 1$) in $f \circ g$ is a polynomial of the form

$$f_k + g_k + Q_k(f_2, \ldots, f_{k-1}, g_2, \ldots, g_{k-1})$$

that is isobaric of weight $k-1$. Moreover, we see that each monomial occurring in $Q_k(f_2, \ldots, f_{k-1}, g_2, \ldots, g_{k-1})$ involves both coefficients f_i and g_j (because $Q_k = 0$ in case either $f = \mathbb{1}$ or $g = \mathbb{1}$). Hence the mth iterate $f^{\circ m}$ takes the form $X + \sum_{k=2}^{\infty} P_k(m) X^k$, where $P_k(m)$ is a polynomial in the coefficients f_i that is isobaric of weight $k-1$.

We have $P_2(m) = m f_2$, of degree 1 in m, and we proceed by induction on k. Suppose that $P_r(m)$ has degree at most $r-1$ in m for $2 \leqslant r < k$. We have

$$P_k(m) = f_k + P_k(m-1) + Q_k(f_2, \ldots, f_{k-1}, P_{k-1}(2), \ldots, P_{k-1}(m-1)).$$

But since each monomial occurring in $Q_k(f_2, \ldots, f_{k-1}, g_2, \ldots, g_{k-1})$ involves coefficients f_i and g_j, the total weight contributed by the coefficients g_j is *less* than $k-1$, so

$$R_k(m) = f_k + Q_k(f_2, \ldots, f_{k-1}, P_{k-1}(2), \ldots, P_{k-1}(m-1))$$

has degree at most $k-2$ in m. Let us write $R_k(m) = d_{k-2} m^{d-2} + \cdots + d_1 m + d_0$. Then $P_k(m) = P_k(m-1) + R_k(m)$ yields

$$P_k(m) = R_k(m) + R_k(m-1) + \cdots = d_{m-2} \sum_{j=0}^{m} j^{m-2} + \cdots + d_1 \sum_{j=0}^{m} j + d_0 \sum_{j=0}^{m} 1,$$

which is a polynomial of degree at most $k-1$ in m. $\qquad\square$

This allows us to define the 'formal iterate':

$$f^{\circ t} = X + \sum_{n=2}^{\infty} P_n(t; f_2, \ldots, f_n) X^n,$$

whenever $t \in F$.

We draw some conclusions about iterates from Lemma 10.15.

Proposition 10.16 *Let* $f \in \mathscr{G}_1$, $h \in \mathscr{G}$, *and* $t \in F$.

(i) *If* $h \circ f = f \circ h$, *then* $h \circ f^{\circ t} = f^{\circ t} \circ h$.

(ii) *If* h *reverses* f, *then it also reverses* $f^{\circ t}$.

Proof First we prove (i). Suppose h commutes with f. The equation $h \circ f^{\circ t} = f^{\circ t} \circ h$ amounts to a sequence of identities between the coefficients of the two sides. Each of these identities is a polynomial in t (with coefficients that depend on the coefficients of f and h). We know that the identities for which t is a positive integer are true; therefore the identities hold for all values of t.

Next we prove (ii). Suppose h reverses f. Then h reverses $f^{\circ n}$ for each integer n, by Proposition 2.2. Thus the identity $f^{\circ t} \circ h \circ f^{\circ t} = h$ holds for each integer t. As with (i), this equation is equivalent to a sequence of identities, each polynomial in t, and hence it holds identically. $\square$

Corollary 10.17 *If* $f, g \in \mathscr{G}_1$, *and* $f \circ h = h \circ f$, *then* $f^{\circ t} \circ h^{\circ u} = h^{\circ u} \circ f^{\circ t}$ *whenever* $t, u \in F$. $\square$

A particular case of this corollary is worth highlighting.

Corollary 10.18 *If* $f \in \mathscr{G}_1$, *then* $f^{\circ t} \in C_f(\mathscr{G}_1)$, *for each* $t \in F$. $\square$

In fact, we have the following stronger result of this type.

Proposition 10.19 *If* $f \in \mathscr{G}_1$, *then*

$$f^{\circ (t+u)} = f^{\circ t} \circ f^{\circ u},$$

whenever $t, u \in F$.

Proof Arguing much as before, we observe first that the equation holds for all integers t and u, and hence it holds whenever $t, u \in F$. $\square$

This proposition says that the map $t \mapsto f^{\circ t}$ is a group isomorphism from $(F, +)$ into $\mathscr{G}_1$.

10.4.4 Classification of centralisers

The following theorem describes the centraliser of an element in special form. We use the homomorphism Ψ from Subsection 10.4.2.

Theorem 10.20 *Suppose that an element f of $\mathcal{G}_1$ is given by*

$$f = X + f_{p+1}X^{p+1} + \sum_{k=2}^{\infty} f_{kp+1}X^{kp+1}, \qquad (10.5)$$

with $f_{p+1} \in F^{\times}$. Then Ψ maps $C_f(\mathcal{G})$ isomorphically onto $S_p \times F$.

Proof Let $\omega \in S_p$ and $\alpha \in F$. Let $t = \omega\alpha/f_{p+1}$. The iterate $f^{\circ m}$ takes the form

$$f^{\circ m} = X + \sum_{k=1}^{\infty} P_{kp+1}(m;f)X^{kp+1} = X + mf_{p+1}X^{p+1} + \cdots,$$

so

$$f^{\circ t} = X + \sum_{k=1}^{\infty} P_{kp+1}(t;f)X^{kp+1} = X + tf_{p+1}X^{p+1} + \cdots.$$

Now $\omega X \in C_f(\mathcal{G})$, by Proposition 10.9, so $\omega f^{\circ t} \in C_f(\mathcal{G})$, and we calculate

$$\Psi(\omega f^{\circ t}) = \Psi(\omega X)\Psi(f^{\circ t}) = (\omega, \alpha).$$

Thus Ψ maps $C_f(\mathcal{G})$ onto $S_p \times F$.

To see that Ψ is injective on $C_f(\mathcal{G})$, fix an element g of $C_f(\mathcal{G})$ with $\Psi(g) = (1,0)$. We have to show that $g = \mathbb{1}$. If $g \neq \mathbb{1}$, then

$$g \equiv X + g_n X^n \pmod{X^{n+1}},$$

with $g_n \neq 0$ and $n > p+1$. We then find, on comparing the coefficients of X^{p+n} in $f \circ g$ and $g \circ f$, that $ng_n f_{p+1} = (p+1)g_n f_{p+1}$, which is impossible. Thus Ψ is injective. $\qquad\square$

Corollary 10.21 *When f is an element of $\mathcal{G}_1$ of the form (10.5), then the centraliser $C_f(\mathcal{G})$ is an abelian group comprised of all series of the form*

$$\omega^j f^{\circ t},$$

where $\omega \in S_p$, $t \in F$, and j is an integer. $\qquad\square$

Note that there are just $|S_p|$ values of ω^j, so the set of all formal iterates has finite index $|S_p|$ in the centraliser. We remark also that the set of powers of ω is the torsion subgroup of the centraliser (isomorphic to S_p), since $f^{\circ t}$ has infinite order whenever $t \neq 0$.

Using Theorem 10.20 we can describe the centralisers of more general formal series.

Corollary 10.22 *Let f be an element of $\mathscr{G}_1$ given by*

$$f = X + aX^{p+1} + \cdots,$$

where $a \in F^\times$. Then the centraliser $C_f(\mathscr{G})$ is an abelian group isomorphic to $S_p \times F$, which consists of all elements of the form

$$\sigma^{\circ j} \circ f^{\circ t},$$

where σ is an element of $\mathscr{G}$ of order $|S_p|$, $t \in F$, and j is an integer. The torsion subgroup of $C_f(\mathscr{G})$ is the set of powers of σ.

Proof Let h be a formal series in $\mathscr{G}$ that conjugates f to the normal form

$$g = X + aX^{p+1} + bX^{2p+1}.$$

The elements of $C_g(\mathscr{G})$ have the form $\omega^j g^{\circ t}$, so taking $\sigma = h \circ (\omega X) \circ h^{\circ -1}$, we see that the elements of $C_f(\mathscr{G})$ have the form $\sigma^{\circ j} \circ f^{\circ t}$, as stated. $\square$

This corollary itself has two corollaries.

Corollary 10.23 *Suppose that f is a nontrivial element of $\mathscr{G}_1$. Then $C_f(\mathscr{G}_1)$ consists of all the formal iterates $f^{\circ t}$, for $t \in F$.* $\square$

The next corollary is about centralisers of elements of $\mathscr{G}$ that do not necessarily belong to $\mathscr{G}_1$.

Corollary 10.24 *Suppose that f is an element of $\mathscr{G}$. If $m(f) \in S_p$ and $f^{\circ p} \neq \mathbb{1}$, then the group $C_f(\mathscr{G})$ is abelian.*

Proof To see why this is so, observe that $C_f(\mathscr{G})$ is a subgroup of $C_{f^{\circ p}}(\mathscr{G})$, and $f^{\circ p} \in \mathscr{G}_1$. $\square$

10.4.5 A cautionary note

Given an element f of $\mathscr{G}_1$, with invariant $p(f) = p$, the centraliser $C_f(\mathscr{G})$ is isomorphic to $S_p \times F$, but it does not always lie in $\mathscr{H}_p$. Let us construct an example to justify this. Consider the involution $-X$, which commutes with the odd series $k = X + X^7$. Let $h = X + X^4$. Then

$$g = h^{\circ -1} \circ (-X) \circ h = -X - 2X^4 + \cdots$$

commutes with

$$f = h^{\circ -1} \circ k \circ h = X + X^7 + \cdots.$$

So $f \in \mathscr{G}_1$, with invariant $p(f) = 6$, and g commutes with f but does not belong to $\mathscr{H}_6$.

However, the usual coefficient-comparison method gives the following proposition.

Proposition 10.25   *Suppose that f is an element of $\mathscr{G}_1$ with invariant $p(f) =$
p. Then $C_f(\mathscr{G}_1) \subset \mathscr{K}_p$.* $\square$

Thus we may apply Ψ to $C_f(\mathscr{G}_1)$. A simpler version of the argument used to
prove Theorem 10.20 gives the following observation.

Proposition 10.26   *Suppose that f is an element of $\mathscr{G}_1$. Then Ψ maps $C_f(\mathscr{G}_1)$
isomorphically onto $\{1\} \times F$.* $\square$

Given an element f of $\mathscr{G}_1$ with invariant $p(f) = p$, it is possible to define a
modified version Ψ' of Ψ on the whole of $C_f(\mathscr{G})$, to give an explicit isomorphism onto $S_p \times F$. For an element g of $C_f(\mathscr{G})$, define $\lambda = m(g)$. Then $\lambda \in S_p$
and $g^{\circ p} \in \mathscr{G}_1$, and $g^{\circ p}$ takes the form $X + bX^{p+1} + \cdots$. Define

$$\Psi'(g) = \left(\lambda, \frac{b}{p} \right).$$

Then one can see that Ψ' does the trick.

10.5 Products of involutions modulo X^4

Recall that I^n denotes the set of those elements of $\mathscr{G}$ that can be expressed as
a product of n involutions, and I_k^n denotes the set of those elements of $\mathscr{G}/J_k$,
where

$$J_k = \{ f \in \mathscr{G} : f \equiv X \quad (\mathrm{mod}\ X^{k+1}) \},$$

that can be expressed as a product of n involutions.

We have noted that membership in I^n, for $n = 1$, translates into an infinite
chain of identities for the coefficients. We shall see that for $n = 2$ the condition
simplifies considerably, and that for $n = 4$ it simplifies radically.

The key to this is what happens modulo X^4. The situation is very simple in
this case, as the next proposition demonstrates. In this proposition, we use the
most natural notation for $\mathscr{G}/J_k$, where, for example, we write

$$X + aX^2 + bX^3$$

for the coset in $\mathscr{G}/J_3$ corresponding to that element.

Proposition 10.27

(i) $I_3 = \{X\} \cup \{-X + aX^2 - a^2X^3 : a \in F\}$

(ii) *An element g of $\mathcal{G}/J_3$ is the product of two proper involutions if and only if*

$$g = X + aX^2 + a^2X^3$$

for some constant a in F.

(iii) *The product of any odd number of proper involutions in $\mathcal{G}/J_3$ is a proper involution.*

(iv) *We have $I_3^\infty = I_3^2$; that is, each product of involutions modulo X^4, is the product of two.*

Proof Referring to formula (10.2), we deduce that for $\varepsilon_i = \pm 1$, $i = 1, 2$, we have

$$(\varepsilon_1 X + cX^2 + \varepsilon_1 c^2 X^3) \circ (\varepsilon_2 X + dX^2 + \varepsilon_2 d^2 X^3)$$
$$\equiv \varepsilon_1\varepsilon_2 X + (c + \varepsilon_1 d)X^2 + \varepsilon_1\varepsilon_2(c + \varepsilon_1 d)^2 X^3 \pmod{X^4}.$$

The lemma follows directly from this equation. $\square$

10.6 Strongly-reversible elements

We start with some more facts about involutions.

Lemma 10.28 *If a formal series $\sum_{n=1}^{\infty} a_n X^n$ is a proper involution, then the first (if any) integer greater than or equal to 2 with $a_n \neq 0$ is even.*

Proof Let $\tau = -X + a_n X^n + \mathrm{O}(X^{n+1})$, with n odd and $a_n \neq 0$. Then we calculate

$$\tau^{\circ 2} = -\left(-X + a_n X^n + \mathrm{O}(X^{n+1})\right) + a_n \left(-X + a_n X^n + \mathrm{O}(X^{n+1})\right)^n + \mathrm{O}(X^{n+1})$$
$$= X - 2a_n X^n + \mathrm{O}(X^{n+1}),$$

so τ is not an involution. $\square$

Corollary 10.29 *If τ is a proper involution in $\mathcal{G}$, then the only involutions that commute with τ are $\mathbb{1}$ and τ.*

Proof By Lemma 10.28, the only odd involutions in $\mathcal{G}$ are $\pm X$. The centraliser of $-X$ in $\mathcal{G}$ is the group of elements of $\mathcal{G}$ with all even coefficients zero. Thus the only involutions commuting with $-X$ are $\pm X$, and the result follows on conjugating $-X$ to τ. $\square$

This shows that $\mathcal{G}$ has a rich supply of noncommuting involutions, and I^2 is a substantially larger set than I.

Lemma 10.30   *Let τ be a proper involution in $\mathcal{G}$ given by*

$$\tau = -X + \sum_{n=2m}^{\infty} a_n X^n, \qquad (10.6)$$

with $a_{2m} \neq 0$. Then $a_n = 0$ for each odd integer $n < 4m - 1$, and $a_{4m-1} = ma_{2m}^2$.

Proof   This is seen by comparing coefficients in $\tau^{\circ 2} = X$.   □

Corollary 10.31   *Let τ be an involution given by (10.6). The conjugacy invariants associated to $-\tau$ are $p = 2m - 1$, $[a_{p+1}]_p$, and $\alpha = m$.*

Corollary 10.32   *Suppose that an element g of $\mathcal{G}$, with invariants $p = 2m - 1$ and α, can be expressed as a product of two proper involutions. Then $\alpha = m$.*

Proof   Let the proper involutions be τ_1 and τ_2, with $g = \tau_1 \circ \tau_2$. We can conjugate τ_1 to $-X$, and then $\tau_2 = -g$ is a proper involution. It follows that $\alpha = m$.   □

Lemma 10.33

(i) *For each positive integer m and $a \in F$, there is an involution of the form*

$$-X - aX^{2m} - ma^2 X^{4m-1} + O(X^{4m}).$$

(ii) *Thus there is a product of two proper involutions conjugate to*

$$X + aX^{2m} + ma^2 X^{4m-1},$$

with conjugacy invariants $p = 2m - 1$, $[a]_p$, and $\alpha = m$.

Proof   In fact, the linear fractional transformations

$$\frac{X}{aX - 1} = -X - aX^2 - a^2 X^3 - \cdots$$

are involutions, for each $a \in F$. Taking $p = 2m - 1$, replacing a by ap, and 'conjugating with X^p' gives

$$\frac{-X}{(1 - apX^p)^{1/p}} = -X - aX^{2m} - ma^2 X^{4m-1} + \cdots,$$

so this is formally involutive. This proves (i).

(Note that this argument does not require the field F to have pth roots of all its elements. When $F = \mathbb{R}$, and a is real, the left-hand side of the above displayed equation is an involutive *function* (since p is odd), and is represented by the series near 0. Hence the right-hand side is a formally-involutive series. This fact will not change when a becomes an element of an arbitrary field, since it amounts to a collection of polynomial identities.)

Following the series from (i) with $-X$, and applying Proposition 10.8, we get (ii).   □

Another way to generate examples as in (i) is to form conjugations

$$(X + \lambda X^{2m})^{\circ -1} \circ (-X) \circ (X + \lambda X^{2m}),$$

with $\lambda \in F$.

We saw earlier that there are no polynomial involutions except $\pm X$, but now we see that the composition of two involutions may be a polynomial, and that any degree except 2 is possible.

Corollary 10.34 *Suppose that an element g of $\mathcal{G}_1$ has conjugation invariants $p = 2m - 1$ and $\alpha = m$ for some positive integer m. Then g is the product of two distinct proper involutions.*

Proof From Lemma 10.33, g is conjugate to a product of distinct proper involutions, and conjugation preserves the set $I^2 \setminus I$.   □

Combining this corollary and Corollary 10.32 we obtain a characterization of $I^2(\mathcal{G})$.

Theorem 10.35 *Let $g \in \mathcal{G}$. Then g is the product of two proper involutions if and only if g is conjugate to $X + aX^{2m} + ma^2 X^{4m-1}$ for some positive integer m and element a of F.*   □

Note, in particular, that if

$$g \not\equiv X \pmod{X^4},$$

but the coset gJ_3 belongs to I_3^2 in the quotient group $\mathcal{G}/J_3$, then $g \in I^2$ (because in those circumstances its invariants are $p = 1$ and $\alpha = 1$).

10.7 Products of involutions

In the next theorem and proof, given an element g of $\mathcal{G}$, we write '$g \bmod X^4$' to refer to the image of g in $\mathcal{G}/J_3$.

Theorem 10.36

(i) *The product of any odd number of proper involutions is the product of three. A series g in $\mathcal{G}$ is the product of three proper involutions if and only if $g \bmod X^4$ is a proper involution mod X^4, and this happens if and only if*

$$g \equiv -X + aX^2 - a^2 X^3 \pmod{X^4},$$

for some $a \in F$.

(ii) *The product of any even number of proper involutions is the product of four. A series g in $\mathscr{G}$ is such a product if and only if g mod X^4 is the product of two proper involutions mod X^4, and this happens if and only if*

$$g \equiv X + aX^2 + a^2X^3 \quad (\text{mod } X^4),$$

for some $a \in F$. Thus

$$I^\infty = I^4 = \{g \in \mathscr{G} : g \text{ mod } X^4 \text{ belongs to } I_3^\infty\}.$$

Proof First we prove (i). Let $g \in \mathscr{G}$ and $g \equiv -X$ (mod X^4). Thus

$$g = -X + \sum_{n=4}^{\infty} a_n X^n,$$

where $a_n \in F$. Take $\tau_1 = \sum_{n=1}^{\infty} -X^n$. Then τ_1 is an involution and obviously

$$\tau_1 \circ g = X - X^2 + X^3 + \sum_{n=4}^{\infty} b_n X^n,$$

where $b_n \in F$. The note at the end of the previous section tells us that $\tau_1 \circ g$ is a product $\tau_2 \circ \tau_3$ of involutions, hence $g = \tau_1 \circ \tau_2 \circ \tau_3$.

Now let ψ be any product of an odd number of proper involutions. Then ψ mod X^4 belongs to I_3 (by Proposition 10.27), and hence ψ mod X^4 is conjugate in $\mathscr{G}/J_3$ to $-X$ mod X^4. Thus there is some cubic k in $\mathscr{G}$ such that

$$k^{\circ-1} \circ \psi \circ k \equiv -X \quad (\text{mod } X^4).$$

Therefore $k^{\circ-1} \circ \psi \circ k$ is the product of three involutions, and hence so is ψ. The rest follows from Proposition 10.27.

Next we prove (ii). From (i), it is clear that any product of proper involutions is the product of three or four. Thus $I^\infty = I^4$.

Obviously

$$I^\infty \subset \{g \in \mathscr{G} : g \text{ mod } X^4 \text{ belongs to } I_3^\infty\}.$$

To see the converse, let g mod X^4 belong to I_3^∞. If the multiplier is -1, then we have just seen that g is the product of three involutions. If the multiplier is $+1$, then $-g$ has multiplier -1 and $-g$ mod X^4 belongs to I_3^∞, so $-g$ is the product of three involutions, and hence g is the product of four. The rest follows from Proposition 10.27. $\square$

10.8 Reversible series

We move on to reversible series and products of them. The results depend on an algebraic aspect of the coefficient field F. Recall that $M(F)$ denotes the collection of those positive integers for which F has a pth root of -1.

Proposition 10.37 *Suppose that ω is an element of the field F that satisfies $\omega^p = -1$. Let μ be either -1 or 1. Then, for each λ in F, the series*

$$g_\lambda = \frac{\mu X}{(1+\lambda X^p)^{\frac{1}{p}}}$$

is reversed by ωX.

Proof In fact $(g_\lambda)^{\circ-1} = g_{-\lambda}$, and a short calculation shows that $g_\lambda \circ (\omega X) = \omega g_{-\lambda}$. $\square$

This gives us a supply of reversible series whenever $p \in M(F)$. It turns out that these are all the reversibles, up to conjugacy.

Theorem 10.38 *Let $f \in \mathcal{H}$, and let μ be the multiplier of f (either -1 or 1). The following are equivalent:*

(i) *f is reversible in $\mathcal{G}$*

(ii) *f is conjugate to*

$$\frac{\mu X}{(1+\lambda X^p)^{1/p}},$$

for some integer p in $M(F)$ and $\lambda \in F$

(iii) *f is conjugate to*

$$\mu X + aX^{p+1} + \mu \frac{(p+1)a^2}{2} X^{2p+1},$$

for some integer p in $M(F)$ and $a \in F$.

Condition (iii) obviously provides the basis for an explicit algorithm to determine whether or not $f \in R$. Condition (ii) is interesting because it embeds f in a one-parameter subgroup of $\mathcal{G}$, and $f^{\circ-1}$ is obtained explicitly by replacing λ by $-\lambda$.

Proof First we show that (ii) is equivalent to (iii). Expanding the series in (ii), we get

$$\frac{\mu X}{(1+\lambda X^p)^{1/p}} = \mu X + aX^{p+1} + bX^{2p+1} + \cdots,$$

where $a = \mu\lambda/p$ and $b = \mu(p+1)a^2/2$. Thus the canonical conjugate polynomial form of this series is as in (iii). Hence (ii) and (iii) are equivalent.

That (ii) implies (i) follows from Proposition 10.37. We complete the proof by showing that (i) implies (iii).

Consider first a nontrivial reversible element f of $\mathscr{G}$ with multiplier 1, and let $p = p(f)$ and $\alpha = \alpha(f)$, that is,

$$f \equiv X + aX^{p+1} + \alpha a^2 X^{2p+1} \quad (\text{mod } X^{2p+2}),$$

with $a \neq 0$.

We calculate

$$f^{\circ-1} \equiv X - aX^{p+1} + (p+1-\alpha)a^2 X^{2p+1} \quad (\text{mod } X^{2p+2}).$$

Thus $[a]_p = [-a]_p$, which implies that $p \in M(F)$, and $\alpha = p+1-\alpha$, so $\alpha = (p+1)/2$. Thus condition (iii) holds, and we are done, in case $m(f) = 1$.

Now suppose that f is a reversible element of $\mathscr{G}$ with multiplier -1. Let h be an element of $\mathscr{G}$ that reverses f. Conjugating if need be, we may assume that f is an involution or takes the form

$$f = -X + aX^{p+1} + bX^{2p+1},$$

with p an even positive integer, and $a \neq 0$.

In the former case, f is conjugate to $-X$, and we are done. In the latter, $f^{\circ 2}$ is also reversed by h, and we calculate that

$$f^{\circ 2} \equiv X - 2aX^{p+1} + ((p+1)a^2 - 2b) X^{2p+1} \quad (\text{mod } X^{2p+2}).$$

Thus, by the first part,

$$(p+1)a^2 - 2b = \frac{(p+1)(-2a)^2}{2},$$

$$b = -\frac{(p+1)a^2}{2}.$$

Therefore f is conjugate to

$$-X + aX^{p+1} - \frac{(p+1)a^2}{2}X^{2p+1},$$

as required. $\square$

Recall that the series

$$f = \frac{X}{(1 + \lambda X^p)^{1/p}},$$

where $p \in M(F)$, is reversed by ωX, whenever $\omega^p = -1$. Therefore f can be reversed by an element of order $2p$, but not by any element of order p. In the next section we shall see that much more than this is true.

10.9 Reversers

A remarkable thing happens for the power series groups.

Theorem 10.39 *Suppose that f is a noninvolutive element of $\mathcal{G}$, reversed by another element h. Let $p = p(f)$. Then h has finite even order $2s$, where p/s is an odd integer.*

Proof By Theorem 10.38, it suffices to consider the case when

$$f = \frac{\mu X}{(1 + \lambda X^p)^{\frac{1}{p}}},$$

where p is an integer in $M(F)$, $\lambda \in F$, and $\mu = \pm 1$. Notice that if $\mu = -1$ then p is even, because if p is odd then f is an involution. Now, we already know that f is reversed by the series $\sigma = \omega X$ of order $2p$, where ω is an element of F with $\omega^p = -1$, and (using Corollary 10.21) the centraliser of f consists of elements of the form $\sigma^{\circ 2j} \circ f^{\circ t}$, where j is an integer, $t \in F$, and

$$f^{\circ t} = \frac{X}{(1 + tX^p)^{\frac{1}{p}}}.$$

Since $\sigma^{\circ -1} \circ h$ commutes with f we see that $h = \sigma^{\circ 2j+1} \circ f^{\circ t}$. The series σ reverses each $f^{\circ t}$, therefore

$$h^{\circ 2p} = \left(\sigma^{\circ 2j+1} \circ f^{\circ t} \circ \sigma^{\circ 2j+1} \circ f^{\circ t} \right)^p = \left(\sigma^{\circ 2j+1} \circ f^{\circ t} \circ (f^{\circ t})^{-1} \circ \sigma^{\circ 2j+1} \right)^p = \mathbb{1}.$$

Therefore the order of h divides $2p$. It must have even order $2s$ (because odd powers of h also reverse f, so $\mathbb{1}$ cannot be an odd power), where s divides p. Since $h^{\circ 2s} = \sigma^{\circ 2(2j+1)s}$, and σ has order $2p$, it follows that p divides $(2j+1)s$. That is, $(2j+1)s = kp$ for some integer k, or $2j + 1 = kp/s$. Hence p/s is odd. $\square$

The next two corollaries concern the relationship between reversible and strongly-reversible elements in $\mathcal{G}$. Recall from Theorem 10.35 that the invariant $p(f)$ is odd if f is a product of two proper involutions.

Corollary 10.40 *Suppose that $f \in \mathcal{H}$ and $p(f)$ is odd. Then f is reversible if and only if it is strongly reversible.*

Proof   Suppose that f is reversible, and not an involution. Theorem 10.39 tells us that f is reversed by an element h of order $2s$, where s is odd. Therefore f is reversed by $h^{\circ s}$, which is an involution.   □

Corollary 10.41   *The set of reversible elements coincides with the set of strongly-reversible elements in $\mathscr{G}$ if and only if F has no primitive fourth root of 1 (that is, F has no square root of -1).*

Proof   Suppose first that F has no primitive fourth root of 1. Let f be a noninvolutive reversible element of $\mathscr{G}$. Theorem 10.39 says that f is reversed by an element h of order $2s$, where p/s is odd. Since $m(h)^{2s} = \pm 1$, s cannot be even, else F has a primitive fourth root of unity. Therefore p is odd, so f is strongly reversible by Corollary 10.40.

Conversely, if ω is a primitive fourth root of 1, then the series

$$\frac{X}{(1+X^2)^{1/2}}$$

is reversible by ωX, but it is not strongly reversible.   □

We close this section with a result that will be useful when we come to study the diffeomorphism group of the real line.

Corollary 10.42   *If f is a nontrivial element of $\mathscr{G}$ with multiplier 1 that is reversed by an element h of $\mathscr{G}$ with multiplier -1, then h is an involution.*

Proof   Since f has infinite order, Theorem 10.39 implies that h has finite order. By Proposition 10.3, h is conjugate to its leading term, $-X$, and hence has order 2.   □

10.10   Products of reversible series

In this section we establish the following theorem about products of reversible series.

Theorem 10.43

(i)   *Suppose F has no square root of -1. Then*

$$I^\infty = R^2 = R^\infty \neq \mathscr{H},$$

and $f \in R^\infty$ if and only if

$$f \equiv \mu X + aX^2 + \mu a^2 X^3 \pmod{X^4},$$

for some $\mu = \pm 1$ and $a \in F$.

(ii) *Suppose F has a square root of -1. Then*

$$I^\infty \neq R^2 = R^\infty = \mathscr{H}.$$

Proof of Theorem 10.43   First we prove (i). By Corollary 10.41 we have $R = I^2$, and the rest of assertion (i) follows from Theorem 10.36.

Next we prove (ii). Suppose that F has a square root of -1. Fix an element f of $\mathscr{H}$, and let $\mu = m(f)$ and $p = p(f)$. Then up to conjugacy,

$$f \equiv \mu X + aX^{p+1} + bX^{2p+1} \pmod{X^{2p+2}},$$

where p is even if $\mu = -1$. We split the remainder of the argument into three cases.

In the first case we assume that $p = 1$. By Theorem 10.38, the series

$$g = \mu X + aX^2 + \mu a^2 X^3 \quad \text{and} \quad h = X + cX^3 + \frac{3c^2}{2}X^5$$

are reversible, so by a suitable choice of c we obtain

$$g \circ h \equiv f \pmod{X^4}.$$

Therefore $f \in R^2$.

In the second case we assume that $p \geqslant 2$. In this case, $f \equiv \pm X \pmod{X^4}$, so, by Theorem 10.36, $f \in I^4 \subset R^2$.

In the third case we assume that $p = 2$, so

$$f \equiv \mu X + aX^3 + bX^5 \pmod{X^6}.$$

Take $g = \mu X + aX^3 + X^4 + cX^5$. We can choose $c \in F$ to make $g \in R$, by Proposition 10.8(ii) and Theorem 10.38. Thus

$$g^{\circ -1} \circ f \equiv X - \mu X^4 + dX^5 \pmod{X^6},$$

for some $d \in F$. Applying Proposition 10.8 and Theorem 10.38 again, we may choose $e \in F$ such that

$$h = X - \mu X^4 + dX^5 + eX^7$$

belongs to R. Then

$$h^{\circ -1} \circ g^{\circ -1} \circ f \equiv X \pmod{X^6}.$$

Therefore $f \equiv g \circ h \pmod{X^6}$, so f is conjugate to $g \circ h$, and so $f \in R^2$.

Thus we have shown that in all cases $R^2 = \mathscr{H}$, and the rest of the assertions follow from Theorem 10.36. $\square$

Corollary 10.44   *The composition of any number of reversible series is the composition of two.* $\square$

Notes

Sources

Material for this chapter was drawn in part from the work of Baker, Cartan, Chen, Kasner, Liverpool, and Lubin. See [24, 49, 52, 123, 144, 171, 172, 189].

Theorem 10.7 is apparently due to Kasner (Voronin attributes the formal normal form $z + z^p + \lambda z^{2p+1}$ in the complex case to an N. Venkov, without a reference. Kasner's work was published in 1915–1916. This normal form has since been rediscovered by many.)

The authors would also like to thank Alexander Stanoyevich and John Murray for useful conversations about the material.

Group automorphisms

A *field automorphism* of $\mathscr{G}$ is an automorphism of $\mathscr{G}$ as a group that arises from an automorphism of the underlying field F. That is, given an automorphism ϕ of F, the corresponding field automorphism of $\mathscr{G}$ is given by

$$a_1 X + a_2 X^2 + \cdots \quad \mapsto \quad \phi(a_1)X + \phi(a_2)X^2 + \cdots .$$

Muckenhoupt [185, Theorem 1] proved that every automorphism of $\mathscr{G}$ is the composition of an inner automorphism and a field automorphism. For simplicity, we now assume that $F = \mathbb{C}$. The only *continuous* field automorphisms of $\mathbb{C}$ are the identity map and complex conjugation. All other field automorphisms are called *wild automorphisms*. It is known (see, for example, [249]) that if ϕ is a wild automorphism of $\mathbb{C}$, then ϕ fixes pointwise a dense subset of $\mathbb{R}$, but $\phi(\mathbb{R})$ is dense in $\mathbb{C}$.

If we discard the wild automorphisms, then we are led to consider the group $\mathscr{G}^*$ consisting of series of one of the types

$$\sum_{n=1}^{\infty} a_n z^n \quad \text{or} \quad \sum_{n=1}^{\infty} a_n \bar{z}^n,$$

where $a_1 \neq 0$. Clearly, this group contains $\mathscr{G}$ as an index-two subgroup. It has more involutions than just those in $\mathscr{G}$, as it has 'anticonformal reflections' such as $\bar{z}$. Kasner [144] showed that the subgroup of $\mathscr{G}^*$ generated by involutions consists of the collection of those series with $|a_n| = 1$, called *unimodular series*. In fact, he showed that each unimodular series can be expressed as a product of four involutions in $\mathscr{G}^*$.

Algebra automorphisms

The group $\mathscr{G}$ of this chapter is isomorphic to the group of algebra automorphisms of the algebra $F[[X]]$ of formal power series. An element $g \in \mathscr{G}$ acts on $F[[X]]$ by sending a series $f \in F[[X]]$ to $f \circ g$. Given an algebra automorphism $\Phi : F[[X]] \to F[[X]]$, the corresponding g is just $\Phi(\mathbb{1})$.

Products of involutions modulo X^{k+1}

For the quotient groups modulo X^{k+1} we have the following theorem, proven in [189].

Theorem 10.45

$$I_k^\infty \subset \begin{cases} I_k^2 & \text{if } 1 \leqslant k \leqslant 6 \\ I_k^4 & \text{if } k \geqslant 7 \end{cases}$$

$$-I_k^\infty \subset \begin{cases} I_k & \text{if } 1 \leqslant k \leqslant 4 \\ I_k^3 & \text{if } k \geqslant 5 \end{cases}$$

$$I_k^\infty = \begin{cases} I_k^2 & \text{if } 1 \leqslant k \leqslant 4 \\ I_k^3 & \text{if } 5 \leqslant k \leqslant 6 \\ I_k^4 & \text{if } k \geqslant 7 \end{cases}$$

Open problems

Other coefficient rings

One may consider formally-invertible series over any commutative ring A with identity, and there is a good deal to be done to answer the corresponding problems in that generality. Lubin [172] provides much of the necessary foundation for this study in case A is an integral domain of characteristic zero. The questions for the Nottingham groups (in which A is a finite field) appear to be unexplored so far, although the elements of finite order are well understood (at least when the order of A is odd), thanks to Klopsch [46, page 16]. The corresponding problems for convergent series over a general complete metric field also appear to be open.

For the group $\mathbb{Z}[[X]]^{\circ-1}$ of series with integral coefficients that have an inverse with integral coefficients, it is known that all reversible elements are strongly reversible. Of course, reversible series in this group are reversible modulo p for each prime p.

Formal power series in several variables

There are some recent results about formal maps in several variables [194, 193], but much remains to be done. It has been established that when the coefficient field is $\mathbb{C}$, for power series in n variables, where $n \geqslant 2$, we have $I^k = R^m = R^\infty$ for some numbers k and m, which may depend on n. Sharp values of k and m are not known.

The so-called *generic* reversibles have been classified for the two-variables case, but there is not yet a complete solution to the reversibility problem even in this case.

It is not known whether each reversible formal map can be reversed by an element of finite order.

There are no results for formal maps over fields that are not algebraically closed or that have positive characteristic.

11

Real diffeomorphisms

A *diffeomorphism* of the real line $\mathbb{R}$ is an infinitely-differentiable homeomorphism of $\mathbb{R}$, onto $\mathbb{R}$, with nonvanishing derivative. In this chapter, we consider the group $\text{Diffeo}(\mathbb{R})$ of all diffeomorphisms of $\mathbb{R}$, and the subgroup $\text{Diffeo}^+(\mathbb{R})$ of orientation-preserving (order-preserving) diffeomorphisms.

The set of diffeomorphisms of a real interval J onto itself will be denoted by $\text{Diffeo}(J)$. (When J has an endpoint as a member, the derivatives at this point should be interpreted as the appropriate one-sided derivatives.) Similarly, we denote the set of order-preserving diffeomorphisms of J by $\text{Diffeo}^+(J)$. Notice that if J is a half-open interval $[a,b)$, then each map f in $\text{Diffeo}^+(J)$ fixes a, and $\text{Diffeo}(J) = \text{Diffeo}^+(J)$. When J is an interval such as $[a,b)$, we write $\text{Diffeo}[a,b)$, rather than $\text{Diffeo}([a,b))$, for simplicity, and we follow similar conventions for other groups and intervals.

Given a diffeomorphism f, and a real number a, we use the notation $T_a f$ to denote the *truncated* Taylor series

$$T_a f = \sum_{n=1}^{\infty} \frac{f^{(n)}(a)}{n!} X^n,$$

regarded as a (formally-invertible) formal power series in the indeterminate X. Here $f^{(n)}(a)$ denotes the nth derivative of f at a. If g is a diffeomorphism, then one can check that

$$T_a(fg) = (T_{g(a)}f)(T_a g),$$

where the product on the right-hand side is a formal composition of formally-invertible formal power series, the group operation that we studied in the previous chapter. In particular, we see that $T_{f(a)}(f^{-1}) = (T_a f)^{-1}$.

While it is known how to tackle the conjugacy problem in the real homeomorphism group, the same problem in the real diffeomorphism group is much trickier, and less well understood. To explain the difference between the two

216

groups, consider, for example, two order-preserving homeomorphisms of $[0, \infty)$ that both fix 0 and satisfy $f(x) < x$ and $g(x) < x$ for $x > 0$. As we saw in Chapter 8, f and g are conjugate in the group $\text{Homeo}^+[0, \infty)$. On the other hand, if f and g are diffeomorphisms then they need not be conjugate in $\text{Diffeo}^+[0, \infty)$; indeed, $f'(0) = g'(0)$ is a necessary condition for conjugacy in $\text{Diffeo}^+[0, \infty)$. Conversely, if $f'(0) = g'(0)$ *and* $f'(0) \neq 1$, then Sternberg's linearization theorem [220] says that f and g are conjugate in $\text{Diffeo}^+[0, \infty)$. When $f'(0) = 1$, we can use the truncated Taylor series $T_0 f$ of f at 0 to help us determine whether f and g are conjugate as diffeomorphisms. If $f = hgh^{-1}$ for some h in $\text{Diffeo}^+[0, \infty)$, then $T_0 f = (T_0 h)(T_0 g)(T_0 h)^{-1}$. Takens [224] proved that, provided $T_0 f$ is not the identity power series, f and g are conjugate in $\text{Diffeo}^+[0, \infty)$ if and only if $T_0 f$ and $T_0 g$ are conjugate as formal power series. The difficult case, which Takens omitted, is when $T_0 f$ and $T_0 g$ are both the identity series (see the survey [190] for more about this tricky case).

Fortunately, the reversibility problem in $\text{Diffeo}(\mathbb{R})$ can be handled without full understanding of the conjugacy problem. In this chapter we prove only a limited number of results about conjugacy (such as Corollary 11.7, which says that all fixed-point-free diffeomorphisms are conjugate) but we give a reasonably complete account of reversibility.

As usual, we will use the notation $\mathbb{1}$ for the identity element of our group, which in this case is the identity diffeomorphism. In contrast, we denote the identity in the group of formally-invertible formal power series with real coefficients by X, where

$$X = X + 0X^2 + 0X^3 + \cdots.$$

We inherit some concepts and notation from Chapter 8, such as the *signature* of a diffeomorphism g, and the fixed-point set $\text{fix}(g)$ of g. We write $\text{Diffeo}^-(\mathbb{R})$ for the set of orientation-reversing (order-reversing) diffeomorphisms of $\mathbb{R}$. This is the second coset of the subgroup $\text{Diffeo}^+(\mathbb{R})$ of $\text{Diffeo}(\mathbb{R})$. We denote the collection of all infinitely-differentiable real-valued functions on $\mathbb{R}$ by C^∞.

Like the groups $\text{Homeo}(\mathbb{R})$ and $\text{Homeo}(\mathbb{S})$, the group of diffeomorphisms of the real line has trivial centre (by Proposition 2.19), and the only automorphisms of $\text{Diffeo}(\mathbb{R})$ are inner automorphisms [180].

11.1 Involutions

We learned from Proposition 8.1 that the only homeomorphisms of $\mathbb{R}$ of finite order are orientation-reversing involutions and the identity map. Since $\text{Diffeo}(\mathbb{R})$ is a subgroup of $\text{Homeo}(\mathbb{R})$, the same is true of diffeomorphisms;

that is, the only nontrivial diffeomorphisms of finite order are orientation-reversing involutions. There are many nontrivial involutions in $\mathrm{Diffeo}(\mathbb{R})$, but, as in the homeomorphism group and the formal series group, all are conjugate.

Proposition 11.1   *Each nontrivial involution in* $\mathrm{Diffeo}(\mathbb{R})$ *is conjugate to the map* $x \mapsto -x$.

Proof   We can use the same proof as that of Proposition 8.2. That is, if τ is an orientation-reversing involution, then we define $h(x) = x - \tau(x)$. The map h is an orientation-preserving diffeomorphism and $h\tau(x) = -h(x)$ for each real number x. $\square$

11.2 Fixed-point-free maps

In this section we shall show that all fixed-point-free diffeomorphisms are conjugate. To prove this fact, we need an interpolation result for diffeomorphisms, Proposition 11.5, to follow.

We begin with a well-known theorem. There are many simple proofs – see, for example, [182]. In this theorem we denote by $\mathbb{R}[[X]]$ the ring of formal power series in the variable X over $\mathbb{R}$.

Theorem 11.2 (É. Borel)   *Given any formal power series S in $\mathbb{R}[[X]]$ and any point $a \in \mathbb{R}$, there exists an infinitely-differentiable function $f : \mathbb{R} \to \mathbb{R}$ with Taylor series at a equal to S.* $\square$

The next lemma should also be familiar.

Lemma 11.3   *There exists a function f in C^∞ such that $f(x) = 0$ when $x \leqslant 0$ and $f(x) > 0$ when $x > 0$.*

Proof   The function

$$f(x) = \begin{cases} 0 & \text{if } x \leqslant 0, \\ \exp\left(-\frac{1}{x}\right) & \text{if } x > 0. \end{cases}$$

has the required properties. $\square$

Corollary 11.4   *Given real numbers $a < b$, there exists a function g in C^∞ such that*

(i) $g(x) = 0$ *when* $x \leqslant a$
(ii) $g(x) = 1$ *whenever* $x \geqslant b$
(iii) g *is strictly increasing on the open interval* (a, b).

Proof Let f be the function defined in Lemma 11.3. Define h to be the function in C^∞ given by

$$h(x) = \int_0^x f(t)f(1-t)\,dt.$$

Then the function g given by $g(x) = h(x)/h(1)$ has the required properties. $\square$

Now we can state our interpolation result for diffeomorphisms.

Proposition 11.5 *Let $a_1 < a_2 < \cdots < a_n$ and $b_1 < b_2 < \cdots < b_n$ be real numbers, and let $P_1, P_2, \ldots, P_n$ be formally-invertible formal power series with positive multipliers. Then there is a diffeomorphism f in $\mathrm{Diffeo}^+(\mathbb{R})$ with $f(a_i) = b_i$ and $T_{a_i}f = P_i$, for $i = 1, 2, \ldots, n$.*

Proof It suffices to deal with the cases $n = 1$ and $n = 2$; the more general case can then be handled using an inductive argument.

For the case $n = 1$, choose (by Borel's theorem) a function g in C^∞ with $T_{a_1}g = P_1{}'$, the term-by-term derivative of P_1. Then $g(a_1) > 0$. Choose $\delta > 0$ such that g is positive on the closed interval $(a_1 - 2\delta, a_1 + 2\delta)$. Now we construct what is known as a partition of unity, in this case into three functions. Corollary 11.4 allows us to choose h_1 and h_2 in C^∞ such that both take only values between 0 and 1, and

$$h_1(x) = \begin{cases} 1 & \text{if } x < a_1 - 2\delta, \\ 0 & \text{if } x > a_1 - \delta, \end{cases} \qquad h_2(x) = \begin{cases} 0 & \text{if } x < a_1 + \delta, \\ 1 & \text{if } x > a_1 + 2\delta. \end{cases}$$

Let $h_3(x) = 1 - h_1(x) - h_2(x)$. Then $h_3 \in C^\infty$ and

$$h_3(x) = \begin{cases} 1 & \text{if } a_1 - \delta < x < a_1 + \delta, \\ 0 & \text{if } |x - a_1| > 2\delta. \end{cases}$$

Let f_1 be given by $f_1(x) = h_1(x) + h_2(x) + h_3(x)g(x)$. Then f_1 agrees with g near a_1, and is everywhere positive. Now define

$$f(x) = b_1 + \int_{a_1}^x f_1(t)\,dt.$$

The diffeomorphism f satisfies $f(a_1) = b_1$ and $T_{a_1}f = P_1$.

The case $n = 2$ may be proved in a similar way. One uses a partition of unity into five maps h_i instead of three, and adjusts the middle one to make sure that

$$\int_{a_1}^{a_2} f_1(t)\,dt = b_2 - b_1.$$ $\square$

Proposition 11.6 *Each fixed-point-free member of $\mathrm{Diffeo}^+(\mathbb{R})$ is conjugate either to the map $x \mapsto x+1$ or the map $x \mapsto x-1$.*

Proof Let f be a fixed-point-free member of $\mathrm{Diffeo}^+(\mathbb{R})$. We may suppose that $f(x) > x$ for all real numbers x. Choose any increasing diffeomorphism $k : [0, f(0)] \to [0, 1]$ such that $T_0 k = X$ and $T_{f(0)} k = (T_0 f)^{-1}$.

Define $h : \mathbb{R} \to \mathbb{R}$ by

$$h(x) = n + k f^{-n}(x), \quad x \in [f^n(0), f^{n+1}(0)],$$

for each integer n. Then h is well defined at the endpoints $f^n(0)$, and is a homeomorphism of $\mathbb{R}$ onto itself, mapping $[f^n(0), f^{n+1}(0)]$ onto $[n, n+1]$ for each integer n. Moreover, $h(f(x)) = h(x) + 1$ for all real x, so that h conjugates f to the translation $x \mapsto x + 1$.

Finally, the condition on the Taylor series ensures that the left-hand and right-hand Taylor series agree at each point $f^n(0)$, so h is a diffeomorphism. $\square$

We note that $x \mapsto x + 1$ and $x \mapsto x - 1$ are not conjugate in $\mathrm{Diffeo}^+(\mathbb{R})$, but they are conjugate by the map $x \mapsto -x$ in $\mathrm{Diffeo}(\mathbb{R})$.

Corollary 11.7 *Each fixed-point-free member of* $\mathrm{Diffeo}(\mathbb{R})$ *is conjugate to the map* $x \mapsto x + 1$. $\square$

11.3 Centralisers

The centraliser of a real homeomorphism is usually enormous, but this is not so for real diffeomorphisms. Kopell was the first to realise this. Essentially, for a map f in $\mathrm{Diffeo}[a, b)$ that fixes only a, the centraliser of f is just a one-parameter group. In this section we prove this result and some other facts about centralisers in $\mathrm{Diffeo}[a, b)$.

Before proving Kopell's observation, we need a preliminary observation.

Proposition 11.8 *Let* $a < b$ *be real numbers. Suppose that* g *is a* C^1 *order-preserving homeomorphism of the interval* $[a, b]$ *that fixes* a *and* b. *Suppose there exists* $M > 0$ *such that*

$$(g^n)'(x) \leqslant M, \quad x \in [a, b],$$

for each positive integer n. *Then* $g(x) = x$ *for* x *in* $[a, b]$.

Proof Suppose, in order to reach a contradiction, that $g(x) \neq x$ for some real x in $[a, b]$. Let (c, d) be one of the connected components of the open set

$$\{x \in [a, b] : g(x) \neq x\}.$$

Choose two points $a_0 < b_0$ of (c, d). Let $a_n = g^{-n}(a_0)$ and $b_n = g^{-n}(b_0)$.

There are two cases to consider. In the first case, $g(x) > x$ throughout (c,d). Then $c < a_n < b_n$ and $b_n \to c$ as $n \to \infty$. It follows that $b_n - a_n \to 0$. Therefore

$$0 < b_0 - a_0 = g^n(b_n) - g^n(a_n) \leqslant M(b_n - a_n) \to 0,$$

which is a contradiction.

In the second case, $g(x) < x$ throughout (c,d), and we obtain another contradiction in a similar way. $\square$

Theorem 11.9 (Kopell's lemma)   *Suppose that f and g are C^2 order-preserving homeomorphisms of an interval $[a,b)$ onto itself with nonvanishing derivatives such that $fg = gf$. If f has no fixed points in (a,b), but g has a fixed point in (a,b), then g is the identity map.*

Proof   By conjugating, we may take $a = 0$ and $b = +\infty$. Replacing either or both of f and g by their inverses, if need be, we may assume that $f(x) < x$ and $g(x) < x$ on $(0, +\infty)$. Suppose g fixes the point p, and define $p_n = f^n(p)$. The sequence $p_1, p_2, \ldots$ decreases strictly monotonically to 0. One sees inductively that g fixes each point p_n. Thus by the mean-value theorem there exists a point q_n between p_n and p_{n+1} with $g'(q_n) = 1$, and we conclude by taking limits that $g'(0) = 1$.

Let

$$\delta = \int_0^p \left| \frac{f''(t)}{f'(t)} \right| dt$$

(the variation of $\log f'$ on the interval $[0,p]$). Then whenever $u, v \in [f(p), p]$, we have

$$\left| \log \frac{(f^n)'(v)}{(f^n)'(u)} \right| \leqslant \sum_{i=1}^{n} \left| \log f'(f^{i-1}(v)) - \log f'(f^{i-1}(u)) \right| \leqslant \delta.$$

Choose any point x in $[f(p), p]$. Let $u = x$ and $v = g(x)$. Since v is also equal to $f^{-n} g f^n(x)$ (because f and g commute) we obtain

$$g'(x) = \frac{(f^n)'(x)}{(f^n)'(f^{-n} g f^n(x))} g'(f^n(x))$$

$$= \frac{(f^n)'(x)}{(f^n)'(g(x))} g'(f^n(x)).$$

Therefore

$$g'(x) \leqslant e^\delta g'(f^n(x)),$$

and letting $n \to \infty$ we see that $g'(x) \leqslant e^\delta$.

We can apply this same argument to an iterate g^n, where $n \in \mathbb{N}$, in place of g. This gives

$$(g^n)'(x) \leqslant e^\delta$$

for each positive integer n and $x \in [f(p), p]$. Then Proposition 11.8 shows that $g(x) = x$ for x in $[f(p), p]$, and since f and g commute we deduce that g is the identity map on $[0, +\infty)$. $\qquad\square$

Corollary 11.10 *Suppose that an element f of* Diffeo$[a, b)$ *fixes the point a, but fixes no other points, and suppose that elements h and k of the centraliser of f in* Diffeo$[a, b)$ *have the same value at a point in (a, b). Then $h = k$.*

Proof   This follows immediately from Kopell's lemma, applied with $g = hk^{-1}$. $\qquad\square$

This corollary shows that the centraliser of such a diffeomorphism f is at most a one-parameter group. In fact, the elements g of the centraliser C_f are parametrised by the values $g(c)$ at any one point $c \in (a, b)$. One can exploit this to show that the centraliser is abelian. Kopell actually showed that, generically, the centraliser consists simply of the infinite cyclic group generated by f.

Proposition 11.11 *Suppose that f and g are elements of* Diffeo$^+[a, b)$ *that both fix a and commute. If $T_a f = X$ and f is not the identity in a neighbourhood of a, then $T_a g = X$ also.*

Proof   Suppose that

$$T_a g = X + \alpha X^{p+1} \quad (\mathrm{mod}\ X^{p+2}),$$

with $\alpha \neq 0$. Then a is an isolated fixed point of g. If a is not an isolated fixed point of f, then Kopell's lemma tells us that f is the identity near a, a contradiction. Thus a is an isolated fixed point of f too. By changing b, if need be, we can assume that both f and g are fixed-point free on (a, b). Furthermore, we can assume that $f(x) < x$ and $g(x) < x$ on (a, b), because if either condition is violated we can replace the function by its inverse.

   Choose a point c in (a, b) with $f(c) < g(c)$; this is possible by Taylor's theorem. Since $g^n(c) \to 0$ as $n \to \infty$, we can choose a positive integer n with $g^n(c) < f(c)$. Note that

$$T_a g^n = X + n\alpha X^{p+1} \quad (\mathrm{mod}\ X^{p+2}).$$

None of the equations $g^n(x) = f(x)$ has a solution in (a, b). This is because such a solution would be a fixed point of $f^{-1} g^n$, and this map commutes with

g, so Kopell's lemma would yield g^n identically equal to f on (a,x), which is false, as g^n and f have different Taylor series at a. It follows that

$$g^n(x) < f(x) < g(x), \quad x \in (a,c).$$

But this implies that $T_a f$ has a nonzero coefficient of X^{p+1}, a final contradiction. $\qquad\square$

11.4 Reversibility of order-preserving diffeomorphisms

The next proposition shows that there are far fewer reversible maps in the group or order-preserving diffeomorphisms than there are in the group of order-preserving homeomorphisms. The proof uses Kopell's lemma, and the fact that the square of a reverser of a diffeomorphism lies in the centraliser of the diffeomorphism.

Proposition 11.12 *Suppose that f and h are order-preserving diffeomorphisms of $\mathbb{R}$ such that $hfh^{-1} = f^{-1}$. If h has a fixed point, then f is the identity map.*

Proof If h is the identity map, then f is an order-preserving involution, and therefore f is also the identity map. Suppose then that h is not the identity map, but that it nevertheless has a fixed point. Choose a component (a,b) in the complement of fix(h). One of a or b is a real number (that is, we cannot have both $a = -\infty$ and $b = +\infty$). Let us assume that a is a real number; the other case can be dealt with similarly. By Lemma 8.19, f fixes (a,b) as a set. The map f cannot be free of fixed points on (a,b). To see this, suppose, by switching f and f^{-1} if necessary, that $f(x) > x$ for each real number x in (a,b). Then

$$x > f^{-1}(x) = hf(h^{-1}(x)) > hh^{-1}(x) = x,$$

which is a contradiction. Since f has a fixed point in (a,b), Kopell's lemma, Lemma 11.9, applied to the maps f and h^2 shows that f coincides with the identity map on (a,b). We already know that f coincides with the identity map on fix(h); thus f is the identity map. $\qquad\square$

We are now in a position to prove the first main result of the chapter, characterising the reversible elements of Diffeo$^+(\mathbb{R})$.

Theorem 11.13 *Each reversible element of* Diffeo$^+(\mathbb{R})$ *is conjugate to a map f that fixes each integer and satisfies*

$$f(x+1) = f^{-1}(x) + 1, \quad x \in \mathbb{R}.$$

Proof Suppose that f is a reversible element of $\mathrm{Diffeo}^+(\mathbb{R})$. We may assume that f is not the identity, since that case is trivial.

By Proposition 11.12, only a fixed-point-free map h can satisfy the equation $hfh^{-1} = f^{-1}$. All such maps h are, by Proposition 11.6, conjugate to either $x \mapsto x+1$ or $x \mapsto x-1$. Applying the same conjugacy to f yields a function that is reversed by $x \mapsto x+1$. Such a map must have a fixed point in each interval of length 1, and by conjugating by a translation we can ensure that 0 is fixed, and hence each integer is fixed. $\square$

The converse of this theorem is obviously true, so from this we obtain an explicit way to construct all the reversible elements of $\mathrm{Diffeo}^+(\mathbb{R})$.

Corollary 11.14 *Let $f \in \mathrm{Diffeo}^+(\mathbb{R})$. The following are equivalent.*

(i) *The map f is reversible in $\mathrm{Diffeo}^+(\mathbb{R})$.*

(ii) *There exist*

  (a) *a formally-invertible formal power series P*

  (b) *an order-preserving diffeomorphism ϕ of $[0,1]$ to itself, with*

$$T_0\phi = P, \qquad T_1\phi = P^{-1}$$

  (c) *a map h in $\mathrm{Diffeo}^+(\mathbb{R})$, such that $f = hgh^{-1}$, where*

$$g(x) = \begin{cases} \phi(x-2n)+2n & \text{if } 2n \leqslant x < 2n+1, \\ \phi^{-1}(x-2n-1)+2n+1 & \text{if } 2n+1 \leqslant x < 2n+2, \end{cases}$$

  for each integer n. $\square$

Each map g of part (ii) commutes with $x \mapsto x+2$, and is reversed by $x \mapsto x+1$. Therefore g is the lift under the covering map $x \mapsto \exp(\pi ix)$ of a reversible orientation-preserving diffeomorphism of the unit circle that is reversed by the rotation by π.

Let us now consider products of reversible maps in $\mathrm{Diffeo}^+(\mathbb{R})$. We will show that each orientation-preserving diffeomorphism can be expressed as a product of four reversible elements of $\mathrm{Diffeo}^+(\mathbb{R})$. We do not know whether two or three reversible diffeomorphisms will suffice.

Proposition 11.15 *Each fixed-point-free element of $\mathrm{Diffeo}^+(\mathbb{R})$ can be expressed as a composite of two reversible elements of $\mathrm{Diffeo}^+(\mathbb{R})$.*

Proof By Proposition 11.6 it suffices to find a single fixed-point-free map that can be expressed as a composite of two reversible diffeomorphisms. Let f be a reversible order-preserving diffeomorphism such that $f(x+1) = f^{-1}(x) + 1$

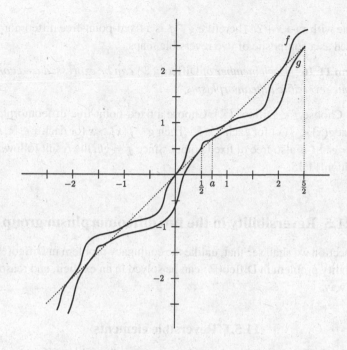

Figure 11.1 The graphs of two reversible diffeomorphisms

for each real number x, and $f(y) > y$ for each element y of $(0,1)$. The graph of such a map f is shown in Figure 11.1.

Let a be an element from the interval $\left(\frac{1}{2}, f\left(\frac{1}{2}\right)\right)$. Notice that every order-preserving diffeomorphism h of $\left[\frac{1}{2}, a\right]$ satisfies $h(x) < f(x)$ for $x \in \left[\frac{1}{2}, a\right]$. Choose an order-preserving diffeomorphism g of $\left[a, \frac{5}{2}\right]$ such that $T_a g = T_{\frac{5}{2}} g = X$, and such that $g(x) < f(x)$ for each $x \in \left[a, \frac{5}{2}\right]$. (This construction is possible by Proposition 11.5.)

Next, choose an order-preserving diffeomorphism k from $\left[\frac{1}{2}, a\right]$ to $\left[a, \frac{5}{2}\right]$ such that $T_{\frac{1}{2}} k = T_a k = X$.

We extend the definition of g to $\mathbb{R}$ by defining $g(x) = k^{-1} g^{-1} k(x)$ for $x \in \left[\frac{1}{2}, a\right]$, and $g(x+2) = g(x) + 2$ for all $x \in \mathbb{R}$. We extend the definition of k by defining $k(x) = k^{-1}(x) + 2$ for $x \in \left[a, \frac{5}{2}\right]$ and $k(x+2) = k(x) + 2$ for all $x \in \mathbb{R}$. The resulting maps g and k are both order-preserving diffeomorphisms. Moreover, one can check that the equation $g(x) = k^{-1} g^{-1} k(x)$ is satisfied for points x in $\left[\frac{1}{2}, \frac{5}{2}\right]$. Since both maps commute with $x \mapsto x+2$, this equation is satisfied throughout $\mathbb{R}$. Finally, we have defined g such that $f(x) > g(x)$ for elements x of $\left[\frac{1}{2}, \frac{5}{2}\right]$, and in fact $f(x) > g(x)$ everywhere, again, because both maps commute with $x \mapsto x + 2$. Therefore $g^{-1}f$ is a fixed-point-free diffeomorphism expressed as a composite of two reversible maps.   □

Theorem 11.16 *Each member of* $\mathrm{Diffeo}^+(\mathbb{R})$ *can be expressed as a composite of four reversible diffeomorphisms.*

Proof Choose $f \in \mathrm{Diffeo}^+(\mathbb{R})$. Choose a fixed-point-free diffeomorphism g such that $g(x) < f(x)$ for each $x \in \mathbb{R}$. Then $g^{-1}f(x) > x$ for each $x \in \mathbb{R}$, so the map $h = g^{-1}f$ is also free of fixed points. Since $f = gh$, the result follows from Proposition 11.15.   □

11.5 Reversibility in the full diffeomorphism group

In this section we shall see that, unlike the conjugacy problem in $\mathrm{Diffeo}(\mathbb{R})$, the reversibility problem in $\mathrm{Diffeo}(\mathbb{R})$ can be solved in an explicit, and reasonably simple way.

11.5.1 Reversible elements

Let us begin by classifying the order-reversing reversible diffeomorphisms. The next result shows again how much tighter the situation is for diffeomorphisms, as opposed to homeomorphisms.

Theorem 11.17 *An order-reversing member of* $\mathrm{Diffeo}(\mathbb{R})$ *is reversible in* $\mathrm{Diffeo}(\mathbb{R})$ *if and only if it is an involution.*

Proof Involutions are all reversible by the identity map. Conversely, suppose that $f \in \mathrm{Diffeo}^-(\mathbb{R})$, $h \in \mathrm{Diffeo}(\mathbb{R})$, and $hfh^{-1} = f^{-1}$. By replacing h with hf if necessary, we can assume that h preserves order. Corollary 2.18 tells us that h fixes the unique fixed point of f (this is easily verified without referring to that corollary). Finally, $hf^2h^{-1} = f^{-2}$, and f^2 preserves order, therefore Proposition 11.12 applies to show that f^2 is the identity map, as required.   □

We now turn to order-preserving reversible diffeomorphisms. In the previous section we considered order-preserving diffeomorphisms that are reversed by other order-preserving diffeomorphisms. That leaves order-preserving diffeomorphisms that are reversed by order-*reversing* diffeomorphisms.

Proposition 11.18 *Fixed-point-free diffeomorphisms are strongly reversible.*

Proof A fixed-point-free diffeomorphism is, by Proposition 11.6, conjugate in the group $\mathrm{Diffeo}(\mathbb{R})$ to $x \mapsto x + 1$, and this map is reversed by the involution $x \mapsto -x$.   □

Theorem 11.19 *If an element of* $\mathrm{Diffeo}^+(\mathbb{R})$ *is reversed by an element of* $\mathrm{Diffeo}^-(\mathbb{R})$, *then it is strongly reversible.*

Proof Let $f \in \mathrm{Diffeo}^+(\mathbb{R})$, $h \in \mathrm{Diffeo}^-(\mathbb{R})$, and $hfh^{-1} = f^{-1}$. We wish to show that f is strongly reversible. Given Proposition 11.18, we can assume that f has a fixed point. By conjugation, we can assume that the fixed point of h is 0. We define an involutive homeomorphism k by

$$k(x) \doteq \begin{cases} h^{-1}(x) & \text{if } x \leqslant 0, \\ h(x) & \text{if } x > 0. \end{cases}$$

Suppose that 0 is a fixed point of f; we consider the alternative that 0 is not a fixed point of f later. Clearly, $kfk = f^{-1}$; however, k may not be a diffeomorphism. Note that $f'(0) = 1$, because $f'(0) = (f^{-1})'(0)$. We consider three cases.

First, suppose that f coincides with the identity on a neighbourhood of 0. In this case we have freedom to adjust the definition of k near 0 so that it is an involutive diffeomorphism, without disturbing the validity of the equation $kfk = f^{-1}$.

Second, suppose that f does not coincide with the identity near 0, but $T_0 f = X$. Since h^2 commutes with f, it follows from Proposition 11.11 that $T_0 h$ is an involution, so that k is already a diffeomorphism.

Third, suppose that 0 is a fixed point of f and $T_0 f \neq X$. From the equation

$$(T_0 h)(T_0 f)(T_0 h)^{-1} = T_0 f^{-1}$$

we see that the formally-invertible formal powers series $T_0 f$ is reversed by $T_0 h$. By Corollary 10.42, $T_0 h$ is an involution, and again we see that k is a diffeomorphism.

Suppose now that 0 is not a fixed point of f; instead 0 lies inside a component (a,b) of $\mathbb{R} \setminus \mathrm{fix}(f)$. This component (a,b) is not the whole real line, because f has a fixed point. Lemma 2.17 tells us that h permutes the fixed points of f, so it also permutes the components of $\mathbb{R} \setminus \mathrm{fix}(f)$. Since $0 \in (a,b)$, we deduce that h fixes (a,b) as a set. Therefore (a,b) is a finite-width interval, and h interchanges a and b. By Kopell's lemma, Lemma 11.9, $h^2(x) = x$ for each real x in $[a,b]$. This implies that h and h^{-1} coincide inside (a,b), so that k is a diffeomorphism, and $kfk^{-1} = f^{-1}$. $\qquad\square$

Since all nontrivial involutions in $\mathrm{Diffeo}(\mathbb{R})$ are conjugate to $x \mapsto -x$ we have the following explicit method for constructing all the strongly-reversible elements of $\mathrm{Diffeo}(\mathbb{R})$ that are not involutions.

Corollary 11.20 *Let* $f \in \mathrm{Diffeo}^+(\mathbb{R})$. *The following are equivalent.*

(i) *The map f is reversible in* Diffeo$(\mathbb{R})$ *by an order-reversing diffeomorphism.*

(ii) *There exist*

   (a) *a point p and an order-preserving diffeomorphism $\phi : [p,\infty) \to [-p,\infty)$ such that $T_p\phi$ is strongly reversible in the group of formally-invertible formal power series by the power series $-X$*

   (b) *a map h in* Diffeo$(\mathbb{R})$ *such that $f = hgh^{-1}$, where*

$$g(x) = \begin{cases} -\phi^{-1}(-x) & \text{if } x \leqslant p, \\ \phi(x) & \text{if } x > p. \end{cases}$$

$\square$

Theorem 11.19 shows that elements of Diffeo$^+(\mathbb{R})$ that are reversed by order-*reversing* diffeomorphisms are strongly reversible in Diffeo$(\mathbb{R})$. There are, however, elements of Diffeo$^+(\mathbb{R})$ that are reversed by order-*preserving* diffeomorphisms that are not strongly reversible in Diffeo$(\mathbb{R})$. In fact, for order-preserving diffeomorphisms, just as for order-preserving homeomorphisms, the properties of being reversible and strongly reversible in Diffeo$(\mathbb{R})$ are logically independent. To demonstrate this, we give examples of order-preserving diffeomorphisms that are

(a) reversible by an order-preserving map but not by an order-reversing map
(b) reversible by an order-reversing map but not by an order-preserving map
(c) reversible by both order-preserving and order-reversing maps.

We should also find an order-preserving diffeomorphism that is not reversible at all; there are plenty of such maps.

First we construct a diffeomorphism f of type (a). We choose f such that $\text{fix}(f) = \mathbb{Z}$. To specify f up to topological conjugacy, we have only to describe the signature of f on $\mathbb{R} \setminus \mathbb{Z}$. On the intervals $(0,1), (1,2), \ldots, (5,6)$ we choose f such that $s(f)$ takes the values

$$+1, +1, +1, -1, -1, +1,$$

respectively. We define f elsewhere using the equation

$$f(x+6) = f^{-1}(x) + 6.$$

We can choose f to be a diffeomorphism. A portion of the graph of such a function is shown in Figure 11.2. By definition, f is reversible by the translation $x \mapsto x + 6$. However, it is not reversible by an orientation-reversing map even in the group Homeo$(\mathbb{R})$ because the pattern of signs is different read forwards than it is read backwards.

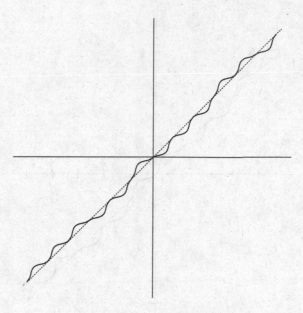

Figure 11.2 The graph of a reversible diffeomorphism that is not strongly reversible

Examples of type (b) are readily constructed. For instance, choose a nontrivial strongly-reversible diffeomorphism that coincides with the identity map outside a compact set; Theorem 11.13 tells us that such a map cannot be reversible in $\mathrm{Diffeo}^+(\mathbb{R})$.

Finally, we supply a nontrivial example of type (c). Choose any nontrivial function ϕ in $\mathrm{Diffeo}^+[0,1]$, with $T_0\phi = T_1\phi = X$, and define in turn

$$
\begin{aligned}
\tau(x) &= -\phi(x), & 0 \leqslant x \leqslant 1, \\
\tau(x) &= \tau^{-1}(x), & -1 \leqslant x < 0, \\
\tau(x+2) &= -\tau(-x) - 2, & -1 < x \leqslant 1,
\end{aligned}
$$

and extend τ to $\mathbb{R}$ by requiring that

$$
\tau(x+4) = \tau(x) - 4, \qquad x \in \mathbb{R}.
$$

A graph of τ is shown in Figure 11.3. Then τ is an involutive element of $\mathrm{Diffeo}^-(\mathbb{R})$, so that $f = -\tau$ is an element of $\mathrm{Diffeo}^+(\mathbb{R})$ that is strongly reversible in $\mathrm{Diffeo}(\mathbb{R})$. On the other hand, $\tau(x+2) = -\tau(-x) - 2$ for all real numbers x, so

$$
f(x+2) = -\tau(x+2) = \tau(-x) + 2 = f^{-1}(x) + 2.
$$

Hence f is also reversible in $\mathrm{Diffeo}^+(\mathbb{R})$.

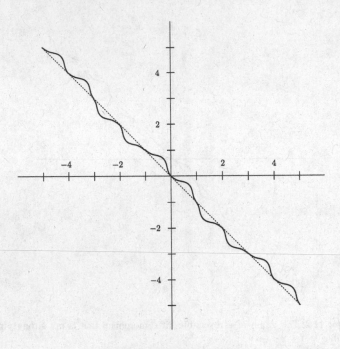

Figure 11.3 The graph of a diffeomorphism τ such that $-\tau$ is reversible by both order-preserving and order-reversing maps

11.6 Products of involutions and reversible elements

We finish with a selection of results that are similar to the equivalent results for real homeomorphisms.

Proposition 11.21   *Each element of* $\mathrm{Diffeo}^-(\mathbb{R})$ *can be expressed as a composite of three involutions in* $\mathrm{Diffeo}(\mathbb{R})$.

Proof   Given an element f of $\mathrm{Diffeo}^-(\mathbb{R})$, choose an involution τ such that $\tau(x) > f(x)$ for each real number x. Then $\tau f(x) > x$. The map τf is strongly reversible, by Proposition 11.18. Therefore f can be expressed as a composite of three involutions.   □

Theorem 11.22   *Each element of* $\mathrm{Diffeo}^+(\mathbb{R})$ *can be expressed as a composite of four involutions in* $\mathrm{Diffeo}(\mathbb{R})$.

Proof   Given an element f in $\mathrm{Diffeo}^+(\mathbb{R})$, the map $-f$, an order-reversing diffeomorphism, can be expressed as a composite of three involutions. The result follows immediately.   □

Corollary 11.23 *Each element of* Diffeo($\mathbb{R}$) *can be expressed as a composite of two reversible maps.*

Notes
Sources

Proposition 11.6 is due to Sternberg, [220]. Theorem 11.9 and Proposition 11.11 are due to Kopell, [155] who elucidated the structure of centralisers in the group Diffeo($\mathbb{R}$). The proof we give is due to Navas [187]. An account of the full conjugacy problem in Diffeo($\mathbb{R}$) can be found in [190].

The study of reversibility in Diffeo($\mathbb{R}$) was initiated by Calica [45]. The results from Section 11.4 to the end appeared in [192]. The authors are happy to acknowledge the valuable assistance of É. Ghys, who drew our attention to the importance of Kopell's lemma.

Higher dimensions

There is little work on reversibility in homeomorphism and diffeomorphism groups of higher-dimensional Euclidean spaces and manifolds. We recall the theorem due to Brouwer and Kerékjártó [56], mentioned in the Notes of Chapter 9, which says that a periodic homeomorphism of the closed unit disc, the plane, or the two-dimensional sphere is conjugate to an orthogonal map. In particular, involutions are conjugate to reflections or rotations by π.

In the plane, it is no longer true that fixed-point-free elements are reversible. Fixed-point-free elements remain useful for factorisation though, as the following proposition demonstrates.

Proposition 11.24 *Let $f \in$* Homeo($\mathbb{R}^d$). *If there exists $M > 0$ such that*

$$|f(x) - x| \leqslant M, \quad x \in \mathbb{R}^d,$$

then $f = gh$, where g and h are fixed-point-free homeomorphisms of $\mathbb{R}^d$.

Proof Let $e_1 = (1, 0, \ldots, 0)$, an element of $\mathbb{R}^d$. The maps $g(x) = x - 2Me_1$ and $h(x) = f(x) + 2Me_1$ have the required properties. $\square$

Note that if f is smooth, or piecewise-linear, then so are the maps g and h constructed in the proof. In dimension one, this observation shows that in many groups of maps, each element f such that $f(x) - x$ is bounded may be factored as the product of four involutions, because (often) fixed-point-free maps are conjugate to translations, and hence to the product of two reflections.

However, once we go to two dimensions, it is no longer true that fixed-point-free maps are conjugate to translations in $\text{Homeo}(\mathbb{R}^2)$. Instead, the plane is divided up into so-called Reeb components (or Brouwer cells), in each of which the map is conjugate to a translation of the whole plane. A conjugacy from f to f^{-1} would permute the Reeb components of f. It is not difficult (once you think of doing it) to construct an example in which no such conjugacy is possible. This was first pointed out by Mather. There is a beautiful theory, due to Haefliger, which classifies the nonsingular oriented plane foliations by using simply-connected oriented nonhausdorff one-dimensional manifolds, and the example is most readily constructed by producing such a manifold that does not admit an orientation-reversing homeomorphism.

Polynomial automorphisms

Another group of interest is the group of polynomial automorphisms of the plane with polynomial inverses. Each element g of this group is a bijection of $\mathbb{R}^2$ of the form $g(x,y) = (p(x,y), q(x,y))$, where p and q are polynomials in both variables x and y, and g^{-1} must also be a bijection of the same form. This group includes the well known family of Hénon quadratic maps

$$h_{c,\delta} : (x,y) \mapsto (y, y^2 + c - \delta x), \qquad \delta \neq 0.$$

Notice that δ is the determinant of the Jacobian of $h_{c,\delta}$, and

$$h_{c,\delta}^{-1}(x,y) = \left(\tfrac{1}{\delta} \left(x^2 + c - y \right), x \right).$$

Those Henon maps $h_{c,\delta}$ that preserve area (that is, $|\delta| = 1$) are strongly reversible; for example, $h_{c,1}$ is reversed by $(x,y) \mapsto (y,x)$ and $h_{c,-1}$ is reversed by $(x,y) \mapsto -(y,x)$. Reversibility in groups of polynomial automorphisms has been studied by Gómez and Meiss [103, 104], Baake and Roberts [17, 18], and Jordan, Jordan, and Jordan [142].

Open problems

Composites of reversibles

We saw in Theorem 11.16 that each element of $\text{Diffeo}^+(\mathbb{R})$ can be expressed as a composite of four reversible elements. We do not know whether each element can be expressed as a composite of three, or even two, reversible elements.

Circle diffeomorphisms

The reversibility problem in the group of diffeomorphisms of the circle is open. It is a classic result of Denjoy that an orientation-preserving circle diffeomorphism with irrational rotation number is topologically conjugate to a rotation. This conjugacy need not be smooth though. The reader interested in this problem should consult Navas [187].

Other degrees of differentiability

One could investigate reversibility in the group of n-times continuously differentiable homeomorphisms of the real line, or in the group of real analytic homeomorphisms of the real line.

Higher dimensions

Even in two dimensions, the theory of reversibility is in its infancy, although much is known about special groups of plane diffeomorphisms. An interesting, as yet unexplored case, is that of the full group of quasiconformal mappings on $\mathbb{R}^d$, $d > 1$.

An important class of reversible maps is the family of Taylor–Chirikov *standard maps* [173]. These arose in the first place from a problem in modelling plasma containment. They are reversible homeomorphisms of $\mathbb{R}^2$. They exhibit extremely complex global behaviour, even when they are real-analytic, and this makes it unlikely that any simple-minded general classification of reversibles is possible. However, it is not inconceivable that a reasonable classification of real-analytic map germs might be possible.

12

Biholomorphic germs

This chapter is about the reversible elements in the group G of invertible biholomorphic germs in one variable and some of its subgroups. The theory of reversibility for formally-reversible formal power series in one variable has already been dealt with in Chapter 10. In the present chapter, we shall denote the group of these formal series by $\mathcal{G}$. The group G is a subgroup of $\mathcal{G}$, which implies that reversible biholomorphic germs are formally reversible.

A priori, biholomorphic conjugacy is a much finer relation than formal conjugacy, so one expects that the formal conjugacy class of an element of G will split into many distinct biholomorphic conjugacy classes. This is often the case, and indeed we shall see (in Section 12.6) that there exist germs that are formally reversible, but not biholomorphically reversible.

Let us discuss the groups to be studied in more detail. A *germ* at 0 is an equivalence class of functions under the relation that regards two functions as equivalent if they agree on some neighbourhood of 0. Let S denote the set of those invertible complex holomorphic maps defined on a neighbourhood of 0 that fix 0. The group G of *biholomorphic germs* consists of the equivalence classes of S under the equivalence relation just described. Thus an element of G is represented by some function f, holomorphic on some neighbourhood of 0 (which depends on f) with $f(0) = 0$ and $f'(0) \neq 0$. Two such functions represent the same germ if they agree on some neighbourhood of 0. The group operation is composition and the identity is the germ of the identity function $z \mapsto z$, which we denote by $\mathbb{1}$.

Next we introduce notation for some important subgroups of G and for some quantities that were used in Chapter 10 to distinguish the conjugacy classes of formal power series. Each element of G can be represented by a convergent complex power series with no constant term and with a nonzero z coefficient. The *multiplier map* $m : az + \cdots \mapsto a$ is a homomorphism from G onto the multiplicative group $\mathbb{C}^{\times}$ of the complex field. Obviously, since $\mathbb{C}^{\times}$ is abelian, the value $m(f)$ depends only on the conjugacy class of f in G. Indeed, as we already know from Chapter 10, $m(f)$ is a formal conjugacy invariant.

Let

$$H = \{f \in G : m(f) = \exp(i\pi q), \text{ for some } q \in \mathbb{Q}\},$$
$$H_0 = \{f \in G : m(f) = \pm 1\},$$
$$G_1 = \ker m.$$

These normal subgroups satisfy $G_1 \lhd H_0 \lhd H \lhd G$.

Formally-reversible elements of $\mathcal{G}$ must have multiplier ± 1. Consequently, in any subgroup of G, the reversible elements must lie in H_0.

For each integer $p > 1$, we define

$$G_p = \{f \in G_1 : f^{(k)}(0) = 0 \text{ whenever } 2 \leqslant k \leqslant p\}, \quad A_p = G_p \setminus G_{p+1}.$$

Then G_1 is the disjoint union of $\{\mathbb{1}\}$ and the sets A_p. Given a nontrivial element f of G_1 we denote by $p(f)$ the unique integer p such that $f \in A_p$. The natural number $p(f)$ is a formal conjugacy invariant of f (with respect to conjugation in G), so that each G_p is a normal subgroup of G.

Given an element f of G_p, we may write $f(z) = z + f_{p+1}z^{p+1} + \mathrm{O}(z^{p+2})$. The map $f \mapsto f_{p+1}$ is a group homomorphism from G_p onto $(\mathbb{C}, +)$. Thus f_{p+1} is a conjugacy invariant of f in G_p. It is even invariant under conjugation in G_1, but it is not invariant under conjugation in G. Each element f of A_p may be conjugated to the form $z + z^{p+1} + a(f)z^{2p+1} + \mathrm{O}(z^{2p+2})$, and then the complex number $a(f)$ is a conjugacy invariant of f in G.

Also, it is always true that for f in G_1, $p(f) = p(f^{-1})$, and we saw in Theorem 10.38 that, for purely formal reasons, the condition $a(f) = a(f^{-1})$ is equivalent to $a(f) = (p(f) + 1)/2$.

The invariants $p(f)$ and $a(f)$ classify the elements of G_1 up to formal conjugacy. The complete biholomorphic conjugacy classification requires additional invariants, and these have been provided by the equivalence class of the EV data $\Phi(f)$ of Écalle–Voronin theory, which is reviewed briefly in Section 12.3 below.

12.1 Elements of finite order

The elements of finite order in $\mathcal{G}$ were classified in Proposition 10.3. A similar result, with a similar proof, applies in the group of biholomorphic germs.

Proposition 12.1 *Suppose that an element g of G has finite order s. Then the multiplier $m(g)$ is an sth root of unity, and g is conjugate in H to $z \mapsto m(g)z$.*

Proof Since $g^s = 1$ it follows that $m(g)^s = 1$, so $m(g)$ is an sth root of unity. Let $\beta = m(g)$ and

$$h(z) = \frac{1}{s}\left(z + \frac{g(z)}{\beta} + \cdots + \frac{g^{s-1}(z)}{\beta^{s-1}}\right).$$

Then $h \in H$ and $h(g(z)) = \beta h(z)$. Therefore $hgh^{-1}(z) = \beta z$. $\square$

Corollary 12.2 *Every nontrivial involution in G is conjugate to the germ $z \mapsto -z$.* $\square$

12.2 Conjugacy

The most basic tool in understanding the conjugacy classes of biholomorphic germs is the multiplier function m. If a biholomorphic germ f is reversible, then $m(f) = m(f^{-1})$. Since m is a homomorphism it follows that

$$m(f)^2 = m(f)m(f^{-1}) = m(ff^{-1}) = 1,$$

so $m(f) = \pm 1$. For this reason we focus our attention on biholomorphic germs with multiplier ± 1; however, let us briefly review the conjugacy classification when $m(f) \neq \pm 1$ (which can be found in more detail in [48]). If $|m(f)| \neq 1$ then f is conjugate to the map $z \mapsto m(f)z$. This is also true if $m(f) = e^{2\pi i\theta}$ and θ is Diophantine (which, informally, means that θ can only be poorly approximated by rational numbers). When θ is neither rational nor Diophantine f may or may not be conjugate to $z \mapsto m(f)z$. The next theorem handles the case when $m(f)$ is a root of unity.

Theorem 12.3 *Suppose that biholomorphic germs f and g have the same multiplier λ, a primitive nth root of unity. Then f and g are conjugate in G if and only if f^n and g^n are conjugate in G.*

Proof It is clear that if f and g are conjugate, then so are f^n and g^n. For the converse, suppose there is an element h of G with $hf^nh^{-1} = g^n$. Let $f_0 = hfh^{-1}$, so that $f_0^n = g^n$. It suffices to show that f_0 and g are conjugate.

Let $k = f_0^n$. If k is the identity, then f_0 and g are periodic with the same multiplier, so, by Proposition 12.1, they are conjugate. If k is not the identity, then, by Corollary 10.22, the centraliser of k in $\mathscr{G}$ is abelian. It follows that the centraliser of k in the subgroup G of $\mathscr{G}$ is abelian. Both f_0 and g belong to this centraliser, hence $(f_0^{-1}g)^n = f_0^{-n}g^n = 1$. But $f^{-1}g$ has multiplier 1, which implies that $f_0^{-1}g = 1$, and $f_0 = g$. $\square$

In fact, we have shown that if $hf^n h^{-1} = g^n$ then $hfh^{-1} = g$ (so the same conjugating function can be used for f^n and g^n, and f and g).

Theorem 12.3 reduces the conjugacy classification of biholomorphic germs with multiplier -1 to the classification of germs with multiplier 1. It follows that the classification of reversible elements is also reduced to the multiplier 1 case.

Corollary 12.4 *A biholomorphic germ f with multiplier -1 is reversible in G if and only if f^2 is reversible in G.* $\square$

To deal with conjugacy of germs with multiplier 1, we turn to a theory developed independently by Écalle and Voronin.

12.3 Écalle–Voronin theory

Let Q denote the collection of those holomorphic functions h defined on a half-plane $\text{Im}[\zeta] > k$ (where k depends on h) such that (i) the function $h(\zeta) - \zeta$ is bounded, and (ii) $h(\zeta + 1) = h(\zeta) + 1$ for all ζ. Let p be a positive integer. *Écalle–Voronin p-data* (or just *EV data*) consist of an ordered $2p$-tuple $\Phi = (\Phi_1, \ldots, \Phi_{2p})$, where $\Phi_1(\zeta), -\Phi_2(-\zeta), \Phi_3(\zeta), \ldots, -\Phi_{2p}(-\zeta) \in Q$.

EV p-data Φ and q-data Ψ are *equivalent* if $p = q$ and there exists an integer k and complex constants $c_1, \ldots, c_{2p}$, such that for each j we have

$$\Phi_{j+2k}(\zeta + c_j) = \Psi_j(\zeta) + c_{j+1}$$

for all values ζ in a half-plane on which both $\Phi_{j+2k}(\zeta + c_j)$ and $\Psi_j(\zeta)$ are defined. (We define Φ_j, Ψ_j, and c_j for all integers j by making them periodic in j, with period $2p$.)

There is a procedure, which we shall outline shortly, that associates Écalle–Voronin data $\Phi(f) = (\Phi_1, \ldots, \Phi_{2p})$ to each element f of G_1. (As usual, G_1 is the group of biholomorphic germs with multiplier 1. Also, A_p is the subset of G_1 of germs f with $p(f) = p$.) One then has the following two fundamental results.

Theorem 12.5 *Let $f, g \in G_1$. Then f is conjugate to g in G if and only if $\Phi(f)$ is equivalent to $\Phi(g)$.* $\square$

Theorem 12.6 *Given any EV data Φ, there exists a function f in A_p with equivalent EV data.* $\square$

These theorems imply that the equivalence classes of EV data are in bijective correspondence with the conjugacy classes in G of the elements of G_1. One says that EV data provide *functional moduli* for these conjugacy classes.

We are not going to prove these two theorems (proofs can be found in [2, pages 7–19]). We shall describe the main ideas in the construction of EV data, without complete justification, but with enough detail to allow the reader to follow the applications we need.

To begin with, for an integer p, consider the special (reversible) germ g given by

$$g(z) = \frac{z}{(1+z^p)^{\frac{1}{p}}},$$

where $|z| < 1$. For each integer j, let T_j be the sector

$$T_j = \left\{ z \in \mathbb{C}^\times : \left| \arg z - \frac{j\pi}{p} \right| < \frac{\pi}{p} \right\}.$$

These sectors are shown for $p = 5$ in Figure 12.1 (of course, there is repetition, such as $T_0 = T_{10}$).

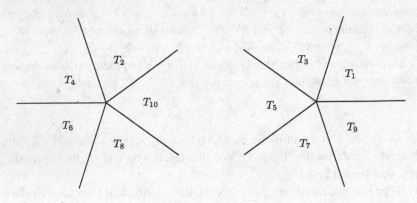

Figure 12.1 The sectors $T_1, \ldots, T_{10}$ for $p = 5$

The transformation $F(z) = 1/z^p$ maps T_{2j} bijectively onto the slit plane $\mathbb{C} \setminus (-\infty, 0]$, and, for points ζ in $\mathbb{C} \setminus (-\infty, 0]$, we have $FgF^{-1}(\zeta) = \zeta + 1$. The transformation $-F$ maps T_{2j-1} bijectively onto the slit plane $\mathbb{C} \setminus (-\infty, 0]$, and $(-F)g(-F)^{-1}(\zeta) = \zeta - 1$ for ζ in $\mathbb{C} \setminus [0, +\infty)$.

For $R > 0$, let

$$S^+(R) = \{ \zeta \in \mathbb{C} : \mathrm{Re}[\zeta] \leqslant R \text{ and } |\mathrm{Im}[\zeta]| \leqslant R \}$$

and $B^+(R) = \mathbb{C} \setminus S^+(R)$. Let P_{2j} be the intersection of T_{2j} and the preimage of

$B^+(R)$ under F; that is

$$P_{2j}(R) = \left\{ z \in \mathbb{C} : \left| \arg z - \frac{2j\pi}{p} \right| < \frac{\pi}{p} \text{ and } \frac{1}{z^p} \in B^+(R) \right\}.$$

The translation $\tau(\zeta) = \zeta + 1$ maps the set $B^+(R)$ to itself, and hence g maps P_{2j} into itself. We call P_{2j} an *attracting petal* for g. One such petal (for $p = 5$) is shown in Figure 12.2.

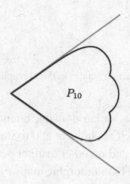

Figure 12.2 The attracting petal P_{10}

Similarly, let

$$S^-(R) = \{ \zeta \in \mathbb{C} : \text{Re}[\zeta] \geqslant -R \text{ and } |\text{Im}[\zeta]| \leqslant R \}$$

and $B^-(R) = \mathbb{C} \setminus §^-(R)$, and let

$$P_{2j-1}(R) = \left\{ z \in \mathbb{C} : \left| \arg z - \frac{(2j-1)\pi}{p} \right| < \frac{\pi}{p} \text{ and } \frac{1}{z^p} \in B^-(R) \right\}.$$

The set $B^-(R)$ is mapped into itself by $\tau^{-1}(\zeta) = \zeta - 1$, and hence g^{-1} maps P_{2j-1} into itself. We call P_{2j-1} a *repelling petal* for g. Full collections of attracting and repelling petals are shown in Figure 12.3, for $p = 5$.

Now take a general map f in G_1, and let p be its invariant $p(f)$. Then f may be conjugated in G to the form

$$f(z) = z - \frac{z^{p+1}}{p} + \cdots$$

on some neighbourhood of 0. We assume that f takes this actual form. Then f is a perturbation of g, agreeing with g up to terms of degree $p + 1$. The basic idea of Écalle and Voronin was that there are corresponding perturbations $\tilde{P}_j$ of the petals P_j, and conjugations F_j of f on P_j to $\tau^{\pm 1}$ on $B^\pm(R)$. Let us be more precise.

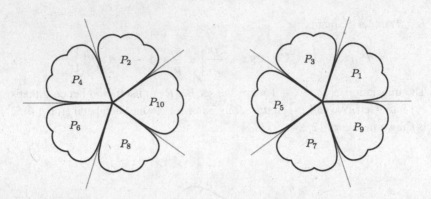

Figure 12.3 Attracting petals (left) and repelling petals (right)

○ For $j = 1, \dots, 2p$, $\tilde{P}_j$ is a Jordan domain, bounded, like P_j, by three real-analytic arcs, two of which are tangent at 0 to the sides of the sector T_j.

○ The petal $\tilde{P}_j$ meets $\tilde{P}_{j\pm1}$, and no other distinct petals.

○ For $j = 1, \dots, p$, there are biholomorphic maps $F_{2j} : \tilde{P}_{2j} \to B^+(R)$ that satisfy

$$F_{2j}(f(z)) = F_{2j}(z) + 1. \tag{12.1}$$

○ For $j = 1, \dots, p$, there are biholomorphic maps $F_{2j-1} : \tilde{P}_{2j-1} \to B^-(R)$ that satisfy

$$F_{2j-1}(f^{-1}(z)) = F_{2j-1}(z) - 1. \tag{12.2}$$

○ Each F_j takes the form

$$F_j(z) = \frac{1}{z^p} + s_j(z),$$

for z in $\tilde{P}_j$, where s_j is holomorphic in $\tilde{P}_j$.

We shall not discuss the detailed (and rather deep) proof that these perturbed petals and maps exist (see [2], but note that the notation there is not quite the same). Instead we take for granted the existence of $\tilde{P}_j$, F_j, and R, and proceed to construct the EV data for f.

The attracting and repelling petals are superimposed in Figure 12.4, and the intersections of consecutive petals $\tilde{P}_j$ and $\tilde{P}_{j+1}$ are shaded.

Let $H^+(k)$ and $H^-(k)$ denote the half-planes consisting of those numbers with imaginary part greater than k or less than $-k$, respectively. On $\tilde{P}_{2j} \cap \tilde{P}_{2j+1}$, the maps F_{2j} and F_{2j+1} are both defined, and there is positive number S (where $S > R$) such that the map

$$\Phi_{2j+1} = F_{2j+1} \circ F_{2j}^{-1}$$

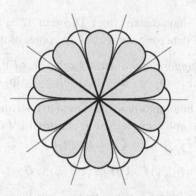

Figure 12.4 Attracting and repelling petals superimposed

is holomorphic on $H^+(S)$. Likewise, on $\tilde{P}_{2j-1} \cap \tilde{P}_{2j}$, the maps F_{2j-1} and F_{2j} are both defined, and there is a positive number T (where $T > R$) such that the map

$$\Phi_{2j} = F_{2j} \circ F_{2j-1}^{-1}$$

is holomorphic on $H^-(T)$. For instance, in the special case when $f = g$, we have $F_j(z) = 1/z^p$ for all j, so Φ_j is the identity for each j.

Let $H^+(S)$ be the half-plane on which Φ_{2j+1} is defined. Given $\zeta \in H^+(S)$, let $z = F_{2j}^{-1}(\zeta)$. Then (12.1) and (12.2) give

$$
\begin{aligned}
\Phi_{2j+1}(\zeta+1) &= F_{2j+1} \circ F_{2j}^{-1}(F_{2j}(z)+1) \\
&= F_{2j+1}(f(z)) \\
&= F_{2j+1}(z)+1 \\
&= \Phi_{2j+1}(\zeta)+1.
\end{aligned}
$$

Further,

$$\lim_{y \uparrow \infty} \Phi_{2j+1}(\zeta+iy) - (\zeta+iy) = 0$$

uniformly in ζ on compact subsets of $H^+(S)$, so Φ_{2j+1} is bounded on $H^+(S)$. The maps $-\Phi_{2j}(-\zeta)$ have similar periodic and bounded properties, therefore $(\Phi_1, \ldots, \Phi_{2p})$ is valid EV data.

We conclude with two consequences of the EV data construction.

Proposition 12.7 *A germ f in G_1 with EV data $(\Phi_1, \ldots, \Phi_{2p})$ is conjugate in G to the special map g (with $p = p(f)$) if and only if there exist constants c_j such that $\Phi_j(\zeta) = \zeta + c_j$ for each j.*

Proof This follows immediately from Theorem 12.5, because, as we observed earlier, the EV data for g consists of $2p$ copies of the identity map. $\square$

Proposition 12.8 *Suppose that a germ f in G_1 has EV data $(\Phi_1, \ldots, \Phi_{2p})$. Then f^{-1} has EV data $(-\Phi_2-, -\Phi_3-, \ldots, -\Phi_{2p}-, -\Phi_1-)$.*

Proof Let F_j be the biholomorphic maps used to construct EV data for f and let G_j be the biholomorphic maps used to construct EV data for f^{-1}. On the petal $\tilde{P}_{2j-1}$, $F_{2j-1} f^{-1} F_{2j-1}^{-1}(\zeta) = \zeta - 1$. Let $\eta(\zeta) = -\zeta$. Then

$$(\eta F_{2j-1}) f^{-1} (\eta F_{2j-1})^{-1}(\zeta) = \zeta + 1.$$

A similar calculation with ηF_{2j} shows that we can choose $G_j = \eta F_{j-1}$. Since $\Phi_j = F_j \circ F_{j-1}^{-1}$, the result follows immediately. $\square$

This proposition can be used to test whether a germ is reversible.

12.4 Roots

We can use the EV data to characterise the existence of composition roots of biholomorphic germs.

Theorem 12.9 *Given g in G_1, and a positive integer m, there exists an element f of G_1 with $f^m = g$ if and only if each holomorphic function in the EV data $\Phi(g)$ has period $1/m$.*

In view of Theorem 12.6, this implies that generic elements of G_1 have no mth roots at all, because generic holomorphic maps of period 1 do not have a smaller period.

Proof Suppose first that there is a germ f with $f^m = g$. After conjugating, we may assume that $f(z) = z - \dfrac{z^{p+1}}{p} + \cdots$. Let $\Phi(f) = (\Phi_1, \ldots, \Phi_{2p})$ be the EV data of f. Observe that

$$f^m(z) = z - \frac{mz^{p+1}}{p} + \cdots.$$

Let $\alpha = m^{1/p}$ and let $g_0(z) = \alpha f^m(z/\alpha)$. Then

$$g_0(z) = z - \frac{z^{p+1}}{p} + \cdots.$$

Let F_j be the conjugating maps that go with f, and let

$$F_j^*(z) = \frac{1}{m} F_j(z/\alpha).$$

Now, (12.1) gives

$$F_{2j-1}(f^m(z)) = F_{2j-1}(z) + m.$$

Thus

$$F_{2j-1}^*(g_0(z)) = \frac{1}{m} F_{2j-1}(f^m(z/\alpha)) = \frac{1}{m}(F_{2j-1}(z/\alpha) + m) = F_{2j-1}^*(z) + 1.$$

Similarly,

$$F_{2j}^*(g_0^{-1}(z)) = F_{2j}^*(z) - 1.$$

It follows that the maps F_j^* give rise to EV data Φ_j^* for g_0; that is, $\Phi_j^* = F_j^* \circ (F_{j-1}^*)^{-1}$ for $j = 1, \ldots, 2p$. A straightforward calculation gives

$$\Phi_j^*(\zeta) = \frac{1}{m} \Phi_j(m\zeta),$$

so Φ_j^* has period $1/m$. Since g is conjugate to g_0, we see from Theorem 12.5 that its EV data also has period $1/m$.

Conversely, suppose that the EV data $\Phi(g)$ has period $1/m$. Define $\Phi^* = (\Phi_1^*, \ldots, \Phi_{2p}^*)$ by

$$\Phi_j^*(\zeta) = m\Phi_j(\zeta/m).$$

Each Φ_j^* has period 1, and clearly Φ^* enjoys all the other properties of EV data. By Theorem 12.6, there exists a germ f with $\Phi(f) = \Phi^*$. From the first part of this proof we see that f^m is conjugate to a germ with EV data Φ, so, by Theorem 12.5, f^m is conjugate to g. Therefore $hf^mh^{-1} = g$ for some germ h, which implies that hfh^{-1} is an mth root of g. $\square$

12.5 Centralisers and flows

By Corollary 10.22, the formal centraliser of a nontrivial element f of G_1 is always the inner direct product of the group $\{f^t : t \in \mathbb{C}\}$ of formal iterates of f and a finite cyclic group of order $p(f)$. Since there are no nontrivial germs of finite order in G_1, it follows that $C_f(G_1)$ is isomorphic to an additive subgroup of $\mathbb{C}$. It consists of those formal iterates f^t that have positive radius of convergence. On the face of it, any additive subgroup of $\mathbb{C}$ that contains $\mathbb{Z}$ may occur. However, a remarkable result of Baker and Liverpool [24, 171] (see also [73]) shows that the only possible subgroups that arise are $\mathbb{C}$ itself and an infinite cyclic group $\frac{1}{d}\mathbb{Z}$ (the set of all integer multiples of $\frac{1}{d}$), for some positive integer d.

Theorem 12.10 (Baker–Liverpool) *Suppose $f \in G_1$. Let*

$$\Lambda = \{t \in \mathbb{C} : f^t \in G_1\}.$$

Then either $\Lambda = \mathbb{C}$ and f is conjugate in G_1 to the map $z/(1+z^p)^{1/p}$, with $p = p(f)$, or otherwise $\Lambda = \frac{1}{d}\mathbb{Z}$ for some positive integer d. In the latter case, d is the largest integer such that f has a dth root in G_1. □

A *flow* in G_1 is a continuous group homomorphism $t \mapsto f_t$ from $(\mathbb{C}, +)$ to G_1. A germ f in G_1 is *flowable* if there exists a flow (f_t) with $f_1 = f$. If all formal iterates f^t have positive radius of convergence, then the map $t \mapsto f^t$ is a flow, with image $C_f(G_1)$.

The next proposition shows that all flows in G_1 have the form $t \mapsto f^t$ for some germ f, which implies that each flowable germ is flowable in a unique way. In fact, this is a purely formal result, so we work in the group $\mathscr{G}_1$ of formally-invertible formal power series with multiplier 1. (A flow in $\mathscr{G}_1$ is a continuous group homomorphism from $(\mathbb{C}, +)$ to $\mathscr{G}_1$.)

Proposition 12.11 *If $t \mapsto f_t$ is a flow in $\mathscr{G}_1$, and $f = f_1$, then $f_t = f^t$ for each t in $\mathbb{C}$.*

Proof Each f_t commutes with f, and hence belongs to $C_f(\mathscr{G}_1)$. But this centraliser consists of formal iterates of f, so $f_t = f^{u_t}$ for some element u_t of $\mathbb{C}$, and we have $f_t = x + u_t x^{p+1} + \cdots$. The map $t \mapsto u_t$ is an additive homomorphism from $\mathbb{C}$ to itself, so $u_t = t u_1 = t$ for each rational t. Therefore, by continuity, $f_t = f^t$ for each t in $\mathbb{C}$. □

Corollary 12.12 *A nontrivial element f of G_1 is flowable if and only if $C_f(G_1) = C_f(\mathscr{G}_1)$, the set of all formal iterates f^t, $t \in \mathbb{C}$.* □

Given a germ f that is not flowable, we denote by $d(f)$ the largest positive integer d such that f has a dth root (so that $C_f(G_1)$ consists of iterates of $f^{1/d}$). Combining Theorem 12.10 with the description of the formal centraliser in Corollary 10.22, we obtain the following theorem.

Theorem 12.13 *Suppose a germ f in G_1 is not flowable. Let $p = p(f)$ and $d = d(f)$. Then $C_f(G)$ is abelian, and there exist positive integers q and k with $k|q$ and $q|p$, and elements τ and ω in $C_f(G)$, such that*

(i) *$C_f(G)/C_f(G_1)$ is cyclic of order q*
(ii) *$C_f(G)$ is generated by τ and $f^{1/d}$*
(iii) *ω has finite order k*
(iv) *$\tau^{\frac{q}{k}} = \omega f^{1/d}$*
(v) *$C_f(G) = \langle \tau \rangle \times \langle \omega \rangle$.*

Proof   The group $C_f(G)$ is abelian because it is contained in the abelian group $C_f(\mathcal{G})$. From the formal theory we know that if an element g of G commutes with f then its multiplier $m(g)$ is a pth root of unity. The kernel of m restricted to $C_f(G)$ is $C_f(G_1)$. Therefore $C_f(G)/C_f(G_1)$ is a subgroup of the cyclic group of order p; so it is a cyclic group of order q, where $q|p$. It follows that $C_f(G)$ is generated by $f^{1/d}$ and a germ τ_0, where $\tau_0^q \in C_f(G_1)$ and $\tau_0^s \notin C_f(G_1)$ for any positive integer s less than q. Next, from Corollary 10.22 we know that the abelian group $C_f(\mathcal{G})$ is generated by σ and f^t ($t \in \mathbb{C}$), where $\langle \sigma \rangle$ is a cyclic group of order p. Let l be the smallest positive integer such that τ_0^l lies in $\langle \sigma, f^{1/d} \rangle$. Then $l|q$, and we define $k = q/l$. There are integers i and j with $\tau_0 = \sigma^i f^{j/(dl)}$, where j and l are coprime. Choose integers a and b such that $aj + bl = 1$; make this choice such that a is also coprime to k, and hence a is coprime to q. Now define $\tau = \tau_0^a f^{b/d}$. Since a is coprime to q, τ and $f^{1/d}$ generate $C_f(G)$. Let $\omega = \sigma^{ial}$; this has order k. Also

$$\tau^l = \tau_0^{al} f^{\frac{bl}{d}} = \sigma^{ial} f^{\frac{aj+bl}{d}} = \omega f^{\frac{1}{d}}.$$

Therefore τ and ω generate $C_f(G)$, and since they commute, $C_f(G) = \langle \tau \rangle \times \langle \omega \rangle$. $\square$

12.6 Reversible and strongly-reversible elements

We begin by describing a standard form for reversible germs, up to conjugacy. Recall that H is the group of germs whose multipliers are roots of unity.

Theorem 12.14   *Let f be a germ in G_1, with $p = p(f)$. Then f is reversible in G if and only if there is an element h of H such that hfh^{-1} is reversed by $z \mapsto \exp[(\pi i/s)z]$, for some integer s such that p/s is an odd integer. Equivalently, f is reversible in G if and only if there is an element h of H with*

$$hfh^{-1}(z) = z + z^{p+1} + \sum_{k=1}^{\infty} c_k z^{sk+p+1}, \tag{12.3}$$

where p/s is an odd integer, and

$$hf^{-1}h^{-1}(z) = z - z^{p+1} + \sum_{k=1}^{\infty} (-1)^k c_k z^{sk+p+1}. \tag{12.4}$$

This result allows us to understand reversibility in G: one reverses a germ f essentially by rotating it (using a rotation modulo conjugacy), so as to swap the attracting and repelling petals of its Leau flower (which is shown in Figure 12.4).

Proof Since every reverser of G is of finite order $2s$ with p/s odd (by Theorem 10.39) and every element of finite order $2s$ is conjugate to

$$\beta(z) = \exp[(\pi i/s)z]$$

(by Proposition 12.1), the first assertion is immediate. The second assertion follows, because (12.3) and (12.4) are standard forms for an element reversed by β. $\qquad\square$

We also have a stronger version of Corollary 12.4.

Corollary 12.15 *Let $f \in G$. Then f is reversible if and only if f^2 is reversible.*

Proof It is true in any group that if f is reversible then f^2 is reversible. For the converse, in which f^2 is reversible, there are two cases: $m(f) = \pm 1$. The case $m(f) = -1$ has been dealt with already in Corollary 12.4, so we may assume that $m(f) = 1$. In this case we appeal to the formal result Theorem 10.39, which shows that each reverser of f^2 reverses every element of the formal flow $(f^2)^t$. In particular, it reverses $(f^2)^{1/2} = f$. $\qquad\square$

Theorem 10.39 enables us to characterise those reversible germs that are strongly reversible.

Theorem 12.16 *Let $f \in G$. Then f is strongly reversible if and only if it is reversible and either it is an involution or $p(f)$ is odd.*

Proof Since involutions are strongly reversible, we may assume that f is not an involution. Let $p = p(f)$.

Suppose first that f is strongly reversible. Then $m(f) = 1$. Theorem 10.39 tells us that any reverser of f has order $2s$, where p/s is odd. In this case $s = 1$, so p is odd. Conversely, suppose that f is reversible and p is odd. Then f is reversed by an element h of order $2p$, so it is also reversed by h^p, which is an involution. Therefore f is strongly reversible. $\qquad\square$

The following summarises our conclusions about reversible elements in subgroups of G. Recall that G_1 consists of germs with multiplier 1, H_0 consists of germs with multiplier ± 1, and H consists of germs with multiplier a root of unity.

Corollary 12.17 *We have*

$$\{\mathbb{1}\} = R(G_1) \subset R(H_0) = I^2(G) \subset R(H) = R(G) \subset H_0,$$

and the three inclusions are proper.

12.7 The order of a reverser

Flowable reversible germs in the class A_p are special – all are conjugate to $z/(1+z^p)^{1/p}$, and each reverser has order $2s$, where p/s is an odd integer.

In the nonflowable case, we can relate the possible orders for reversers to the centraliser generators τ and ω, and the natural numbers d, q, and k of Theorem 12.13. The numbers d, q, and k are uniquely determined by f: the $\frac{1}{d}$th power of f is the smallest positive power whose series converges, q is the index of $C_f(G_1)$ in $C_f(G)$, and k is the order of the (cyclic) torsion subgroup of $C_f(G)$.

Theorem 12.18 *Suppose that f is a reversible member of G_1 that is not flowable. Let $p = p(f)$, and let τ and ω, and d, q, and k, be the germs and integers from Theorem 12.13.*

(i) *If h reverses f, then h commutes with ω, and h reverses $f^{m/d}$ for any integer m.*

(ii) *We have $k = q$, and $\frac{p}{q}$ is odd.*

(iii) *We have*

$$\{h^2 : hfh^{-1} = f^{-1}\} = \{\omega^l : l \text{ is odd}\},$$

and we always have

$$\{\operatorname{ord}(h) : hfh^{-1} = f^{-1}\} = \{2r \in \mathbb{N} : r|k \text{ and } \tfrac{k}{r} \text{ is odd}\}.$$

Notice that the set $\{\omega^l : l \text{ is odd}\}$ is the collection $\langle\omega\rangle$ of all powers of ω if k is odd, and if k is even then it is the coset $\omega\langle\omega^2\rangle$ of $\langle\omega\rangle$.

Proof   First we prove (i). Notice that h and h^{-1} both reverse f, and ω commutes with f. It follows that $h\omega h^{-1}$ commutes with f, has order k, and has the same multiplier as ω. Therefore it equals ω. The final assertion from ((i)) follows from the fact that it holds formally, by Proposition 10.16.

Next we prove (ii). If h reverses f, then, by Theorem 10.39, it has order $2s$, where p/s is odd. Now h^2 commutes with f, so we must have $h^{2k} = 1$. It follows that s divides k, and hence p/k is odd. Let us now show that $q = k$. The commutator $h\tau h^{-1}\tau^{-1}$ commutes with f and has multiplier 1, so $h\tau h^{-1}\tau^{-1} = f^{\frac{n}{d}}$ for some integer n. Raise both sides of this identity to the power q and we get $f^{-\frac{2k}{d}} = f^{\frac{qn}{d}}$. It follows that $-2k = qn$, or $-2 = \frac{q}{k}n$, so that q/k is either 1 or 2. It cannot be 2, as p/k is odd (and q divides p). Hence $q = k$.

Last we prove (iii). Suppose first that $hfh^{-1} = f^{-1}$. Then h has finite order, and h^2 commutes with f, so $h^2 = \omega^l$ for some integer l. Notice that $\omega^j h$ is also a reverser of f for any integer j, and $\omega^{2j}h^2 = \omega^{2j+l}$. If k is odd, then each element of $\langle\omega\rangle$ has the form ω^{2j+l} for some integer j, so $\{h^2 : hfh^{-1} = f^{-1}\} =$

$\langle\omega\rangle$. If k is even, then p is even, and, by Theorem 12.16, f is not strongly reversible. It follows that l must be odd (otherwise $\omega^{-l/2}h$ is an involution that reverse f). In this case the collection of squares of reversers $\omega^{2j}h^2$ accounts for only half of $\langle\omega\rangle$, namely $\omega\langle\omega^2\rangle$ – the odd powers of ω.

We conclude that the possible values of ord(h) are the numbers $2\,\text{ord}(\omega^l)$, where l ranges over the odd numbers. Since ω has order k, the order of ω^l is $r = k/u$, where u is the greatest common divisor of l and k. Since l is odd, u must be odd as well. Conversely, suppose that r is a divisor of k and $u = k/r$ is odd. Then, as we have seen, there is a reverser h of f with $h^2 = \omega^u$, which obviously has order r.   □

Corollary 12.19 *If $p = 2^a u$, where u is odd, and a germ f in A_p is nonflowable and reversible in G, then $k = q = 2^a n$ where n divides u. The largest order for a reverser of f is $2k$ and the smallest order is 2^{a+1}.*   □

In the flowable case, this corollary also holds (with, additionally, $q = p$).

12.8 Examples

12.8.1 A reversible germ that is not strongly reversible

Let

$$u(z) = iz, g(z) = z + z^3, \text{ and } f = ugu^{-1}g^{-1}.$$

Then u^2 commutes with g, so

$$ufu^{-1} = u^2gu^{-1}g^{-1}u^{-1} = gug^{-1}u^{-1} = f^{-1}.$$

Therefore f is reversible. However, one calculates that $f(z) = z - 2z^3 + O(z^4)$, so $f \in A_2$. Thus f is not strongly reversible, by Corollary 12.16.

12.8.2 A reversible germ that is not reversible by any germ of order dividing 2^k

Given an even integer p, let $u(z) = e^{2\pi i/p}z$, $g(z) = z + z^{p/2+1}$, and $f = ugu^{-1}g^{-1}$. Then u^2 commutes with g, so

$$ufu^{-1} = u^2gu^{-1}g^{-1}u^{-1} = gug^{-1}u^{-1} = f^{-1}.$$

Therefore f is reversible. A short calculation shows that $f \in A_p$. In case $p = 2^{k+1}$, Theorem 10.39 tells us that no element of order 2^k can reverse f.

12.8.3 A nonflowable reversible germ

Given any positive integer p, define EV data $\Phi = (\Phi_1, \ldots, \Phi_{2p})$ by

$$\Phi_1(\zeta) = \zeta + \exp(-2\pi i\zeta), \qquad \Phi_2(\zeta) = \zeta - \exp(2\pi i\zeta),$$

and $\Phi_{j+2} = \Phi_j$ for all j.

By Theorem 12.6, there is a germ f in A_p with EV data $\Phi(f)$ equivalent to Φ. This germ f is not flowable, by Proposition 12.7, because Φ_1 is not a translation. However, f is reversible, by Proposition 12.8, because $\Phi_2(\zeta) = -\Phi_1(-\zeta)$.

12.8.4 A formally-reversible germ that is not reversible in G

Let $\Phi_1(\zeta) = \zeta + e^{-2\pi i\zeta}$ and $\Phi_2(\zeta) = \zeta$. Suppose that $f = z + az^2 + bz^3 + \cdots$ realizes this EV data. Using the formula at the top of page 19 of [2], one sees that $b/a^2 = 1$. It then follows from Theorem 10.38 that f is formally reversible. In contrast, f is not reversible in G_1, by Proposition 12.8, because $\Phi_2(\zeta) \neq -\Phi_1(\zeta)$.

Notes

Sources

The Écalle–Voronin invariants were defined and explained in Voronin's paper [236]. Essentially the same construction was discovered independently by Écalle [174]. For a clear account (with full details when $p > 1$), see [2, pages 7–19]. The case $p > 1$ was first fully elaborated by Ilyashenko [137]. It appears that the essential idea of the EV data, and Theorem 12.5, were previously discovered by Birkhoff in 1939 [32]. The facts about $C_f(G)$, for $f \in G_1$, were established by Baker and Liverpool [24, 25, 26, 171] (see also [223]).

The main results about reversibility in this chapter were established by Ahern and the first author in [3]. A thorough account of reversibility in the larger group of conformal or anticonformal germs was given by Ahern and Gong [2]. The strongly-reversible elements of G (see Corollary 12.16) were also identified (in terms of EV data) in [2]. The corresponding formal problem was solved by Kasner [143, 144]. Theorem 12.3 goes back to Muckenhoupt [157, Theorem 8.7.6, page 359]. The case $p = 1$ of Corollary 12.16 was given by Voronin [237]. Corollary 12.16 is due to Ahern and Gong [2], who also described the strongly-reversible biholomorphic germs in terms of their EV data.

De Paepe's problem

In 1984, de Paepe asked [64] whether there exists a radius $r > 0$ such that each continuous complex-valued function f on the closed disc $D = \mathbb{B}(0, r)$ may be approximated, uniformly on D, by polynomials of the form $p(z^2, \bar{z}^2 + \bar{z}^3)$, where $p(Z, W) \in \mathbb{C}[Z, W]$ is a polynomial in two variables having complex coefficients.

This question arose naturally following earlier work of Wermer and Minsker and was in some sense the easiest-looking problem at the edge of known theory; however, it turned out that de Paepe (and others) were able to deal with many similar problems, but not with this first example. Also, for some of these examples the approximation result held, and for others it failed, so it was not at all clear what to expect. For a survey of the history, see [65].

Definition 12.20   Let K be a compact set in $\mathbb{C}^n$. The *polynomially convex hull* $\hat{K}$ of K is defined by

$$\hat{K} = \{z \in \mathbb{C}^n : |p(z)| \leqslant \sup_K |p|, \ p \text{ holomorphic polynomial}\}.$$

The compact set K is called *polynomially convex* if $\hat{K} = K$.

The problem of deciding whether a compact set in $\mathbb{C}^n$ is polynomially convex or not is a fundamental and difficult problem in complex analysis (see [4, Chapter 12] and the references therein). The polynomial convexity, or otherwise, of the topological disc

$$K = \{(z^2, \bar{z}^2 + \bar{z}^3) : |z| \leqslant r\}$$

is the main difficulty in de Paepe's problem.

By an *analytic variety attached to K* (see Figure 12.5) we mean a bordered complex manifold (M, B), and a continuous map $\phi : M \cup B \to \mathbb{C}^k$, holomorphic on M, such that $\phi(B) \subset K$. By the maximum modulus principle, such an attached variety is an obstruction to polynomial convexity, unless it lies entirely in K.

De Paepe's problem was solved [191] by showing that the above topological disc K has attached analytic varieties. In fact, this holds for all discs of the form $K = \{(z^2, f(\bar{z})) : |z| \leqslant r\}$, where f is holomorphic on $|z| \leqslant r$, and $f(z) = z^2 + a_3 z^3 + \cdots$, with all coefficients a_n real, and at least one $a_{2n+1} \neq 0$. The attached varieties are images of plane domains in the shape of lunulae (Figure 12.6).

The method of proof uses ideas from complex dynamical systems applied to a strongly-reversible map associated with the disc. Both functions z^2 and f have two-point fibres, and so each gives a holomorphic involution defined

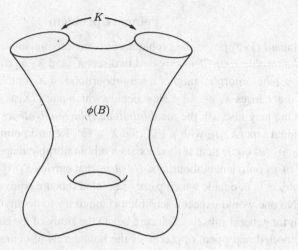

Figure 12.5   Attached variety

Figure 12.6   A lunula

near 0. The composition of these involutions is a biholomorphic map fixing 0, with multiplier 1, and using the theory of one-dimensional complex dynamical systems (see [48, Chapter 2]), it can be conjugated in a Leau petal to the map $z \mapsto z + 1$ on the right half-plane. Using the same conjugation on the map $z \mapsto z + \tau$ gives a semigroup action of the half plane, with orbits that are analytic varieties.

Poincaré problem

Poincaré [196] posed the problem of characterising subsets $X \subset \mathbb{C}^d$ up to *biholomorphic equivalence*: regard two sets X_1 and X_2 as equivalent if there exists a biholomorphic map of a neighbourhood of X_1 onto a neighbourhood of X_2 that carries X_1 onto X_2. Now decide which pairs (X_1, X_2) are equivalent.

One may also ask the *local biholomorphic equivalence* question: Consider pointed sets (X, p), with $p \in X$ and $X \subset \mathbb{C}^d$. Regard pointed sets (X_1, p_1) and (X_2, p_2) as equivalent if there exists a biholomorphic map of a neighbourhood U_1 of p_1 onto a neighbourhood U_2 of p_2 that carries $X_1 \cap U_1$ onto $X_2 \cap U_2$. The problem is to decide which pairs of pointed sets are equivalent.

No one would expect a sensible person to try to do anything with this problem for general subsets . Poincaré began the study of the case when the sets are connected and open, or pieces of the boundaries of connected and open sets. Obviously, any two open sets are locally biholomorphically equivalent, as are any two nonsingular real-analytic curves. He suggested, and Reihhardt proved [94] that the ball and the polydisc in $\mathbb{C}^2$ are globally inequivalent. Subsequent work focused on simple sets of progressively larger topological dimension.

Real-analytic discs are sets that have nonsingular real-analytic parameterisations by maps from the open unit disc. If X is such a set, and $p \in X$, then the tangent plane to X at p is said to be a *complex tangent* if it is a complex line. The presence of a complex tangent at p is a local biholomorphic invariant of the pointed set (X, p). A surface without complex tangents is called *totally real*. Surfaces without complex tangents are all locally equivalent [4]. Thus remaining interesting cases in dimensions less than 3 involve surfaces that have a complex tangent, unions of nonsingular curves or surfaces, and singular curves and surfaces. Reversible maps have come up in connection with several of these cases, two of which are described below.

The case of two curves

Antiholomorphic involutions

Consider sets $X = \Gamma_1 \cup \Gamma_2$, where the $\Gamma_j \subset \mathbb{C}^d$ are nonsingular real-analytic arcs that meet at 0.

When the complex spans of the tangents are distinct, the two arcs may be mapped biholomorphically to a pair of straight lines.

When the Γ_i have tangents at 0 that span a single complex line, then the angle between the real tangents is a biholomorphic invariant.

We shall consider only the case $d = 1$ of two plane arcs. To each arc, we may associate the antiholomorphic reflection across the arc. The association between germs of arcs and germs of antiholomorphic involutions is bijective. Thus the problem of classifying pairs of arc-germs (Γ_1, Γ_2) is the same as the problem of classifying pairs (τ_1, τ_2) of antiholomorphic involutive germs up to conjugation by elements of the group G of holomorphic germs.

Let $\hat{G}$ denote the group of germs at 0 in $\mathbb{C}$ that are either conformal or anticonformal. Then G is an index-two normal subgroup of $\hat{G}$. Let $\bar{G}$ denote the coset of anticonformal germs. We also use $\mathscr{G}$ for the group of invertible formal series in one variable, and we embed it in the larger group $\hat{\mathscr{G}}$ of invertible formal series in z or in $\bar{z}$, where $\bar{z}$ is a formal symbol that obeys the composition rule $\bar{z} \circ \bar{z} = z$.

The indicator of a pair

An antiholomorphic involution is, in these terms, an element of $\bar{G} \cap I(\hat{G})$. Each pair of curve germs determines a strongly-reversible element $f = \tau_1 \circ \tau_2 \in G \cap I^2(\hat{G})$. This f is called the *indicator* of the pair. The conjugacy class of f in G is a biholomorphic invariant of the pair.

A given $f \in G$ is the indicator of some pair if and only if it is reversed by some antiholomorphic involution τ. The problem of classifying pairs (τ_1, τ_2) of antiholomorphic involutions is equivalent to the problem of classifying pairs (f, τ), where τ is an antiholomorphic involution, and τ reverses f.

The indicator of a pair takes the form $f(z) = \lambda z + \cdots$, with $\lambda = \exp(2\alpha i)$, where α is one of the angles between the two arcs. In particular, its multiplier is unimodular.

The conjugacy classification of biholomorphic germs whose multiplier is a root of unity has been discussed in Chapter 12. The other unimodular multipliers $\lambda = \exp(\alpha i)$ divide into two cases. If α is a so-called *Bruno* number [48], then f is conjugate to the irrational rotation λz, and each pair with indicator f is conformally-equivalent to a pair of straight lines. The Bruno numbers may be characterised by a Diophantine condition – a condition on their continued-fraction expansion. If α is not a Bruno number, then Yoccoz [250] showed that there are uncountably many distinct conjugacy classes of biholomorphic f having multiplier λ. Marco even showed [195] that there are indicators f shared by uncountably-many inequivalent pairs of arcs. Despite these deep results, the classification of pairs having indicator with non-Bruno multiplier is not fully understood. Thus we are some way from understanding the biholomorphic classification of pairs of arcs which meet at angles that are not rational multiples of π.

Mutually-tangent pairs

We shall concentrate on the case in which the curves Γ_i are mutually-tangent at 0, that is, on pairs whose indicator $f(z) = z + \cdots$ belongs to the subgroup G_1. (The results in this case can be extended to cover all cases in which the angle between the arcs is a rational multiple of π.)

We know that f has formal invariants $p \in \mathbb{N}$, and $a \in \mathbb{C}$, such that it may be conjugated in $\mathscr{G}$ to the form

$$f(z) = z + z^{p+1} + az^{2p+1} + \cdots.$$

The invariant $p(f)$ associated to a pair (Γ_1, Γ_2) is one less than the order of contact between the tangents to Γ_1 and Γ_2. Up to a biholomorphic change of variables, the arcs take the form $y = 0$ and

$$y = \lambda x^{p+1} + O(|y|^{p+2}),$$

where $\lambda > 0$.

We saw that a given $f \in G_1$ is formally reversible by a formally-conformal series if and only if $a = (p+1)/2$. The difference

$$r(f) = \frac{p+1}{2} - a$$

is called the *residue* of f. (It is actually a residue of something related to f, essentially of branches of $f(\zeta^{-1/p})^{-p}/p$.) Let us now consider reversibility in the larger group $\hat{G}$, beginning with the formal aspects.

We know (by Theorem 10.39) that each conformal reverser of f has finite even order. Considering reversibility in the larger group $\hat{G}$, we have the following result.

Proposition 12.21 *Let $f \in G_1$. Then each anticonformal reverser of f is an involution. Moreover, f is formally reversible by an anticonformal involution if and only if its residue is pure imaginary.* $\square$

This does not tell us how to characterise the conformal germs f that are reversed by anticonformal involutions. It is possible to do this in terms of a symmetry property of the EV data associated to f. We omit the details, which may be found in [2]. See also [186, 237].

Classification of pairs

It turns out that the conjugacy classification of the indicator f in G goes most of the way to classifying the pairs.

Theorem 12.22 (Nakai) *Suppose that $f \in G_1$ is the indicator of a pair (τ_1, τ_2)*

of antiholomorphic involutions, corresponding to a pair (Γ_1, Γ_2) *of mutually-tangent real-analytic arcs. Then there are at most four classes of pairs, under biholomorphic equivalence, that have an indicator function biholomorphically-equivalent to* f.

(i) *If* f *is flowable and* $p(f)$ *is odd, then there is one class.*

(ii) *If* f *is flowable and* $p(f)$ *is even, then there are one or two classes.*

(iii) *If* f *is not flowable and* $p(f)$ *is odd, then there are one or two classes.*

(iv) *If* f *is not flowable and* $p(f)$ *is even, then there are up to four classes.*

$\square$

Trépreau [232] has a nice geometric way of analysing the problem of two tangent arcs, and works it through in detail for some tangent conics. His analysis is based on the repeated reflection of one curve in the other, and the pattern of intersections and singularities in the resulting chain of curves.

Discs with a complex tangent

Consider pointed sets (X, p), where X is a (nonsingular) real-analytic disc, and has an isolated complex tangent at p, that is, the tangent is complex at all points near p, except p.

For this discussion, we let G_2 denote the group of biholomorphic germs in two variables, defined near $0 \in \mathbb{C}^2$.

Without loss in generality, we may take $p = 0$, and X determined by an equation

$$z_2 = F(z_1, \overline{z_1}) = q(z_1, \overline{z_1}) + \mathrm{O}(|z_1|^3),$$

where $F(z, w)$ is a complex-valued analytic function on some neighbourhood of 0 in $\mathbb{C}^2$, and $q(z, w)$ is a quadratic form $az^2 + bzw + cw^2$ with complex coefficients. Assuming that the form is not identically zero, and that $b \neq 0$, an invertible quadratic change of variables reduces the equation of the surface to the form

$$z_2 = F(z_1, \bar{z}_1) = \gamma z_1^2 + z_1 \overline{z_1} + \gamma \overline{z_1}^2 + \mathrm{O}(|z_1|^3), \qquad (12.5)$$

with a nonnegative real γ. This form in this context is known as the *Bishop normal form* [33]. The real number γ is a biholomorphic invariant of the germ of X at 0. The surface is said to be of *elliptic type* if $0 \leqslant \gamma < \frac{1}{2}$. Moser and Webster [184] studied the elliptic case and obtained a complete set of biholomorphic invariants:

Theorem 12.23 (Moser–Webster)   *Let X be a real-analytic surface in $\mathbb{C}^2$ with an isolated complex tangent at 0 having Bishop invariant γ in the range $(0, \frac{1}{2})$. Then there exist $s \in \mathbb{N} \cup \{+\infty\}$ and (when $s < +\infty$) $\delta = \pm 1$ such that the germ of X at 0 is biholomorphically-equivalent to that of the surface S given by*

$$x_2 = z_1 \overline{z_1} + \Gamma(x_2)(z_1^2 + \overline{z_1}^2),$$
$$y_2 = 0,$$

where

$$\Gamma(x_2) = \begin{cases} \gamma + \delta x_2^s & \text{if } s < +\infty, \\ \gamma & \text{if } s = +\infty. \end{cases}$$

The surfaces S determined by distinct (γ, s) are inequivalent, and for $s < +\infty$ the surface determined by $(\gamma, s, +1)$ is not equivalent to that determined by $(\gamma, s, -1)$.   □

This theorem is significant in polynomial approximation theory, but also, and perhaps most interestingly, in demonstrating that these germs are always biholomorphically equivalent to algebraic surface germs.

The elliptic case has been much studied, using fixed-point methods of non-linear functional analysis, beginning with Bishop [33], who established the existence of attached analytic discs, and continuing with results on the regularity properties of the attached discs and a characterisation of the local hull of holomorphy by Hunt, Bedford, and Gaveau, and Kenig and Webster (see [147] and the references therein). See also the work of Huang and Krantz [135, 136].

Theorem 12.23 admits as an immediate corollary that the local hull of holomorphy of M near p is precisely the real analytic 3-manifold with boundary $\tilde{M}$ defined by $x_2 \geqslant z_1 \bar{z}_1 + \Gamma(x_2)(z_1^2 + \bar{z}_1^2)$, $y_2 = 0$. The manifold $\tilde{M}$ is the union of a one parameter family of ellipses, the boundaries of which are the curves on M obtained by setting $x_2 = c > 0$. The idea of the proof of Theorem 12.23 differs from the previous work on the elliptic case in that the analytic discs attached to the surface are not obtained as solutions of a nonlinear functional equation, but as orbits of a complex holomorphic flow. This is where strongly-reversible maps come in, and we now give a little detail about it:

The Moser–Webster method

The *complexification* of the surface X defined by

$$z_2 = F(z_1, \bar{z}_1)$$

is the two-dimensional complex submanifold (four-dimensional real submanifold) $\hat{X} \subset \mathbb{C}^4$ given by

$$z_2 = F(z_1, w_1),$$
$$w_2 = \overline{F}(w_1, z_1),$$

for (z_1, w_1) in some neighbourhood of 0 in $\mathbb{C}^2$, where $\overline{F}(z, w)$ denotes $\overline{F(\bar{z}, \bar{w})}$. The variety $\hat{X}$ is invariant under the linear antiholomorphic involution

$$P(z_1, z_2, w_1, w_2) = (\bar{w}_1, \bar{w}_2, \bar{z}_1, \bar{z}_2)$$

The map $x \mapsto (z, \bar{z})$ embeds X into $\hat{X}$. The image X' is the fixed-point set of $P|\hat{X}$.

Transforming X by a biholomorphic germ $f: (\mathbb{C}^2, 0) \to (\mathbb{C}^2, 0)$ is equivalent to transforming $\hat{X}$ by

$$\begin{aligned} z' &= f(z), \\ w' &= \bar{f}(w). \end{aligned} \tag{12.6}$$

Equation (12.6) defines an action of the group G_2 on the set of germs of two-dimensional varieties at 0, and the problem of classifying germs $X \subset (\mathbb{C}^2, 0)$ under the ordinary action of G_2 is the same as that of classifying the corresponding $\hat{X}$ under this action of G_2.

Let $\pi_i : \mathbb{C}^2 = \mathbb{C}^2 \times \mathbb{C}^2 \to \mathbb{C}^2$ be the ($\mathbb{C}$-linear) projections given by

$$\begin{aligned} \pi_1(z, w) &= z, \\ \pi_2(z, w) &= w, \end{aligned}$$

and let $C : \mathbb{C}^2 \to \mathbb{C}^2$ stand for complex conjugation. We have

$$\begin{aligned} F(z_1, w_1) &= \gamma z_1^2 + z_1 w_1 + \gamma w_1^2 + H(z_1, w_1), \\ \bar{F}(w_1, z_1) &= \gamma z_1^2 + z_1 w_1 + \gamma w_1^2 + \bar{H}(w_1, z_1), \end{aligned}$$

where H and $\bar{H}$ vanish to order 3 at 0. We may use (z_1, w_1) as coordinates on $\hat{X}$. In terms of these coordinates, we may describe the restrictions of the π_i to $\hat{X}$ by

$$\begin{aligned} \pi_1(z_1, w_1) &= (z_1, z_2) = (z_1, F(z_1, w_1)), \\ \pi_2(z_1, w_1) &= (w_1, w_2) = (w_1, \bar{F}(w_1, z_1)). \end{aligned}$$

It follows that, on a sufficiently-small neighbourhood of $0 \in \mathbb{C}^2$, each point has either one or two preimages on $\hat{X}$ under each π_j. Thus (cf. Section 1.3.1) we may define two involutions τ_i on $\hat{X}$ by swapping the preimages. These maps τ_j are each holomorphic on a dense open subset of $\hat{X}$, and extend analytically to all $\hat{X}$.

Let $\phi = \tau_1\tau_2$ be the strongly-reversible map obtained from the two involutions. The function ϕ takes the form

$$\phi(z_1, w_1) = \Phi(z_1, w_1) + O(|(z_1, w_1)|^2),$$

where the linear part has standard matrix in the form

$$\Phi = \begin{pmatrix} \dfrac{1}{\gamma^2} & \dfrac{1}{\gamma} \\ -\dfrac{1}{\gamma} & 1 \end{pmatrix}.$$

The proof proceeds by the simultaneous conjugation of the related series $\tau_1 = T_1 + \cdots$, $\tau_2 = T_2 + \cdots$, and $\phi = \tau_1\tau_2 = \Phi + \cdots$ to canonical forms. This is first done formally and then (not without effort) biholomorphically.

Remark 12.24 The idea of embedding $\phi = \tau_1\tau_2$ into a flow (ϕ_t), and then complexifying the parameter t, first appeared in [184], and was the inspiration for the approach taken to de Paepe's problem in [191].

There has been a great deal of further work on the biholomorphic classification of surfaces. See, for instance [92, 245].

Lower central series

The lower central series of $\mathscr{G}$ is not

$$G \geqslant G_0 \geqslant G_1 \geqslant G_2 \geqslant \cdots \geqslant G_k \geqslant \cdots.$$

The successive commutators go down faster than that. In one variable, for instance, $[\mathscr{G}_1, \mathscr{G}_1] \subset \mathscr{G}_3$. This follows from

$$(x + ax^2 + bx^3)(x + \alpha x^2 + \beta x^3) = x + (a + \alpha)x^2 + (b + \beta + 2a\alpha)x^3,$$
$$(x + ax^2 + bx^3)^{-1} = x - ax^2 + (2a^2 - b)x^3.$$

Open problems

Biholomorphic germs in several variables

The corresponding problem for biholomorphic germs in more than one variable is barely scratched. As already mentioned, there is a little progress on formal reversibility by the first author and Zaitsev [193, 194]. It would be a useful advance if one could prove the conjecture that each reversible germ is reversed by a germ of finite order.

Products of involutions

In one dimension, characterise the classes I^n of products of involutions, for $n = 3, 4, \ldots$. We have seen that for formal power series, $I^\infty = I^4$ (Theorem 10.36). Is there a similar result for biholomorphic germs?

Global reversibility

Consider reversibility in groups of automorphisms of other complex manifolds. For the full automorphism group we have implicitly dealt with compact Riemann surfaces, the ball in $\mathbb{C}^n$, and the polydisc, as the corresponding groups are isomorphic to ones already considered. However, even for these, the point of view of complex analysis suggests focus on various distinguished subgroups.

Generalised de Paepe problem

It is not known whether all discs of the form $K = \{(z^2, f(\bar{z})) : |z| \leqslant r\}$, where f is holomorphic on $|z| \leqslant r$, and $f(z) = z^2 + a_3 z^3 + \cdots$, admit attached analytic varieties if the coefficients a_n are not restricted to be real. This might be accessible to attack by complexification, and the use of biholomorphic involutions and reversibility in $\mathbb{C}^2$, instead of $\mathbb{C}$.

References

[1] Ageev, O. 2005. The homogeneous spectrum problem in ergodic theory. *Invent. Math.*, **160**(2), 417–446.

[2] Ahern, P., and Gong, X. 2005. A complete classification for pairs of real analytic curves in the complex plane with tangential intersection. *J. Dyn. Control Syst.*, **11**(1), 1–71.

[3] Ahern, P., and O'Farrell, A. G. 2009. Reversible biholomorphic germs. *Comput. Methods Funct. Theory*, **9**(2), 473–484.

[4] Alexander, H., and Wermer, J. 1998. *Several complex variables and Banach algebras*. Third edn. Graduate Texts in Mathematics, vol. 35. New York: Springer-Verlag.

[5] Allan, G. R. 2011. *Introduction to Banach spaces and algebras*. Oxford Graduate Texts in Mathematics, vol. 20. Oxford: Oxford University Press. Prepared for publication and with a preface by H. G. Dales.

[6] Anderson, R. D. 1962. On homeomorphisms as products of conjugates of a given homeomorphism and its inverse. Pages 231–234 of: *Topology of 3-manifolds and related topics (Proc. The Univ. of Georgia Institute, 1961)*. Englewood Cliffs, N.J.: Prentice-Hall.

[7] Anzai, H. 1951. On an example of a measure preserving transformation which is not conjugate to its inverse. *Proc. Japan Acad.*, **27**, 517–522.

[8] Arnol'd, V. I. 1984. Reversible systems. Pages 1161–1174 of: *Nonlinear and turbulent processes in physics, Vol. 3 (Kiev, 1983)*. Chur: Harwood Academic Publ.

[9] Arnol'd, V. I. 1988. *Geometrical methods in the theory of ordinary differential equations*. Second edn. Grundlehren der Mathematischen Wissenschaften [Fundamental Principles of Mathematical Sciences], vol. 250. New York: Springer-Verlag. Translated from the Russian by J. M. Szücs.

[10] Arnol'd, V. I. 2006. *Ordinary differential equations*. Universitext. Berlin: Springer-Verlag. Translated from the Russian by Roger Cooke, Second printing of the 1992 edition.

[11] Arnol'd, V. I., and Avez, A. 1968. *Ergodic problems of classical mechanics*. Translated from the French by A. Avez. W. A. Benjamin, Inc., New York-Amsterdam.

[12] Aschbacher, M. 1998. Near subgroups of finite groups. *J. Group Theory*, **1**(2), 113–129.

[13] Aschbacher, M. 2000. *Finite group theory*. Second edn. Cambridge Studies in Advanced Mathematics, vol. 10. Cambridge: Cambridge University Press.

[14] Aschbacher, M., Meierfrankenfeld, U., and Stellmacher, B. 2001. Counting involutions. *Illinois J. Math.*, **45**(3), 1051–1060.

[15] Baake, M., and Roberts, J. A. G. 1997. Reversing symmetry group of Gl$(2, \mathbf{Z})$ and PGl$(2, \mathbf{Z})$ matrices with connections to cat maps and trace maps. *J. Phys. A*, **30**(5), 1549–1573.

[16] Baake, M., and Roberts, J. A. G. 2001. Symmetries and reversing symmetries of toral automorphisms. *Nonlinearity*, **14**(4), R1–R24.

[17] Baake, M., and Roberts, J. A. G. 2003. Symmetries and reversing symmetries of area-preserving polynomial mappings in generalised standard form. *Phys. A*, **317**(1-2), 95–112.

[18] Baake, M., and Roberts, J. A. G. 2005. Symmetries and reversing symmetries of polynomial automorphisms of the plane. *Nonlinearity*, **18**(2), 791–816.

[19] Baake, M., and Roberts, J. A. G. 2006. The structure of reversing symmetry groups. *Bull. Austral. Math. Soc.*, **73**(3), 445–459.

[20] Baake, M., Roberts, J. A. G., and Weiss, A. 2008. Periodic orbits of linear endomorphisms on the 2-torus and its lattices. *Nonlinearity*, **21**(10), 2427–2446.

[21] Baake, M., Neumärker, N., and Roberts, J. A. G. 2013. Orbit structure and (reversing) symmetries of toral endomorphisms on rational lattices. *Discrete Contin. Dyn. Syst.*, **33**(2), 527–553.

[22] Bagiński, C. 1987. On sets of elements of the same order in the alternating group A_n. *Publ. Math. Debrecen*, **34**(3-4), 313–315.

[23] Baker, A. 2002. *Matrix groups*. Springer Undergraduate Mathematics Series. London: Springer-Verlag London Ltd. An introduction to Lie group theory.

[24] Baker, I. N. 1961/1962. Permutable power series and regular iteration. *J. Austral. Math. Soc.*, **2**, 265–294.

[25] Baker, I. N. 1964. Fractional iteration near a fixpoint of multiplier 1. *J. Austral. Math. Soc.*, **4**, 143–148.

[26] Baker, I. N. 1967. Non-embeddable functions with a fixpoint of multiplier 1. *Math. Z.*, **99**, 377–384.

[27] Ballantine, C. S. 1977/78. Products of involutory matrices. I. *Linear and Multilinear Algebra*, **5**(1), 53–62.

[28] Beardon, A. F. 1983. *The geometry of discrete groups*. Graduate Texts in Mathematics, vol. 91. New York: Springer-Verlag.

[29] Bedford, T., Keane, M., and Series, C. (eds). 1991. *Ergodic theory, symbolic dynamics, and hyperbolic spaces*. Oxford Science Publications. New York: The Clarendon Press Oxford University Press. Papers from the Workshop on Hyperbolic Geometry and Ergodic Theory held in Trieste, April 17–28, 1989, Edited by T. Bedford, M. Keane and C. Series.

[30] Bessaga, C., and Pełczyński, A. 1975. *Selected topics in infinite-dimensional topology*. Warsaw: PWN—Polish Scientific Publishers. Monografie Matematyczne, Tom 58. [Mathematical Monographs, Vol. 58].

[31] Birkhoff, G. D. 1915. The restricted problem of three bodies. *Rend. Circ. Mat. Palermo*, **39**, 265–334.

[32] Birkhoff, G. D. 1939. Déformations analytiques et fonctions auto-équivalentes. *Ann. Inst. H. Poincaré*, **9**, 51–122.

[33] Bishop, E. 1965. Differentiable manifolds in complex Euclidean space. *Duke Math. J.*, **32**, 1–21.

[34] Borevich, A. I., and Shafarevich, I. R. 1966. *Number theory*. Translated from the Russian by Newcomb Greenleaf. Pure and Applied Mathematics, Vol. 20. New York: Academic Press.

[35] Botha, J. D. 2009. A unification of some matrix factorization results. *Linear Algebra Appl.*, **431**(10), 1719–1725.

[36] Brauer, R. 1963. Representations of finite groups. Pages 133–175 of: *Lectures on Modern Mathematics, Vol. I*. New York: Wiley.

[37] Brendle, T. E., and Farb, B. 2004. Every mapping class group is generated by 6 involutions. *J. Algebra*, **278**(1), 187–198.

[38] Brin, M. G. 1996. The chameleon groups of Richard J. Thompson: automorphisms and dynamics. *Inst. Hautes Études Sci. Publ. Math.*, **84**, 5–33 (1997).

[39] Brin, M. G., and Squier, C. C. 1985. Groups of piecewise linear homeomorphisms of the real line. *Invent. Math.*, **79**(3), 485–498.

[40] Brin, M. G., and Squier, C. C. 2001. Presentations, conjugacy, roots, and centralizers in groups of piecewise linear homeomorphisms of the real line. *Comm. Algebra*, **29**(10), 4557–4596.

[41] Brucks, K. M., and Bruin, H. 2004. *Topics from one-dimensional dynamics*. London Mathematical Society Student Texts, vol. 62. Cambridge University Press, Cambridge.

[42] Buck, R. C. 1972. On approximation theory and functional equations. *J. Approximation Theory*, **5**, 228–237. Collection of articles dedicated to J. L. Walsh on his 75th birthday, III (Proc. Internat. Conf. Approximation Theory, Related Topics and their Applications, Univ. Maryland, College Park, Md., 1970).

[43] Bullett, S. 1988. Dynamics of quadratic correspondences. *Nonlinearity*, **1**(1), 27–50.

[44] Bünger, F., Knüppel, F., and Nielsen, K. 1997. Products of symmetries in unitary groups. *Linear Algebra Appl.*, **260**, 9–42.

[45] Calica, A. B. 1971. Reversible homeomorphisms of the real line. *Pacific J. Math.*, **39**, 79–87.

[46] Camina, R. 2000. The Nottingham group. Pages 205–221 of: *New horizons in pro-p groups*. Progr. Math., vol. 184. Boston, MA: Birkhäuser Boston.

[47] Cannon, J. W., Floyd, W. J., and Parry, W. R. 1996. Introductory notes on Richard Thompson's groups. *Enseign. Math. (2)*, **42**(3-4), 215–256.

[48] Carleson, L., and Gamelin, T. W. 1993. *Complex dynamics*. Universitext: Tracts in Mathematics. New York: Springer-Verlag.

[49] Cartan, H. 1995. *Elementary theory of analytic functions of one or several complex variables*. New York: Dover Publications Inc. Translated from the French, Reprint of the 1973 edition.

[50] Carter, R. W. 1972. Conjugacy classes in the Weyl group. *Compositio Math.*, **25**, 1–59.

[51] Cassels, J. W. S. 1978. *Rational quadratic forms*. London Mathematical Society Monographs, vol. 13. London: Academic Press Inc. [Harcourt Brace Jovanovich Publishers].

[52] Chen, K-t. 1968. Normal forms of local diffeomorphisms on the real line. *Duke Math. J.*, **35**, 549–555.

[53] Cohn, H. 1978. *A classical invitation to algebraic numbers and class fields*. New York: Springer-Verlag. With two appendices by Olga Taussky: "Artin's 1932 Göttingen lectures on class field theory" and "Connections between algebraic number theory and integral matrices", Universitext.

[54] Cohn, H. 1980. *Advanced number theory*. New York: Dover Publications Inc. Reprint of ıt A second course in number theory, 1962, Dover Books on Advanced Mathematics.

[55] Cohn, P. M. 2003. *Basic algebra*. London: Springer-Verlag London Ltd. Groups, rings and fields.

[56] Constantin, A., and Kolev, B. 1994. The theorem of Kerékjártó on periodic homeomorphisms of the disc and the sphere. *Enseign. Math. (2)*, **40**(3-4), 193–204.

[57] Conway, J. H. 1997. *The sensual (quadratic) form*. Carus Mathematical Monographs, vol. 26. Washington, DC: Mathematical Association of America. With the assistance of Francis Y. C. Fung.

[58] Conway, J. H., and Sloane, N. J. A. 1999. *Sphere packings, lattices and groups*. Third edn. Grundlehren der Mathematischen Wissenschaften [Fundamental Principles of Mathematical Sciences], vol. 290. New York: Springer-Verlag. With additional contributions by E. Bannai, R. E. Borcherds, J. Leech, S. P. Norton, A. M. Odlyzko, R. A. Parker, L. Queen and B. B. Venkov.

[59] Coxeter, H. S. M. 1947. The product of three reflections. *Quart. Appl. Math.*, **5**, 217–222.

[60] Coxeter, H. S. M. 1969. *Introduction to geometry*. Second edn. New York: John Wiley & Sons Inc.

[61] Coxeter, H. S. M. 1974. *Regular complex polytopes*. London: Cambridge University Press.

[62] Curtis, C. W. 1984. *Linear algebra*. Fourth edn. Undergraduate Texts in Mathematics. New York: Springer-Verlag. An introductory approach.

[63] de Melo, W., and van Strien, S. 1993. *One-dimensional dynamics*. Ergebnisse der Mathematik und ihrer Grenzgebiete (3) [Results in Mathematics and Related Areas (3)], vol. 25. Berlin: Springer-Verlag.

[64] de Paepe, P. J. 1986. Approximation on disks. *Proc. Amer. Math. Soc.*, **97**(2), 299–302.

[65] De Paepe, P. J. 2001. Eva Kallin's lemma on polynomial convexity. *Bull. London Math. Soc.*, **33**(1), 1–10.

[66] Devaney, R. L. 1976. Reversible diffeomorphisms and flows. *Trans. Amer. Math. Soc.*, **218**, 89–113.

[67] Dieudonné, J. 1951. On the automorphisms of the classical groups. With a supplement by Loo-Keng Hua. *Mem. Amer. Math. Soc.*, **1951**(2), vi+122.

[68] Dijkstra, J. J., and van Mill, J. 2006. On the group of homeomorphisms of the real line that map the pseudoboundary onto itself. *Canad. J. Math.*, **58**(3), 529–547.

[69] Diliberto, S. P., and Straus, E. G. 1951. On the approximation of a function of several variables by the sum of functions of fewer variables. *Pacific J. Math.*, **1**, 195–210.

[70] Djoković, D. Ž. 1967. Product of two involutions. *Arch. Math. (Basel)*, **18**, 582–584.

[71] Djoković, D. Ž. 1986. Pairs of involutions in the general linear group. *J. Algebra*, **100**(1), 214–223.

[72] Djoković, D. Ž., and Malzan, J. G. 1982. Products of reflections in U(p, q). *Mem. Amer. Math. Soc.*, **37**(259), vi+82.

[73] Écalle, J. 1975. Théorie itérative: introduction à la théorie des invariants holomorphes. *J. Math. Pures Appl. (9)*, **54**, 183–258.

[74] Eisenbud, D., Hirsch, U., and Neumann, W. 1981. Transverse foliations of Seifert bundles and self-homeomorphism of the circle. *Comment. Math. Helv.*, **56**(4), 638–660.

[75] Ellers, E. W. 1977. Bireflectionality in classical groups. *Canad. J. Math.*, **29**(6), 1157–1162.

[76] Ellers, E. W. 1983. Cyclic decomposition of unitary spaces. *J. Geom.*, **21**(2), 101–107.

[77] Ellers, E. W. 1993. The reflection length of a transformation in the unitary group over a finite field. *Linear and Multilinear Algebra*, **35**(1), 11–35.

[78] Ellers, E. W. 1999. Bireflectionality of orthogonal and symplectic groups of characteristic 2. *Arch. Math. (Basel)*, **73**(6), 414–418.

[79] Ellers, E. W. 2004. Conjugacy classes of involutions in the Lorentz group $\Omega(V)$ and in SO(V). *Linear Algebra Appl.*, **383**, 77–83.

[80] Ellers, E. W., and Malzan, J. 1990. Products of reflections in the kernel of the spinorial norm. *Geom. Dedicata*, **36**(2-3), 279–285.

[81] Ellers, E. W., and Nolte, W. 1982. Bireflectionality of orthogonal and symplectic groups. *Arch. Math. (Basel)*, **39**(2), 113–118.

[82] Ellers, E. W., and Villa, O. 2004. The special orthogonal group is trireflectional. *Arch. Math. (Basel)*, **82**(2), 122–127.

[83] Engel, K-J., and Nagel, R. 2000. *One-parameter semigroups for linear evolution equations*. Graduate Texts in Mathematics, vol. 194. New York: Springer-Verlag. With contributions by S. Brendle, M. Campiti, T. Hahn, G. Metafune, G. Nickel, D. Pallara, C. Perazzoli, A. Rhandi, S. Romanelli and R. Schnaubelt.

[84] Falcolini, C. 2002. *Collisions and singularities in the n-body problem*. Lecture Notes in Physics, vol. 590. Berlin: Springer-Verlag. Edited by D. Benest and C. Froeschlé.

[85] Fein, B. 1970. A note on the Brauer-Speiser theorem. *Proc. Amer. Math. Soc.*, **25**, 620–621.

[86] Feit, W. 1967. *Characters of finite groups*. W. A. Benjamin, Inc., New York-Amsterdam.

[87] Feit, W., and Thompson, J. G. 1963. Solvability of groups of odd order. *Pacific J. Math.*, **13**, 775–1029.

[88] Feit, W., and Zuckerman, G. J. 1982. Reality properties of conjugacy classes in spin groups and symplectic groups. Pages 239–253 of: *Algebraists' homage: papers in ring theory and related topics (New Haven, Conn., 1981)*. Contemp. Math., vol. 13. Providence, R.I.: Amer. Math. Soc.

[89] Fine, N. J., and Schweigert, G. E. 1955. On the group of homeomorphisms of an arc. *Ann. of Math. (2)*, **62**, 237–253.

[90] Foguel, T., Kinyon, M. K., and Phillips, J. D. 2006. On twisted subgroups and Bol loops of odd order. *Rocky Mountain J. Math.*, **36**(1), 183–212.

[91] Foreman, M., Rudolph, D. J., and Weiss, B. 2011. The conjugacy problem in ergodic theory. *Ann. of Math. (2)*, **173**(3), 1529–1586.

[92] Forstnerič, F. 1992. A smooth holomorphically convex disc in $\mathbf{C}^2$ that is not locally polynomially convex. *Proc. Amer. Math. Soc.*, **116**(2), 411–415.

[93] Gal′t, A. A. 2010. Strongly real elements in finite simple orthogonal groups. *Sibirsk. Mat. Zh.*, **51**(2), 241–248.

[94] Gamkrelidze, R. V. (ed). 1990. *Several complex variables. I.* Encyclopaedia of Mathematical Sciences, vol. 7. Berlin: Springer-Verlag. Introduction to complex analysis, A translation of Sovremennye problemy matematiki. Fundamentalnye napravleniya, Tom 7, Akad. Nauk SSSR, Vsesoyuz. Inst. Nauchn. i Tekhn. Inform., Moscow, 1985 [MR0850489 (87f:32003)], Translation by P. M. Gauthier, Translation edited by A. G. Vitushkin.

[95] Gauss, C. F. 1986. *Disquisitiones arithmeticae.* New York: Springer-Verlag. Translated and with a preface by Arthur A. Clarke, Revised by William C. Waterhouse, Cornelius Greither and A. W. Grootendorst and with a preface by Waterhouse.

[96] Ghys, É. 2001. Groups acting on the circle. *Enseign. Math. (2)*, **47**(3-4), 329–407.

[97] Ghys, É., and Sergiescu, V. 1980. Stabilité et conjugaison différentiable pour certains feuilletages. *Topology*, **19**(2), 179–197.

[98] Giblin, J., and Markovic, V. 2006. Classification of continuously transitive circle groups. *Geom. Topol.*, **10**, 1319–1346.

[99] Gill, N., and Short, I. 2010. Reversible maps and composites of involutions in groups of piecewise linear homeomorphisms of the real line. *Aequationes Math.*, **79**(1-2), 23–37.

[100] Gill, N., and Singh, A. 2011a. Real and strongly real classes in $\mathrm{PGL}_n(q)$ and quasi-simple covers of $\mathrm{PSL}_n(q)$. *J. Group Theory*, **14**, 461–489.

[101] Gill, N., and Singh, A. 2011b. Real and strongly real classes in $\mathrm{SL}_n(q)$. *J. Group Theory*, **14**, 437–459.

[102] Gill, N., O'Farrell, A. G., and Short, I. 2009. Reversibility in the group of homeomorphisms of the circle. *Bull. Lond. Math. Soc.*, **41**(5), 885–897.

[103] Gómez, A., and Meiss, J. D. 2003. Reversible polynomial automorphisms of the plane: the involutory case. *Phys. Lett. A*, **312**(1-2), 49–58.

[104] Gómez, A., and Meiss, J. D. 2004. Reversors and symmetries for polynomial automorphisms of the complex plane. *Nonlinearity*, **17**(3), 975–1000.

[105] Gongopadhyay, K. 2011. Conjugacy classes in Möbius groups. *Geom. Dedicata*, **151**, 245–258.

[106] Gongopadhyay, K., and Parker, J. R. 2012. Reversible complex hyperbolic isometries. *Preprint.*

[107] Goodson, G., and Lemańczyk, M. 1996. Transformations conjugate to their inverses have even essential values. *Proc. Amer. Math. Soc.*, **124**(9), 2703–2710.

[108] Goodson, G. R. 1996. The structure of ergodic transformations conjugate to their inverses. Pages 369–385 of: *Ergodic theory of $\mathbf{Z}^d$ actions (Warwick, 1993–1994)*. London Math. Soc. Lecture Note Ser., vol. 228. Cambridge: Cambridge Univ. Press.

[109] Goodson, G. R. 1997. The inverse-similarity problem for real orthogonal matrices. *Amer. Math. Monthly*, **104**(3), 223–230.

[110] Goodson, G. R. 1999. Inverse conjugacies and reversing symmetry groups. *Amer. Math. Monthly*, **106**(1), 19–26.

[111] Goodson, G. R. 2000a. Conjugacies between ergodic transformations and their inverses. *Colloq. Math.*, **84/85**(, part 1), 185–193. Dedicated to the memory of Anzelm Iwanik.

[112] Goodson, G. R. 2000b. The converse of the inverse-conjugacy theorem for unitary operators and ergodic dynamical systems. *Proc. Amer. Math. Soc.*, **128**(5), 1381–1388.

[113] Goodson, G. R. 2002. Ergodic dynamical systems conjugate to their composition squares. *Acta Math. Univ. Comenian. (N.S.)*, **71**(2), 201–210.

[114] Goodson, G. R. 2010. Groups having elements conjugate to their squares and connections with dynamical systems. *Applied Mathematics*, **1**, 416–424.

[115] Goodson, G. R., del Junco, A., Lemańczyk, M., and Rudolph, D. J. 1996. Ergodic transformations conjugate to their inverses by involutions. *Ergodic Theory Dynam. Systems*, **16**(1), 97–124.

[116] Goodson, Geoffrey R. 2007. Spectral properties of ergodic dynamical systems conjugate to their composition squares. *Colloq. Math.*, **107**(1), 99–118.

[117] Gorenstein, D. 1968. *Finite groups*. New York: Harper & Row Publishers.

[118] Gow, R. 1975. Real-valued characters of solvable groups. *Bull. London Math. Soc.*, **7**, 132.

[119] Gow, R. 1976. Real-valued characters and the Schur index. *J. Algebra*, **40**(1), 258–270.

[120] Gow, R. 1979. Real-valued and 2-rational group characters. *J. Algebra*, **61**(2), 388–413.

[121] Gow, R. 1981. Products of two involutions in classical groups of characteristic 2. *J. Algebra*, **71**(2), 583–591.

[122] Gow, R. 1988. Commutators in the symplectic group. *Arch. Math. (Basel)*, **50**(3), 204–209.

[123] Graham, D., Keane, S., and O'Farrell, A. G. 2001. Siméadracht amchúlaithe chórais dinimiciúil. *in: R.N. Shorten, T. Ward and T. Lysaght (eds), Proceedings of the Irish Systems and Signals Conference*, 27–31. *Translation available online from AOF*.

[124] Guba, V., and Sapir, M. 1997. Diagram groups. *Mem. Amer. Math. Soc.*, **130**(620), viii+117.

[125] Gustafson, W. H. 1991. On products of involutions. Pages 237–255 of: *Paul Halmos*. New York: Springer.

[126] Gustafson, W. H., Halmos, P. R., and Radjavi, H. 1976. Products of involutions. *Linear Algebra and Appl.*, **13**(1/2), 157–162. Collection of articles dedicated to Olga Taussky Todd.

[127] Halmos, P. R., and von Neumann, J. 1942. Operator methods in classical mechanics. II. *Ann. of Math. (2)*, **43**, 332–350.

[128] Hamkins, J. D. 1998. Every group has a terminating transfinite automorphism tower. *Proc. Amer. Math. Soc.*, **126**(11), 3223–3226.

[129] Higman, G. 1974. *Finitely presented infinite simple groups*. Department of Pure Mathematics, Department of Mathematics, I.A.S. Australian National University, Canberra. Notes on Pure Mathematics, No. 8 (1974).

[130] Hillar, C. J., and Rhea, D. L. 2007. Automorphisms of finite abelian groups. *Amer. Math. Monthly*, **114**(10), 917–923.

[131] Hirsch, M. W., and Smale, S. 1974. *Differential equations, dynamical systems, and linear algebra*. Academic Press [A subsidiary of Harcourt Brace Jovanovich, Publishers], New York-London. Pure and Applied Mathematics, Vol. 60.

[132] Hjorth, G. 2000. *Classification and orbit equivalence relations*. Mathematical Surveys and Monographs, vol. 75. Providence, RI: American Mathematical Society.

[133] Hladnik, M., Omladič, M., and Radjavi, H. 2001. Products of roots of the identity. *Proc. Amer. Math. Soc.*, **129**(2), 459–465.

[134] Hoffman, F., and Paige, E. C. 1970/1971. Products of two involutions in the general linear group. *Indiana Univ. Math. J.*, **20**, 1017–1020.

[135] Huang, X. 1998. On an n-manifold in $\mathbf{C}^n$ near an elliptic complex tangent. *J. Amer. Math. Soc.*, **11**(3), 669–692.

[136] Huang, X. J., and Krantz, S. G. 1995. On a problem of Moser. *Duke Math. J.*, **78**(1), 213–228.

[137] Il'yashenko, Yu. S. 1993. Nonlinear Stokes phenomena. Pages 1–55 of: *Nonlinear Stokes phenomena*. Adv. Soviet Math., vol. 14. Providence, RI: Amer. Math. Soc.

[138] Isaacs, I. M. 1976. *Character theory of finite groups*. New York: Academic Press [Harcourt Brace Jovanovich Publishers]. Pure and Applied Mathematics, No. 69.

[139] Ishibashi, H. 1995. Involutary expressions for elements in $\mathrm{GL}_n(\mathbf{Z})$ and $\mathrm{SL}_n(\mathbf{Z})$. *Linear Algebra Appl.*, **219**, 165–177.

[140] James, G., and Liebeck, M. 2001. *Representations and characters of groups*. Second edn. New York: Cambridge University Press.

[141] Jarczyk, W. 2002. Reversible interval homeomorphisms. *J. Math. Anal. Appl.*, **272**(2), 473–479.

[142] Jordan, C. R., Jordan, D. A., and Jordan, J. H. 2002. Reversible complex Hénon maps. *Experiment. Math.*, **11**(3), 339–347.

[143] Kasner, E. 1915. Conformal classification of analytic arcs or elements: Poincaré's local problem of conformal geometry. *Trans. Amer. Math. Soc.*, **16**(3), 333–349.

[144] Kasner, E. 1916. Infinite Groups Generated by Conformal Transformations of Period Two (Involutions and Symmetries). *Amer. J. Math.*, **38**(2), 177–184.

[145] Katok, A., and Hasselblatt, B. 1995. *Introduction to the modern theory of dynamical systems*. Encyclopedia of Mathematics and its Applications, vol. 54. Cambridge: Cambridge University Press. With a supplementary chapter by Katok and Leonardo Mendoza.

[146] Kaur, D., and Kulsherstha, A. *Strongly real special 2-groups*. To appear.

[147] Kenig, C. E., and Webster, S. M. 1982. The local hull of holomorphy of a surface in the space of two complex variables. *Invent. Math.*, **67**(1), 1–21.

[148] Khavinson, S. Ya. 1995. The annihilator of linear superpositions. *Algebra i Analiz*, **7**(3), 1–42.

[149] Knüppel, F. 1988. Products of involutions in orthogonal groups. Pages 231–247 of: *Combinatorics '86 (Trento, 1986)*. Ann. Discrete Math., vol. 37. Amsterdam: North-Holland.

[150] Knüppel, F., and Nielsen, K. 1987a. On products of two involutions in the orthogonal group of a vector space. *Linear Algebra Appl.*, **94**, 209–216.

[151] Knüppel, F., and Nielsen, K. 1987b. Products of involutions in $O^+(V)$. *Linear Algebra Appl.*, **94**, 217–222.

[152] Knüppel, F., and Nielsen, K. 1991. $SL(V)$ is 4-reflectional. *Geom. Dedicata*, **38**(3), 301–308.

[153] Knüppel, F., and Thomsen, G. 1998. Involutions and commutators in orthogonal groups. *J. Austral. Math. Soc. Ser. A*, **65**(1), 1–36.

[154] Kolesnikov, S. G., and Nuzhin, Ja. N. 2005. On strong reality of finite simple groups. *Acta Appl. Math.*, **85**(1-3), 195–203.

[155] Kopell, N. 1970. Commuting diffeomorphisms. Pages 165–184 of: *Global Analysis (Proc. Sympos. Pure Math., Vol. XIV, Berkeley, Calif., 1968)*. Providence, R.I.: Amer. Math. Soc.

[156] Korkmaz, M. 2005. On stable torsion length of a Dehn twist. *Math. Res. Lett.*, **12**(2-3), 335–339.

[157] Kuczma, M., Choczewski, B., and Ger, R. 1990. *Iterative functional equations*. Encyclopedia of Mathematics and its Applications, vol. 32. Cambridge: Cambridge University Press.

[158] Laffey, T. J. 1997. Lectures on integer matrices. *Unpublished lecture notes*.

[159] Lamb, J. S. W. 1992. Reversing symmetries in dynamical systems. *J. Phys. A*, **25**(4), 925–937.

[160] Lamb, J. S. W. 1995. Resonant driving and k-symmetry. *Phys. Lett. A*, **199**(1-2), 55–60.

[161] Lamb, J. S. W. 1996. Area-preserving dynamics that is not reversible. *Phys. A*, **228**(1-4), 344–365.

[162] Lamb, J. S. W., and Quispel, G. R. W. 1994. Reversing k-symmetries in dynamical systems. *Phys. D*, **73**(4), 277–304.

[163] Lamb, J. S. W., and Quispel, G. R. W. 1995. Cyclic reversing k-symmetry groups. *Nonlinearity*, **8**(6), 1005–1026.

[164] Lamb, J. S. W., and Roberts, J. A. G. 1998. Time-reversal symmetry in dynamical systems: a survey. *Phys. D*, **112**(1-2), 1–39. Time-reversal symmetry in dynamical systems (Coventry, 1996).

[165] Lamb, J. S. W., Roberts, J. A. G., and Capel, H. W. 1993. Conditions for local (reversing) symmetries in dynamical systems. *Phys. A*, **197**(3), 379–422.

[166] Lávička, R., O'Farrell, A. G., and Short, I. 2007. Reversible maps in the group of quaternionic Möbius transformations. *Math. Proc. Cambridge Philos. Soc.*, **143**(1), 57–69.

[167] Lewis Jr., D. C. 1961. Reversible transformations. *Pacific J. Math.*, **11**, 1077–1087.

[168] Liebeck, M. W., O'Brien, E. A., Shalev, Λ., and Tiep, P. H. 2010. The Ore conjecture. *J. Eur. Math. Soc. (JEMS)*, **12**(4), 939–1008.

[169] Liu, K. M. 1988a. Decomposition of matrices into three involutions. *Linear Algebra Appl.*, **111**, 1–24.

[170] Liu, K. M. 1988b. Decomposition of matrices into three involutions. *Linear Algebra Appl.*, **111**, 1–24.

[171] Liverpool, L. S. O. 1974/75. Fractional iteration near a fix point of multiplier 1. *J. London Math. Soc. (2)*, **9**, 599–609.

[172] Lubin, J. 1994. Non-Archimedean dynamical systems. *Compositio Math.*, **94**(3), 321–346.

[173] MacKay, R. S. 1993. *Renormalisation in area-preserving maps.* Advanced Series in Nonlinear Dynamics, vol. 6. River Edge, NJ: World Scientific Publishing Co. Inc.

[174] Malgrange, B. 1982. Travaux d'Écalle et de Martinet-Ramis sur les systèmes dynamiques. Pages 59–73 of: *Bourbaki Seminar, Vol. 1981/1982.* Astérisque, vol. 92. Paris: Soc. Math. France.

[175] Markley, N. G. 1970. Homeomorphisms of the circle without periodic points. *Proc. London Math. Soc. (3)*, **20**, 688–698.

[176] Marshall, D. E., and O'Farrell, A. G. 1979. Uniform approximation by real functions. *Fund. Math.*, **54**, 203–11.

[177] Marshall, D. E., and O'Farrell, A. G. 1983. Approximation by a sum of two algebras. The lightning bolt principle. *J. Funct. Anal.*, **52**(3), 353–368.

[178] Mazurov, V. D., and Khukhro, E. I. (eds). 2014. *The Kourovka notebook.* Eighteenth edn. Novosibirsk: Russian Academy of Sciences Siberian Division Institute of Mathematics. Unsolved problems in group theory, Including archive of solved problems.

[179] McCarthy, P. J., and Stephenson, W. 1985. The classification of the conjugacy classes of the full group of homeomorphisms of an open interval and the general solution of certain functional equations. *Proc. London Math. Soc. (3)*, **51**(1), 95–112.

[180] McCleary, S. H. 1978. Groups of homeomorphisms with manageable automorphism groups. *Comm. Algebra*, **6**(5), 497–528.

[181] Medvedev, V. A. 1992. Refutation of a theorem of Diliberto and Straus. *Mat. Zametki*, **51**(4), 78–80, 142.

[182] Meyerson, M. D. 1981. Every power series is a Taylor series. *Amer. Math. Monthly*, **88**(1), 51–52.

[183] Miller, III, C. F. 1971. *On group-theoretic decision problems and their classification.* Princeton, N.J.: Princeton University Press. Annals of Mathematics Studies, No. 68.

[184] Moser, J. K., and Webster, S. M. 1983. Normal forms for real surfaces in C^2 near complex tangents and hyperbolic surface transformations. *Acta Math.*, **150**(3-4), 255–296.

[185] Muckenhoupt, B. 1961. Automorphisms of formal power series under substitution. *Trans. Amer. Math. Soc.*, **99**, 373–383.

[186] Nakai, I. 1998. The classification of curvilinear angles in the complex plane and the groups of ± holomorphic diffeomorphisms. *Ann. Fac. Sci. Toulouse Math. (6)*, **7**(2), 313–334.

[187] Navas, A. 2007. *Grupos de difeomorfismos del círculo*. Ensaios Matemáticos [Mathematical Surveys], vol. 13. Rio de Janeiro: Sociedade Brasileira de Matemática.

[188] O'Farrell, A. G. 2004. Conjugacy, involutions, and reversibility for real homeomorphisms. *Irish Math. Soc. Bull.*, **54**, 41–52.

[189] O'Farrell, A. G. 2008. Composition of involutive power series, and reversible series. *Comput. Methods Funct. Theory*, **8**(1-2), 173–193.

[190] O'Farrell, A. G., and Roginskaya, M. 2010. Conjugacy of real diffeomorphisms. A survey. *Algebra i Analiz*, **22**(1), 3–56.

[191] O'Farrell, A. G., and Sanabria-Garcia, M.A. 2002. De Paepe's disc has nontrivial polynomial hull. *Bull. LMS*, **34**, 490–494.

[192] O'Farrell, A. G., and Short, I. 2009. Reversibility in the diffeomorphism group of the real line. *Publ. Mat.*, **53**(2), 401–415.

[193] O'Farrell, A. G., and Zaitsev, D. *Formally-reversible maps of* $(\mathbb{C}^2, 0)$. To appear in *Ann. Sc. Norm. Super. Pisa Cl. Sci. (5)*, DOI:10.2422/2036-2145.201201_001.

[194] O'Farrell, A. G., and Zaitsev, D. 2014. Factoring formal maps into reversible or involutive factors. *J. Algebra*, **399**, 657–674.

[195] Pérez Marco, R. 1995. Nonlinearizable holomorphic dynamics having an uncountable number of symmetries. *Invent. Math.*, **119**(1), 67–127.

[196] Poincaré, H. 1907. Les fonctions analytiques de deux variables et la préprésentation conforme. *Rend. Circ. Mat. Palermo*, **23**(1), 185–220.

[197] Poincaré, H. 1996. *Œuvres. Tome VI.* Les Grands Classiques Gauthier-Villars. [Gauthier-Villars Great Classics]. Sceaux: Éditions Jacques Gabay. Géométrie. Analysis situs (topologie). [Geometry. Analysis situs (topology)], Reprint of the 1953 edition.

[198] Quispel, G. R. W., and Capel, H. W. 1989. Local reversibility in dynamical systems. *Phys. Lett. A*, **142**(2-3), 112–116.

[199] Quispel, G. R. W., and Roberts, J. A. G. 1988. Reversible mappings of the plane. *Phys. Lett. A*, **132**(4), 161–163.

[200] Radjavi, H. 1975. Decomposition of matrices into simple involutions. *Linear Algebra and Appl.*, **12**(3), 247–255.

[201] Radjavi, H. 1981. The group generated by involutions. *Proc. Roy. Irish Acad. Sect. A*, **81**(1), 9–12.

[202] Rämö, J. 2011. Strongly real elements of orthogonal groups in even characteristic. *J. Group Theory*, **14**(1), 9–30.

[203] Ratcliffe, J. G. 1994. *Foundations of hyperbolic manifolds.* Graduate Texts in Mathematics, vol. 149. New York: Springer-Verlag.

[204] Roberts, J. A. G., and Capel, H. W. 1992. Area preserving mappings that are not reversible. *Phys. Lett. A*, **162**(3), 243–248.

[205] Roberts, J. A. G., and Quispel, G. R. W. 1992. Chaos and time-reversal symmetry. Order and chaos in reversible dynamical systems. *Phys. Rep.*, **216**(2-3), 63–177.

[206] Robinson, D. J. S. 1996. *A course in the theory of groups.* Second edn. Graduate Texts in Mathematics, vol. 80. New York: Springer-Verlag.

[207] Sarnak, P. 2007. Reciprocal geodesics. Pages 217–237 of: *Analytic number theory*. Clay Math. Proc., vol. 7. Providence, RI: Amer. Math. Soc.

[208] Schreier, J., and Ulam, S. 1933. Über die Permutationsgruppe der natürlichen Zahlenfolge. *Studia Math.*, **4**, 134–141.

[209] Sepanski, M. R. 2007. *Compact Lie groups*. Graduate Texts in Mathematics, vol. 235. New York: Springer.

[210] Series, C. 1985. The geometry of Markoff numbers. *Math. Intelligencer*, **7**(3), 20–29.

[211] Sevryuk, M. B. 1986. *Reversible systems.* Lecture Notes in Mathematics, vol. 1211. Berlin: Springer-Verlag.

[212] Shalev, Aner. 2009. Word maps, conjugacy classes, and a noncommutative Waring-type theorem. *Ann. of Math. (2)*, **170**(3), 1383–1416.

[213] Short, I. 2008. Reversible maps in isometry groups of spherical, Euclidean and hyperbolic space. *Math. Proc. R. Ir. Acad.*, **108**(1), 33–46.

[214] Siegel, C. B., and Moser, J. K. 1995. *Lectures on Celestial Mechanics, reprint of the 1971 edition*. Berlin, Heidelberg: Springer.

[215] Simon, B. 1996. *Representations of finite and compact groups*. Graduate Studies in Mathematics, vol. 10. Providence, RI: American Mathematical Society.

[216] Singh, A., and Thakur, M. 2005. Reality properties of conjugacy classes in G_2. *Israel J. Math.*, **145**, 157–192.

[217] Singh, A., and Thakur, M. 2008. Reality properties of conjugacy classes in algebraic groups. *Israel J. Math.*, **165**, 1–27.

[218] Springer, T. A. 1974. Regular elements of finite reflection groups. *Invent. Math.*, **25**, 159–198.

[219] Stebe, P. F. 1972. Conjugacy separability of groups of integer matrices. *Proc. Amer. Math. Soc.*, **32**, 1–7.

[220] Sternberg, S. 1957. Local C^n transformations of the real line. *Duke Math. J.*, **24**, 97–102.

[221] Sternfeld, Y. 1986. Uniform separation of points and measures and representation by sums of algebras. *Israel J. Math.*, **55**(3), 350–362.

[222] Stewart, I., and Tall, D. 1979. *Algebraic number theory*. London: Chapman and Hall. Chapman and Hall Mathematics Series.

[223] Szekeres, G. 1964. Fractional iteration of entire and rational functions. *J. Austral. Math. Soc.*, **4**, 129–142.

[224] Takens, F. 1973. Normal forms for certain singularities of vectorfields. *Ann. Inst. Fourier (Grenoble)*, **23**(2), 163–195. Colloque International sur l'Analyse et la Topologie Différentielle (Colloques Internationaux du Centre National de la Recherche Scientifique, Strasbourg, 1972).

[225] Taylor, D. E. 1992. *The geometry of the classical groups*. Sigma Series in Pure Mathematics, vol. 9. Berlin: Heldermann Verlag.

[226] Thomas, S. *The automorphism tower problem*. Book in preparation.

[227] Thomas, S. 1985. The automorphism tower problem. *Proc. Amer. Math. Soc.*, **95**(2), 166–168.

[228] Thompson, R. C. 1961. Commutators in the special and general linear groups. *Trans. Amer. Math. Soc.*, **101**, 16–33.

[229] Thompson, R. C. 1962a. Commutators of matrices with coefficients from the field of two elements. *Duke Math. J.*, **29**, 367–373.

[230] Thompson, R. C. 1962b. On matrix commutators. *Portugal. Math.*, **21**, 143–153.

[231] Tiep, P. H., and Zalesski, A. E. 2005. Real conjugacy classes in algebraic groups and finite groups of Lie type. *J. Group Theory*, **8**(3), 291–315.

[232] Trépreau, J-M. 2003. Discrimination analytique des difféomorphismes résonnants de $(\mathbb{C}, 0)$ et réflexion de Schwarz. *Astérisque*, 271–319. Autour de l'analyse microlocale.

[233] Vdovin, E. P., and Gal't, A. A. 2010. Strong reality of finite simple groups. *Sibirsk. Mat. Zh.*, **51**(4), 769–777.

[234] Villa, O. 2003. An example of a bireflectional spin group. *Arch. Math. (Basel)*, **81**(1), 1–4.

[235] Vinroot, C. R. 2004. A factorization in $GSp(V)$. *Linear Multilinear Algebra*, **52**(6), 385–403.

[236] Voronin, S. M. 1981. Analytic classification of germs of conformal mappings $(\mathbf{C}, 0) \to (\mathbf{C}, 0)$. *Funktsional. Anal. i Prilozhen.*, **15**(1), 1–17, 96.

[237] Voronin, S. M. 1982. Analytic classification of pairs of involutions and its applications. *Funktsional. Anal. i Prilozhen.*, **16**(2), 21–29, 96.

[238] Wall, G. E. 1963. On the conjugacy classes in the unitary, symplectic and orthogonal groups. *J. Austral. Math. Soc.*, **3**, 1–62.

[239] Walsh, J. A. 1999. The dynamics of circle homeomorphisms: a hands-on introduction. *Math. Mag.*, **72**(1), 3–13.

[240] Webster, S. M. 1996. Double valued reflection in the complex plane. *Enseign. Math. (2)*, **42**(1-2), 25–48.

[241] Webster, S. M. 1997. A note on extremal discs and double valued reflection. Pages 271–276 of: *Multidimensional complex analysis and partial differential equations (São Carlos, 1995)*. Contemp. Math., vol. 205. Providence, RI: Amer. Math. Soc.

[242] Webster, S. M. 1998. Real ellipsoids and double valued reflection in complex space. *Amer. J. Math.*, **120**(4), 757–809.

[243] Weyl, H. 1997. *The classical groups*. Princeton Landmarks in Mathematics. Princeton, NJ: Princeton University Press. Their invariants and representations, Fifteenth printing, Princeton Paperbacks.

[244] Whittaker, J. V. 1963. On isomorphic groups and homeomorphic spaces. *Ann. of Math. (2)*, **78**, 74–91.

[245] Wiegerinck, J. 1995. Local polynomially convex hulls at degenerated CR singularities of surfaces in $\mathbf{C}^2$. *Indiana Univ. Math. J.*, **44**(3), 897–915.

[246] Wilson, R., Walsh, P., Tripp, J., Suleiman, I., Parker, R., Norton, S., Nickerson, S., Linton, S., Bray, J., and Abbott, R. *ATLAS of finitie group representations – Version 3*. http://brauer.maths.qmul.ac.uk/Atlas/v3/.

[247] Wonenburger, M. J. 1966. Transformations which are products of two involutions. *J. Math. Mech.*, **16**, 327–338.

[248] Xia, Z. 1992. The existence of noncollision singularities in Newtonian systems. *Ann. of Math. (2)*, **135**(3), 411–468.

[249] Yale, P. B. 1966. Automorphisms of the complex numbers. *Math. Magazine*, **39**, 135–141.

[250] Yoccoz, J.-C. 1995. *Petits diviseurs en dimension 1*. Paris: Société Mathématique de France. Astérisque No. 231 (1995).

[251] Young, S. W. 1994. The representation of homeomorphisms on the interval as finite compositions of involutions. *Proc. Amer. Math. Soc.*, **121**(2), 605–610.

List of frequently used symbols

$O(n, \mathbb{R})$	orthogonal group	60
$PSL(2, \mathbb{Z})$	modular group	105
$SL(n, F)$	special linear group	77
$SO(n, \mathbb{R})$	special orthogonal group	68
$SU(n, \mathbb{C})$	special unitary group	71
$Spin(n, \mathbb{C})$	spinor group	88
$Sp(n, \mathbb{C})$	compact symplectic group	84
$U(n, \mathbb{C})$	unitary group	60

Index of names

Subject index

Printed in the United States
By Bookmasters